AF730676

Geology of the Higher Central Himalaya

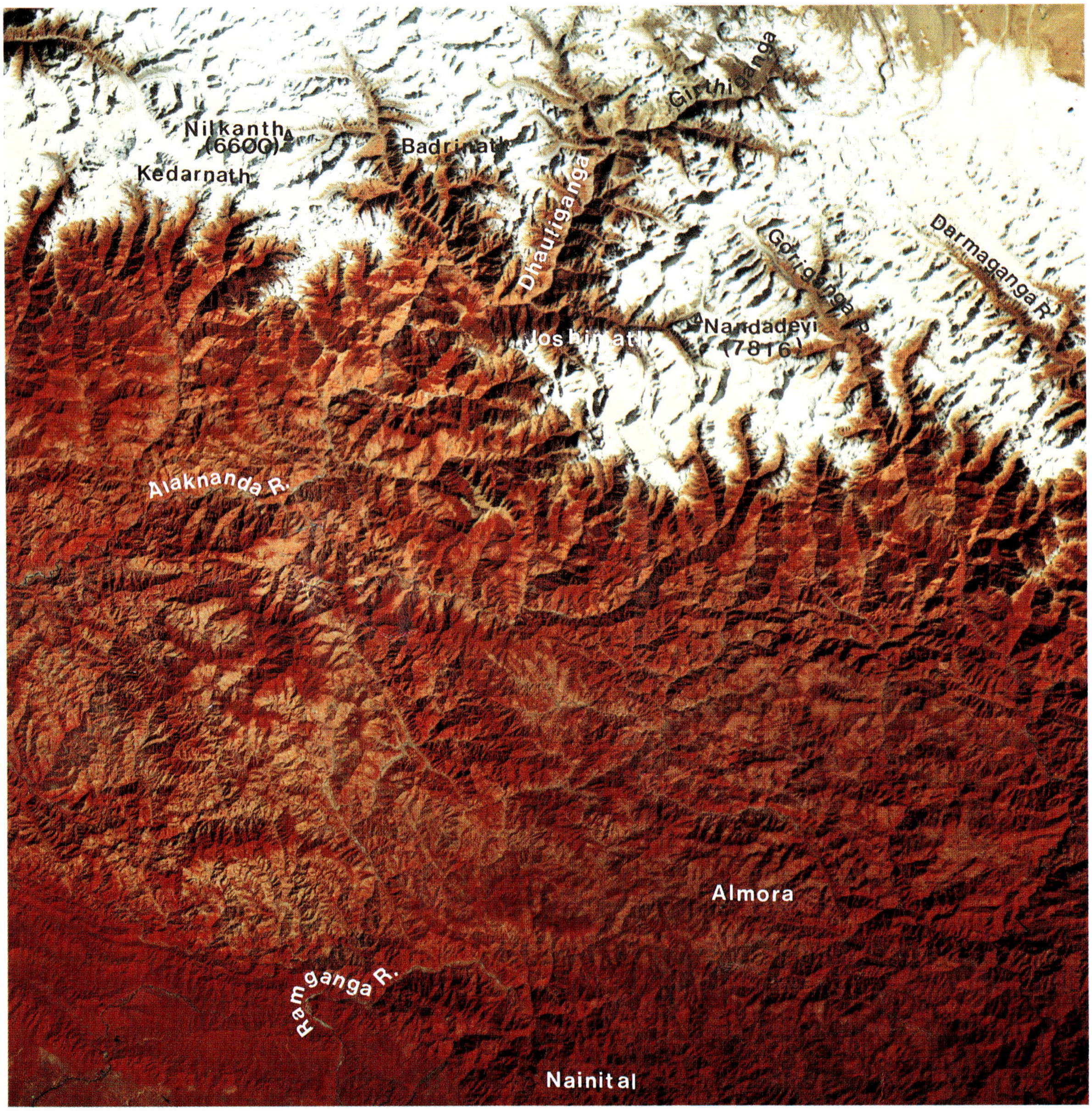

Figure 1. Landsat satellite view of the research area of Higher Central Himalaya (Scale 1:1000,000)

Geology of the Higher Central Himalaya

ANSHU K. SINHA

Wadia Institute of Himalayan Geology
Dehra Dun, India

A Wiley–Interscience Publication

JOHN WILEY & SONS
Chichester · New York · Brisbane · Toronto · Singapore

Copyright © 1989 by John Wiley & Sons Ltd.

All rights reserved.

No part of this book may be reproduced by any means, or transmitted, or translated into a machine language without the written permission of the publisher.

Library of Congress Cataloging-in-Publication Data:
Sinha, Anshu K.
Geology of the higher central Himalaya.

'A Wiley–Interscience publication.'
Bibliography: p.
Includes index.
1. Geology—Himalaya Mountains. I. Title.
QE319.H5S56 1989 555.4 86-9169

ISBN 0 471 91122 4

British Library Cataloguing in Publication Data:
Sinha, Anshu K.
Geology of the Higher Central Himalaya.
1. Geology—Himalaya Mountains
I. Title
555.4 QE319.H5

ISBN 0 471 91122 4

Typeset by Acorn Bookwork, Salisbury, Wiltshire
Printed and bound in Great Britain by Anchor Press, Tiptree, Essex

DEDICATED TO
MY MOTHER
SHRIMATI SAROJINI DEVI SINGH
FOR
HER PROFOUND AFFECTION ENDOWED
ON ME

Contents

Key: ✉ enclosed in slip case

Preface

I was fascinated to study the mysteries of Himalayan mountains after my first visit to Shimla (formerly spelled 'Simla') as a student of geology. After my basic field training in the Caucasus Mountains, I was sent to Himalaya for further research work under the auspices of the Indo-Soviet scientific exchange programme between Friendship University, Moscow, and Panjab University, Chandigarh. This opportunity allowed me to work in the Shimla-Kulu area of Himachal Pradesh in the Lesser Himalayan zone. After joining the Wadia Institute of Himalayan Geology in 1973 I was entrusted with the task of exploring the unknown regions of the Higher Central Himalaya lying behind the lofty high peaks of Nandadevi (7820 m). These inaccessible terrains were used as routes for Indo-Tibetan tradesmen. Some scanty data were available due to the efforts of earlier stalwarts working at the end of the last century and to the traverses taken by late Professor Arnold Heim and Professor Augusto Gansser from Zürich University, Switzerland. The systematic geological surveying work was started in early seventies by the geologists of Wadia Institute, and I joined the expedition team in 1973. Every year from 1973 to 1981 the expeditions were organized and the task was taken up to cover the *terra incognita* in parts. The traverses were always through some passes of over 5000 to 6000 metres and they were not always negotiable, even in the summer time, due to inclemency of the severe weather conditions. A couple of times the expedition team has experienced narrow escapes in snow storm and blizzard.

This monograph is an outgrowth of my research in the Southern Lesser and inaccessible Higher Central Zone of the Himalayan mountain chain. The book comprises ten chapters and a bibliography. Keeping in my mind the expeditional work, some of the factual data, especially incorporating those long traverses and their systematic sketching in the field, I have also tried to give a first hand account dealing with not only the geology but also the other interesting episodes, viz. climatic hazards, negotiability of the terrain, ethnic groups living in this region and human logistic problems. There had occurred some incidents of landslides and heavy downpours when I walked throughout the night covering more than 60 km between Chamoli to Joshimath to deliver important messages for the expedition party. Chapter 3 includes fascinating discoveries made with the help of my colleagues. Coccolith, Dinoflagellates, Scoelecodont, Chitinozoan and other microorganisms changed the basic concept of geological columns from this classical area of Phanerozoic biostratigraphic research. This area was treated as paradise for study of classical stratigraphic sequence, and their correlation throughout the world; especially the 'Triassic of Himalaya' by Karl Diener has attracted attention all over the world for correlation. This chapter has also added to the new geological data available from Southern Lesser Himalaya. Chapter 4 is devoted to the tectonic framework of the Higher Zone, discussing the new concept of division of the Himalaya on the basis of Main Axial Zone and Root-Zone concepts. The long traverses across Higher rugged region and through gorges have been described in Chapter 5. I have tried to give a real and authentic picture of geology in the field. Problems related to structural zonation and synthesis of geological development have been highlighted in Chapter 6. In Chapter 7 the metamorphic and magmatic episodes have been synthesized with geochemical analyses and geochronological dating. Many friendly laboratories besides our own at WIHG extended support for doing geochemical analyses. I have tried to keep those analyses tagged up with geochronological dating and their outcrops of occurrence with the help of diagrams and profiles attached to the book. Sometimes the original field number has been maintained. I hope that in future these valuable data could be utilized by the geologists working in Himalaya for their detailed analytical work and preparation of various models and diagrams. During the course of the expedition in 1973, for the first time the occurrence of barite and polymetallic mineralization was discovered. Successively in 1974, the whole mineralized area was studied for prospecting and exploration. Although the area is situated in a difficult rugged terrain cut into deep gorges by the rapids and mountain rivers, the economic viability of these barite–polymetallic mineralizations should not be underestimated. Rather, barite deposits were found to be economic for commercial exploration.

My synopsis of the geological data presented in this monograph should not be taken as a final statement since every day some new facts are coming to light. Rather, attempts in this monograph should be taken as a basis to prepare researches in the higher, inaccessible, zones of the mighty Himalaya. Naturally, there may be many shortcomings which I would have been able to avoid. The book is a humble attempt to present a synthesis of new data from the Higher Himalayan regions adjoining the Tibetan plateau. I would not hesitate to offer my apologies to the reader for the lapses and omissions inadvertently incorporated into this book due to my ignorance and limited knowledge. I shall always look forward to suggestions for improving this monograph. But I shall feel enlightened and satisfied if the purpose of researchers and students of Himalayan geology is served through this humble work.

Acknowledgements

The initiation of research work in Higher region of Central Himalaya came through the late Professor A. G. Jhingran after I joined the Wadia Institute in 1973. During the course of my expeditions I have received help in many ways from Professor K. S. Valdiya (Kumaun University, Nainital), Professor S. K. Shah (Jammu University, Jammu), Professor S. K. Tandon (Delhi University, Delhi), and Dr V. C. Thakur, Dr A. C. Nanda and the late Dr A. K. Khanna, as my former and present colleagues and members, in the Wadia Institute of Himalayan Geology. Besides them, Dr V. K. Gairola (B.H.U., Varanasi), Dr S. Kumar (Lucknow University, Lucknow) and Dr S. P. Jain (Panjab University, Chandigarh) took part in our other joint expeditions. The Ministry of Defence and the authorities deployed by the Government of India, in these regions, including the Indo-Tibetan Border Police, State Police Department of U.P. and the Civil Authority and other agencies have extended their help and logistic support; without their cooperation it would not have been possible to map and work in this *terra incognita*.

The Survey of India was very generous in providing the genuine help in map preparation and its clearance for the academic publication. For the geochemical analysis I am grateful to the Laboratories of Geological Institute, Institute of Geology of Ore Deposits, Petrography, Mineralogy, and Geochemistry; Academy of Sciences of USSR, Moscow, and the Geological Institute of the Armenian Academy of Sciences, Erevan. Professor G. D. Ashgirei, Head of the Department of Geology in Friendship University, Moscow, has been helpful to me as my guide and teacher throughout, for the researches in the Himalaya. Without his guidance and invitation for DSc thesis my researches would not have culminated in the present form. During my studies in the Department of Dynamic Geology in Moscow State University I had an opportunity to be in touch with the late Professor G. P. Gorshkov and Professor V. E Khain, corresponding member of the Academy of Sciences of USSR. Academician A. L. Yanshin, present Vice-President of the Academy of Sciences, USSR, inspired and helped in many ways to prepare this monograph. I am grateful to my Soviet colleagues especially to Professor Y. G. Safonov, Professor N. N. Trofimov, Professor G. P. Bagdasarian, Dr Y. Choporov, Dr (Mrs) O. B. Dmitrenko, Dr (Mrs) M. E. Raaben, Dr (Mme) Nina Umnova and many others who helped during the course of my preparation of this monograph by supporting laboratory data for palynological identification and doing photographic work. Dr Gerhard Fuchs (Austrian Geological Survey, Vienna) gave me consistent encouragement and helped in supplying geological analyses of barite specimens and has also kindly reviewed my book for the publisher, John Wiley & Sons, on their request. This is a very thoughtful gesture from his side. I express my thanks to Professor Augusto Gansser (Zürich, Switzerland), Professor Brian Windley, Dr Mike Searle (Leicester University, UK), Professor Peter Molnar (MIT, Cambridge, USA); Professor Mike Coward (Imperial College, London, UK); Professor Michel Colchen (Poitier University, France); Dr Patrick LeFort (CNRS, Nancy, France); Dr (Mme) Monique Fort (Paris University, France), and many other friends who helped me in this endeavour.

In WIHG, analytical work was carried out by Drs B. R. Yadav, P. P. Khanna and M. S. Rathi. During the preparation of Mss I have had consistent support from Professor S. P. Nautiyal, Professor I. C. Pande, Shri S. N. Talukdar, Dr S. C. D. Sah (then Director of WIHG) and my colleagues in the Structure & Tectonics Groups of WIHG particularly to Drs K. S. Bist, and S. K. Paul. Thanks are due to Shri G. S. Khattri who prepared the maps and diagrams. Shri D. P Chowdhury, Shri P. P. Dhasmana, and Shri Tapan Banerjee have ably typed the Mss.

Finally, I wish to express my thanks to my publisher John Wiley & Sons for their cooperation throughout the production of the book and particularly to Mrs Helen Bailey, Mme Irene Cooper and their colleagues for their editorial advice and guidance, and the art department for excellent work to present the figures and diagrams.

I believe that it would not be an exaggeration in expressing gratitude to my wife, Dr (Shrimati) Meenakshi Sinha, who in spite of her medical professional career, took all the family responsibilities to help spare me with valuable time for the completion of this endeavour.

Anshu K. Sinha

Dehra Dun
India
28 August 1987

CHAPTER 1

Introduction

1.1 Cultural and Historical Background

He who regards,
with an eye that is equal
Friend and Comrade,
the Foe and the Kin man.
The Vice, the Wicked,
the Men who judge himself
and those who belong
to neither faction.
He is the greatest.

—*Bhagwad Gita*
(Ancient epic of India)

The mighty and magnificent Himalaya, by humbling those that come to them with their magnitude and power, gratifying them with their splendour and beauty, testing their manhood with glaciers and peaks, defying their spirit with their inviolate mysteries, enables man to acquire the qualities which only danger and nature in the raw can sharpen. The might of Himalaya has inspired and evoked the awe and wonder of the people of the Indian subcontinent from time immemorial. The central part of the Great Higher Himalaya, which forms the subject of the geological description in this monograph, comprises the northern region of Garhwal–Kumaun of the state of Uttar Pradesh in the Indian subcontinent (Figures 1, 2 and 3). The name of 'Kumaun' is derived from Kurmanchal (*Kurma* = Tortoise, *Anchal* = Mountain) as described in *Skanda Purana* (the Ancient epic of the Aryans). Lord *Vishnu* had assumed the incarnation of tortoise at Champawat in the Kali valley. It has been cited in the ancient literature that this region of Garhwal–Kumaun in the Central Himalaya bears the name of Uttarakhand (*Uttar* = North, *Khand* = regions). The Aryans had great reverence for this part of Himalaya and every river, glacier, and high mountain peak of this region was sanctified. The sacred shrines of Badrinath, Kedarnath, Gangotri, Yamunotri, Nanda Devi, and further north Kailash–Manasarovar are the most notable throne of Gods for Hindus. *Ved Vyas* attached so much importance to this region of Himalaya that he wrote *Up Purana* on this part of Himalaya. It has been described in *Puranas* that 'There is a region of *Swargya* (Heaven), the seats of the righteous where the wicked did not arrive even after a thousand births. There is not sorrow, nor weariness, nor anxiety, nor hunger, nor apprehension, the inhabitants are exempt from all infirmity and pain, and live in uninterrupted enjoyment. The goddess never sends rain upon them, yet the earth abounds with water. In this region there is no distinction or succession of ages, and time is no more'. The great poet *Kalidas* in his poem '*Purva Megha*' had been inspired by the natural romantic area of Himalaya and imagines that the clouds rest in Himalaya.

In ancient times the notable Himalayan kingdoms which flourished in this region were *Dwigarta*, *Trigarta*, and *Madra*. The kingdom of Brahmpura and Shrugna, in the southeast and northeast of Alaknanda, respectively, and kingdoms of *Badawra* and *Govisana* were confined in the south of Brahmapura in the terai region as described by Huen Tsang, a Chinese traveller (Cunningham, 1963). Although there is no archaeological finding to substantiate the extension of the Mauryan and Gupta empires to this

Figure 2. Panoramic view of Higher Central Himalayan Panchuli group of peaks (heights from left to right 6904 m, 6312 m, 6334 m, 6437 m and 6072 m).

part of the interior Himalaya, scripture has been found on a quartzite boulder of the Jaunsar formation at Kalsi on the left bank of the Yamuna (50 km northwest of Dehra Dun). It is clear that the emperor Ashoka in the third century BC extended his empire up to this region. During the medieval period of the Muslim invasions, a large number of Rajput princes and Brahmin priests had fled to the valleys of Himalaya. In the middle of the 16th century, Rana Balakalyan Chand of the Chand dynasty established his empire in Almora. In the same century, Raja Ajaipal took the incentive to integrate the scattered fifty-two fortresses (*Garh*) and their territories into Garhwal (fort/integrated territory). The races contributing to the culture were notably Kiratash, Khas, Scythian, and Mongoloid.

Further north, the regions known as Bhot Mahals, Painkhanda, and Johar were mostly inhabited by the Bhotiya tribe (Figure 4). This region with Tibet formed one of the ancient culture groups known as *Shaukas*, who were perhaps the descendants of the *Sakas* inhabiting Central Asia (Pande, 1962). The Bhotiyas today usually write their name with their race and the village to which they belong *viz*., Rana, Singh, etc., recalling their belonging to the Rajput clan of Rajasthan. The Bhotiyas are found in every part of the country along Indo-Tibetan border that is sacred to the Hindus. The notable centres of Bhot culture are situated in Mana in the Krishanganga valley, Niti in the Dhauliganga valley, Johar in the Goriganga valley, Darma in the Dharmaganga valley, and Byans in the Kali river valley. From the Tibetan side there had been influx of Lamas and Huniyas known as the Khampas settling in the higher valley (Pant, 1935). The trade route of the Bhotiyas from the Indian side and the Khampas from the Tibetan side were through the passes of Mana (5611 m), Niti (5068 m), Shalshal (4496 m), Kungribingri (5578 m), Lamphya Dhura (5532 m), and Lipulekh (5453 m) (Figure 4).

In 1906, C. A. Sherring, then Deputy Commissioner of Almora, gave a vivid account of the Bhotiyas in his book *Western Tibet and British Borderland*. The passages described in the *Kumaun Himalayas* by M. S. Randhawa (1970, p. 123–125) reads, with modification: '*Bhot*, or more correctly *Bod*, is really the same word as

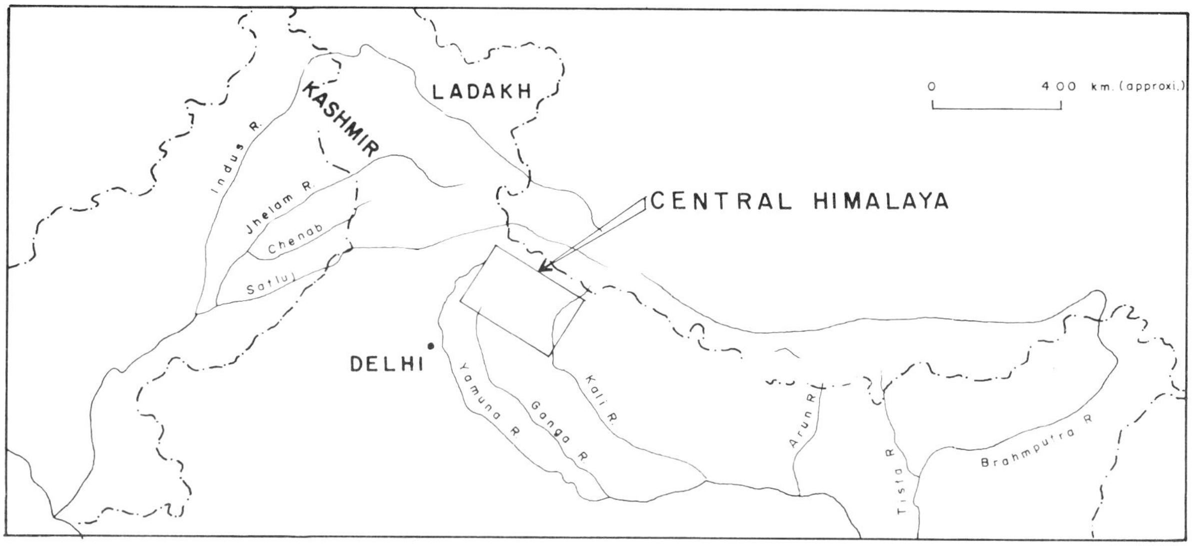

Figure 3. Location map of Central Himalaya.

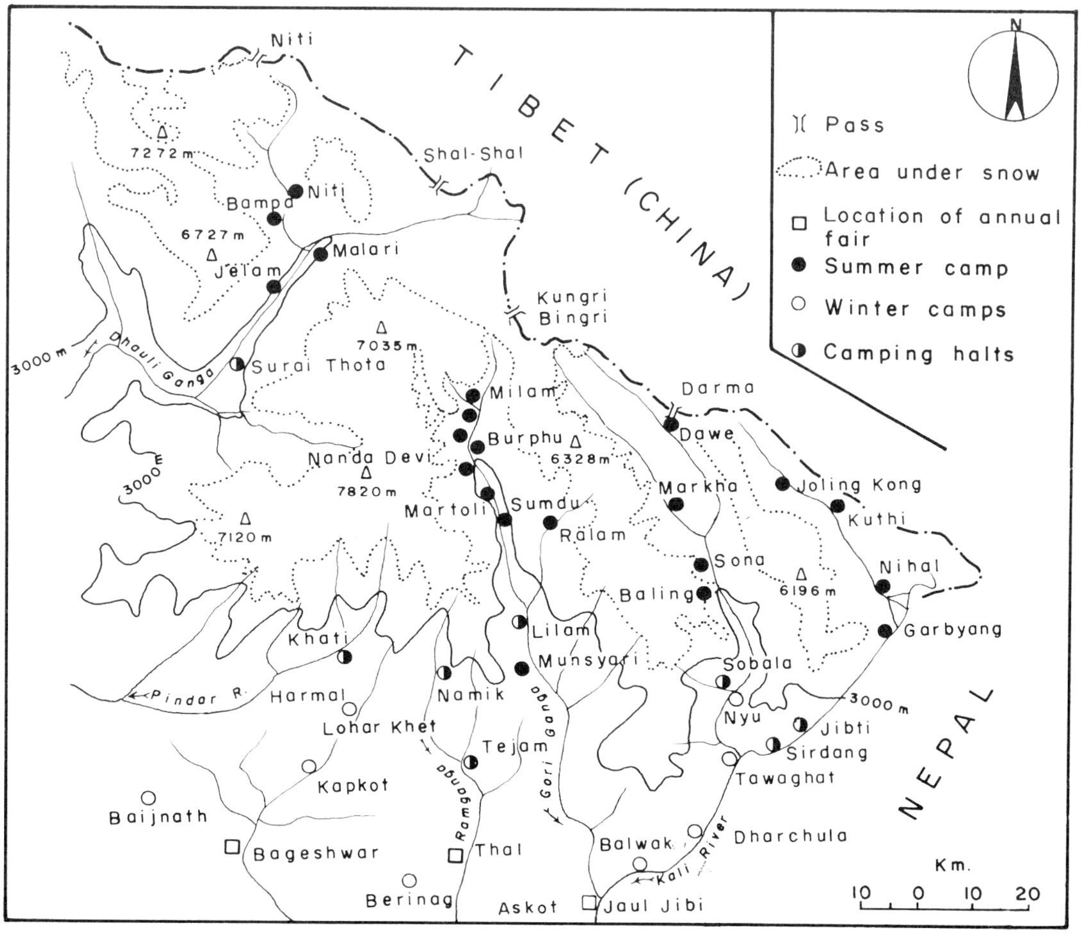

Figure 4. Map showing the inhabitant nature of local population and their seasonal migration (modified after Singh, 1971)

Tibet. The Bhotiyas of our hills are found in that very tract of the country which has been described as being so very sacred to the Hindus. They are to be found at the mouths of all the passes into Tibet. Those by the Mana Pass, near the holy temple of Badrinath, and those by the Niti Pass are known as *Tolchas* and *Marchas*, whereas those by the Unta Dhura Pass (5450 m) in Johar are *Shokas* (otherwise called *Rawats*). Amongst them there are also some Tolchas and Marchas. All these comprise the western division of Bhotiyas, who are in their own estimation superior to all other Bhotiyas, with whom they will not eat or marry, and whose ordinary language they cannot talk, as they have forgotten it. South of the Johar, Bhotiyas are the Jethora Bhotiyas, who do not trade but are cultivators. These have a Tibeto-Burmese dialect of their own (called Rankas or Shokiya and Khun) and consider themselves the first settlers (*Jeth* means elder brother). Lastly there are the eastern Bhotiyas living in the valley of the Darma Ganga, which is subdivided into three patties or subdivisions, *viz.* the Darma, Chandas, and Byans patties. These are much more backward than the others, and their customs are therefore all the more interesting. They are looked down upon by the other Bhotiyas because of their primitive ways, and although all Bhotiyas called themselves Hindus, the acknowledgement of the eastern Bhotiyas Hinduism extends to little more than the name, whereas the eastern Bhotiyas or Jethora, have made some progress in Brahmancial rites and customs. All Bhotiyas everywhere are divided into two castes, *viz.* Rajputs or upper class or Dumras, or low caste'.

During our expedition it was observed that the Bhotiyas in the Darma and Byans valleys are a more romantic people. The Bhotiyas are only Kumaunies who drink alcoholic liquor, brewing it from barley in their homes. 'Rambang' are social gatherings, particularly for all young people to contract marriages. The men and women enjoyed drinking, dancing, and singing at the *Rambang* festival, practising free love and giving an opportunity to the women to select their life-companion. The Bhotiyas are a quite hard-working people but are more superstitious.

Since independence, these tribes have been given more facilities and they have been classified under the schedule tribes categories with the result there has been much improvement in living standards, education, and agricultural facilities, and there is a modern postal system, even in the interior part of the region. As a personal experience, everywhere I found them most sincere and hard working even in the inaccessible and dangerous locations of the expedition.

J. H. Batten, of the British Civil Service (BCS) who retired as Commissioner of Kumaun in 1877 records: 'Kumaun was conquered by the British from the Gurkhas in 1815 and the Honourable Edward Gardner was subsequently promoted to the Residency of Nepal was the first British ruler of Kumaun. Mr. G. W. Traill (after whom the Traill Pass of the Pindari glacier is named) who was then an Assistant under the Resident of Delhi, was first offered the appointment of Commissioner of Kumaun'. Since then the political control and administrative reforms of Kumaun and Garhwal were subject to the British rule until independence. Uttarakhand together with the twin Garhwal and Kumauni cultures are at present represented by eight hill districts, *viz.* Chamoli, Uttarkashi, Pithoragarh, Naini Tal, Almora, Pauri, Tehri and Dehra Dun. The area of this monograph falls in the northern part of the first three districts adjoining the Tibetan border in the north and eastern border with Nepal across Kali River (Figure 4).

1.2 Orography

The Great Himalayan range, forming an enormous curvilinear arch over a length of 2500 km between the eastern and western extremities of Burma and Pakistan (Figure 3), is highly rugged with horned peaks, serrated crests of ridges, cirques, and hanging glacial valleys, and cascades of gushing icy water through deep canyons or dammed as lakes.

The area with which this monograph deals is within the Higher and Tibetan Zone of Central Himalaya, constituting northern Uttarakhand of Garhwal and Kumaun. It falls within the longitudes 79°5′ to 81°0′ and latitudes 30°0′ to 31°1′, excluding the territories across the Tibetan and Nepalese borders, and encompassing an area of approximately 20,000 km² (Figure 3). Geologically this is the area lying north of the Main Central Thrust (MCT), which pass through Helang–Loharkhet–Dharchula, limited in the north by the Tibetan watershed and in the east by the Kali river boundary with Nepal. According to Griesbach (1891), the Central Himalayan region consists of (a) The Southern range (line of great peaks); and (b) The Northern range (line of the watershed) which is rimmed by the Tibetan plateau in the north and the lower Himalaya in the south. Burrard and Hayden (1934) subdivided the Himalaya into three geological zones: (1) *Outer sub-Himalayan zone* composed of sediments for the most part of Tertiary age; (2) *Central* or *Himalayan zone* comprising the ranges known as the lesser Himalaya, a line of high peaks; and (3) *Northern* or *Tibetan zone* lying behind the axis of Great Himalayan range and composed of strata from the Cambrian to the Tertiary. Later, in their classic monograph on Central Himalaya, the late Arnold Heim and Professor Augusto Gansser (1938) called the studied region of the Central Himalayan zone the *High Himalaya* dividing it into (1) The Central High Range (Nampa–Nanda Devi–Badrinath: 15–30 km; and (2) The Northern Ranges or Tethys–Himalaya (20–35 km). Thus, in the present context we shall be dealing with the Central Himalaya in general.

The magnificent group of peaks confined in the central high ranges are the highest peaks of India: Nanda Devi (7820 m), Nanda Devi East (7439 m), Kamet (7758 m), Mana (7274 m), Chaukhambha (7140 m), Trisul (7122 m), Dunagiri (7068 m), Deoban (6979 m), Kalanka (6933 m), Panch Chuli (6905 m) (Figure 2), Changbang (6866 m), Nanda Kot (6863 m), Mrigathuni (6857 m), Devisthan (6680 m), Nanda Khat (6674 m), Penwali Dhar (6665 m), Nilkanth (6597 m), Nanda Ghunti (6314 m), and others (Figures 2 and 4).

1.2.1 Drainage

The above mentioned peaks are separated by the main river valleys of the Alaknanda, Dhauli Ganga, Girthi Ganga, Pindari, Gori Ganga, Darma Ganga, and Kali. The profiles of these valleys are convex with steep valley walls and terraces at different levels indicating the uplifting phases of Himalaya (Figure 1).

Parallel to the southern snowy ranges with the peaks mentioned earlier for the High Central Himalayan ranges, run the northern ranges, sometimes connected by ridges forming the watershed or by broad valleys of great height, over three to four thousand metres. These form a water 'divide' between the drainage systems of the Ganges and the Indus or Sutlej. They are mostly snow free except in the upper reaches of Lissar and Byans and all the passes leading to Tibet are situated on this high land ridge. In the Indian part of this region we have only the drainage pattern of Ganga. The Ganga headwaters rise in the Nanda Devi, Kamet, Badrinath, and Kedarnath group and consisting of two great branches: Alaknanda and Bhagirathi (west of the described area). The Alaknanda is formed by the headwaters of the Dhauli and the Vishnu which are separated by the Kamet peaks. Amongst the prominent streams that join the right bank of the Kali on its Indian side are: Kuti Yangti, Darma (+ Lissar) Ganga and Gori Ganga (+ Gonkha Gad from the Unta Dhura Pass, and the Bamlas and Safed Glaciers). These rivers come down the steep gradient with great volume and strong current. Recently, in 1977, torrential rain caused devastating floods and landslide, which created havoc and led to the death of a complete unit of border guards along with their supply dump and animals at the confluence of Darma and Kali Rivers near the hamlet Tawaghat.

1.2.2 Glaciers

The central higher zone of Garhwal and Kumaun is occupied by a vast area of glaciers and glacial topography (Figure 4). The whole glaciated area could be divided into five groups: (1) Badrinath–Satopath group; (2) Kamet East and North group; (3) Dhauli Ganga Group; (4) Nanda Devi–Pindari–Milam Group; (5) Kala Baland Panch Chulhi–Nama Group. From the earlier works on the Central Himalayan glaciers, reports worth referring to for the details are Strachey (1857), Grinlinton (1912, 1914), and Raina (1963). The Himalayan glaciers provide many features of mountain glaciation which are different to the sheet glaciation features of Greenland and Antarctica. The entire water supply for irrigation and hydroelectric projects related to the northern Indian rivers originating in the Himalaya is dependent not only on the rainfall but on the snowfall and glacial thaw. The data from the Himalayan glaciers for the last seventy years indicates a continuing recession of glacier termini which is a pointer of the negative glacier balances. The Milam glacier between 1849–1957 (108 yr) retreated for 1350 m, whereas the Pindari glacier between 1847–1966 has shown a recession of 2840 m in 121 years. The southern slopes of the Great Himalayan Range and the Nanda Devi–Badrinath group have a snow line of approximately 1200 m lower than the near zone of the inner Great Range and the Kamet group of peaks. U-shaped valleys, truncated spurs with sawn-off faces, aretes, horns, pyramidal and conical peaks, serrated crests of ridges, cirques, glacial troughs, cols, knife-edged precipices, smooth rock walls, steep head-walls, glacial lakes, etc., are the prominent erosional features. The depositional features are represented by talus cones, snow-avalanches, snow-bridges, dead ice, debris deposited by terminal, lateral, medial, group- and surface cover-moraines. The Himalayan glaciers occupy both antecedent and consequent troughs. Alaknanda, Dhauli, Gori and Darma for example are antecedent glacial drainage surviving mountain uplift. During the Pleistocene, these glaciers extended down the valleys, whereas in recent times they have retreated due to ablation. The glacial lakes are formed by the damming of valleys by moraines, which are the remnants of ancient large glaciers, *viz.* Satopath Tal below Chaukhambha, Hemkund in the Bhyundhar valley, Rupkund on the outer slopes of Nanda Ghunti, Pari Tal below the Unta Dhura Pass near the snout of Saphed Glacier, and the Joling Kong lake between Bedang and Kuti village on the watershed between Darma and Kuthi Yangti.

1.2.3 Climate

The southern part of the area of the Main Central Thrust, consisting of crystalline metamorphic rocks and some granite and basic magmatic rocks, experiences heavy rain (37–50 cm) during the monsoon season commencing by the end of June and continuing until the middle of September (Figure 5). The pioneer explorers of the nineteenth century, Hughes and Waagen (1878), record that they took traverses early in August 1873 along the track from Almora to Milam and came back via the Niti Pass. They wrote: 'We were in almost constant rain during our first ten marches, and, however enthusiastic our aspirations in the cause of geology at the outset of each day's journey were, they could not withstand the depressing effect of recurring downpour of water and we hurried to shelter as the most prudent course to pursue. It was not until we reached the Bhotia village of Rilkot, that we passed out of the regime of rain'. And this Rilkot is situated just in the shadow zone. The high rampart of Nanda Devi and its adjoining extension towards the east and west functions as a blocking wall for the profuse and copious clouds of the southwest monsoon, causing a 'shadow zone' behind it (rainfall approximately 15 cm). Thus in the Tethyan zone of Central Himalaya there is no heavy downpour, causing Tibet to be a high altitude desert. The snowfall is unpredictable in the 'shadow zone'. During our 1978 expedition in the Unta Dhura Pass area, the party was stranded for four days between 17 to 21 August at Dung due to continuous heavy snowfall. The area in the 'shadow zone' is usually open from the end of May until the middle of September. Passes like Unta Dhura have perpetual snow specially on the northern slopes. However, the fresh wet snowfalls with gusting stormy winds are more dangerous for the explorers in this area. Usually after 10 am the high passes experience speeding wind from the Tibetan side making it impossible even to stand still without any support. Another hazard is the ultraviolet radiation causing severe sunburn. Worst comes to face the explorer when there is unprecedented snowfall washing away the foot-trail made on the extremely steep slopes of hard rocks and debris of rock fall. In such conditions communication is completely cut off and there exist no supply lines. The ponies carrying the supplies start biting each other's backs and tails in the absence of green grazing pastures. In searching for a trail path, trekking parties have been washed away by the sudden gushing of glaciers, avalanche, and sludge into deep gorge of rivers. Our experience during August 1978 between Unta Dhura-Dung and Milam in the Gonkha Gad-Gori valley is to be taken as lesson for future work in this part of Himalaya.

1.3 Flora and Fauna

1.3.1 Flora

Natural vegetation in Himalaya is divided into four broad groups according to altitude: (1) Subtropical zone (below 1200 m); (2) Temperate zone (1200–1800 m); (3) Subalpine zone (1800–3000 m); and (4) Alpine zone (3000–4500 m). In the central high Himalayan zone, the last three zones are encountered. The temperate forests are generally dominated by Chir pine (*Pinus longifolia*) and Banj oak (*Quercus incana*) and bamboo shrubs. On the way to the Pindar glacier, the beautiful sparaeas of white and crimson colours with wild rose bloom everywhere between the Dhakuri and Khati camping sites.

In the southern fringe of the area crystalline rocks over the Main Central Thrust grows the

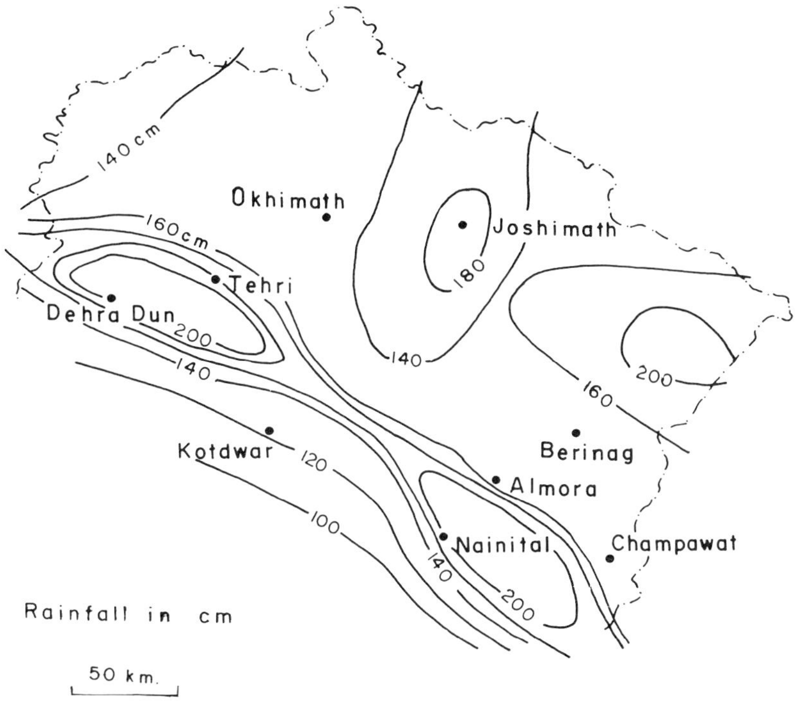

Figure 5. Annual rainfall map of Central Himalaya

Rudraksh tree (*Elaeocarpus ganitrus*). *Rudraksh* (literally meaning 'red-eyed' in *Sanskrit*) is the brick-red coloured dried berry. When the *rudraksha* are dried, the fissures on its hard surface vary from one to fourteen in number, and are known as 'faces' (*Mukha*). An *ek-mukhi* or single-faced is difficult to obtain while the most commonly available ones are the five-faced (*panch-mukhi*). From time immemorial the *rudraksha* have not only been venerated by the Hindus, with the berries being used for rosaries to aid both prayer and meditation, but have been valued for therapeutic effects, and as a gauge to differentiate between hygienic and unhygienic food. Nowadays it is being argued whether the beneficial effects of *rudraksha* in the cases of hypertension, high blood pressure, mental disorders, and coronary malfunctioning could be accepted. Occasionally some species of deciduous forests are observed. Moist conifer forest with deciduous trees are the dominant vegetation of the lower subalpine zone. Amongst the oaks are: *Banj* oak (*Quercus incana*, *Quercus himalayansis*), *Maru* oak (*Quercus dilatata*), and *Kharsu* oak (*Quercus semicaprifolia*). The other most common associations are the *Rhododendron arboreum* and *Abies webbiana* forests. The high alpine forests are mainly composed of silver fir (*Abies pindrow*), blue pine (*Pinus excelsa*), spruce (*Picea morinda*), Cypress (*Cupressus torulosa*), deodar (*Cedras deodar*), birch (*Betula utilis*), *Rhododendron campanulatum*, *Rhododendron lepidotum*, *Salix elegans* and dwarf bamboo (*Arundinaria falcate*). The high level birch forests are usually found in the slopes north of main great Himalayan lands, i.e. Tethyan sedimentary zones, usually between 3000–4000 m. The silver–grey stems and branches of birch are very attractive from a distance. The birch forests are sustained by snow in winter and by rainfall in summer emerging with profuse flowers in early June. The bark of birch, the '*Bhojpatra*', is like paper and was used for writing in the ancient India. In the pastures the shepherds were found to use it as packing paper. The whitish, violetish and light-coloured rhododendrons (*R. anthopogon*, *R. campanulatum* and *Salix elegans*) are found in the Pindar valley. Further up there is a gradual transition to Alpine pastures.

The meadows on the northern slopes of the Martoli and Malari area on the left bank of the Girthi Ganga and in the valley of flowers remain

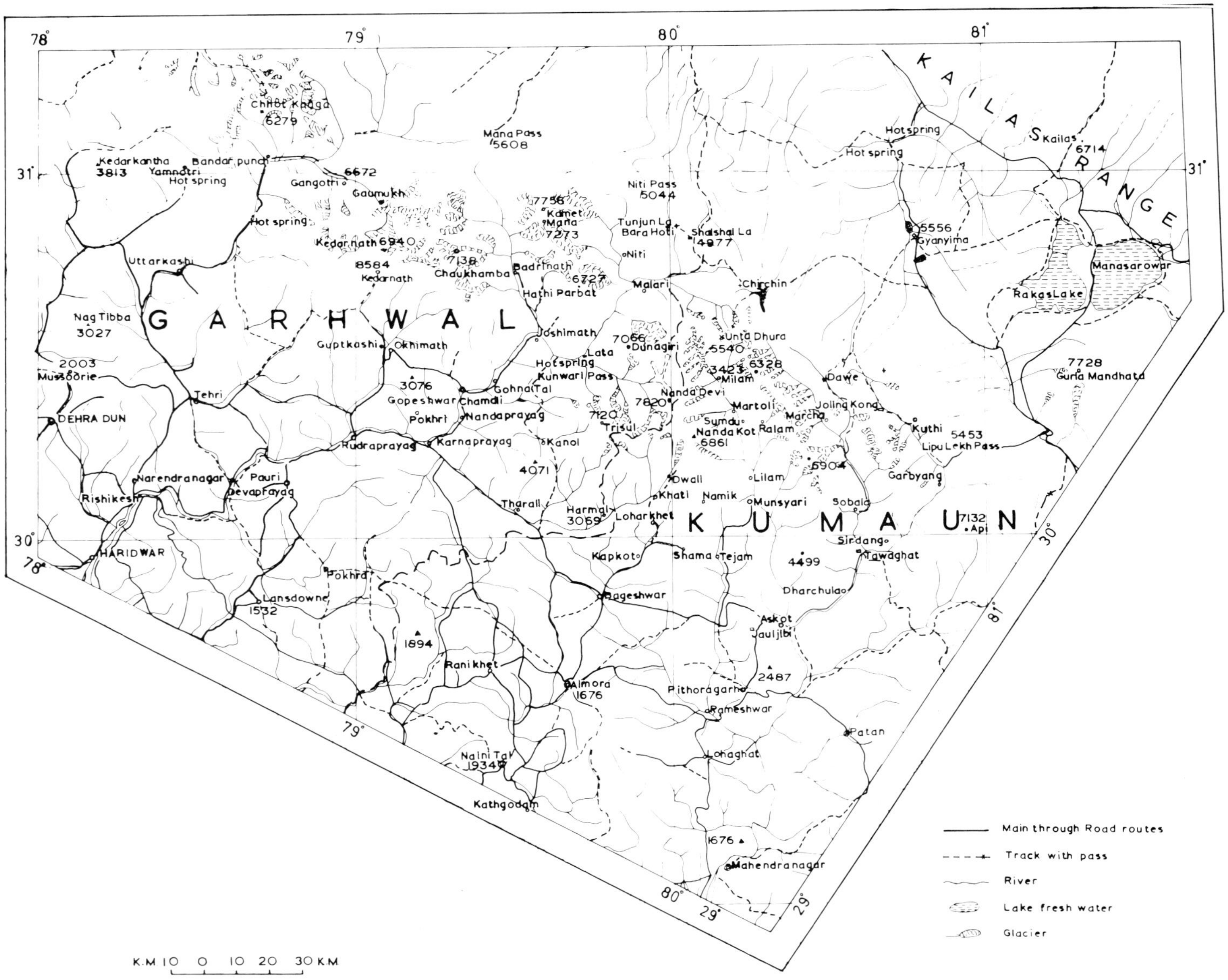

Figure 6. Drainage and communication map of Central Himalaya

covered with snow till the first week of May. With the advent of the melting of the snow, a colourful carpet of purple primulas, blue Corydalis, and yellow Gangea bloom and cover the meadows. In July–August appears the famous blue poppy (*Mecanopsis aculeata*), the queen of Himalayan Alpine meadow.

The valley of flowers situated near Badrinath was brought to the lime light in the world of botanical sciences by Frank Smythe in 1931. He camped in the valley for a month and collected material for the Botanical Gardens in Edinburgh, Scotland. John and Margaret Legge visited the valley in 1939 and, while camping, Margaret unfortunately slipped from a rock slope while picking flowers. The epitaph on her tombstone reads: 'I shall lift my eyes up to the Himalayas, from whence cometh my help'.

Some of the flowers in this eye-feasting valley are: creamy anemones, large purple asters, the rare white androsace, the blue borage, the rosy-petalled cypripedium, blue forgetmenots, green and chequered fritillaries, pink geraniums, purple and dwarf irises, dwarf larkspurs, the indigo-coloured nomocharis, blue and yellow pansies, mauve polemoniums, the blue poppy, white and red potentillas, dwarf rhododendrons, primulas in pink, blue, and deep purple, ranunculus and the *Brahma kamal*, etc.

1.3.2 Fauna

In general there has been adopted three faunal zones in the Himalaya. The wild animals in the Middle and Inner Himalayan zones found over 2000 m to 5500 m are to be referred here only. Notable amongst them are musk deer (*Moschus moschiferus*), barking deer (*Cervulus muntajae*), Himalayan black bear (*Ursus torquatus*), and the goat families, *viz. thar* (*Hemitragus jemlaicus*), *goral* (*Nemorhoedus goral*); and the sheep *bharal* (*Ovis nahura*). In the desert part of Tibetan zone the wild horse Kyang (*Equus hemionus Pallas*) are commonly met with watching the travellers. These pony-like horses are not suitable for load carriage because of their weak backbones. Yak (locally called *Jhaboo*) is the most common animal used for transportation. But their numbers are much less in the Indian part. Most commonly the local breed of ponies are used for transportation. Ponies used on the plains in the military farms have been found to be unsuitable for the rarified air at an altitude over 4500 m, and most of them succumb due to heart failure and shock. The goat and sheep families in the inner zone are nowadays found grazing on the high escarpment grounds, and they were profuse even at the trekking level, but due to indiscriminate poaching they are vanishing fast.

CHAPTER 2

History of Geological Investigations

Captain Herbert (1842), the then Deputy Surveyor General of India, was the first to attempt a systematic examination of the Himalaya, and he divided the Himalayan terrain into four different zones, right of the line of the watershed between the Indus and the Ganga and the low hills of Siwaliks.

The history of geological investigation of this belt adjoining the Tibetan part in the Garhwal Himalaya could be classified into two epochs: *viz.* the pre-independence and post-independence periods of India. During the pre-independence period in the middle of last century Captain Richard Strachey (1857) was the pioneer geologist to give the first account of this part of Indian Himalaya and adjoining Tibetan region. He, in his short paper published in the *Quarterly Journal of the Geological Society*, presented the result of many years of study in the Central Himalaya and the adjoining provinces of 'Hundes'. With the map, it forms by far the most important contribution to our knowledge of the geology of this part of Himalaya. In desending order he distinguished seven stratigraphic units:

(7) Tertiary deposits of Hundes,
(6) Grits, shales and limestones,
(5) Shales (Oxford, spiti shales),
(4) Jurassic series (limestone, etc.),
(3) Trias,
(2) Palaeozoic, and
(1) Azoic slates.

Edward Suess (1885) in his 'Letter on Triassic fossils collected by General Strachey' recognized true Alpine Trias ammonites from the Niti area. His collection of fossils from the adjoining Niti valley were subsequently described by Salter and Blanford (1865). Besides the papers mentioned above, Strachey contributed also for the study of the Pindari glacier (1857). The exhaustive work of the area was known to the world through the publications of C. L. Griesbach (1880, 1891, 1893), who first classified the rocks, recorded the occurrence of a more or less complete Precambrian–Palaeozoic–Mesozoic sequence, and produced broad and generalized geologic maps and profiles of the inaccessible part of Kumaun Himalaya, Sutlej, and Hundes in Tibet, and partly also of the Spiti valley.

Griesbach (1891, p. 2), recalls: 'I began surveying work in the Niti area in 1879, and continued my explorations amongst the snows during intervals when I was not engaged in travelling in the northwest frontier or in Afghanistan'. He was also the first worker to draw the attention to what came to be called subsequently the 'Exotic Block' of Malla Johar and recognized these Chitichun Limestone blocks as 'Klippen'. He recorded (1893, p. 22): 'So far the geological structure of the area northeast of Kungribingri appears simple enough and excepting considerable folding and subsequent crushing the sequence of strata is normal. Nevertheless this area presents features of structure of exceptional interest, features which, so far I am aware, have, up to the present, not been observed in India. There is nothing less than a series of *Klippen Zuge* or *blocs exotiques*, or as we may perhaps call them "detached blocks".'

Diener, in his series of publications (1895a, 1895b, 1895c, 1897a, 1898, 1903, 1906, 1912) added further details to the geology of Malla Johar explaining the 'Klippen' as due to faulting accompanied by intrusions of igneous rocks. He further elaborated his work on fauna collected from exotic blocks and the related formation of the area. His monumental work *The Trias of the Himalaya* (1912) has remained a classic to be used even at the present time and it has formed the backbone for International Geological Correlation Programme (IGCP) No. 4: *Triassic of the Tethys Realm*.

The *exotic blocks* were carefully studied in later years by A. von Kraft (1904). He concluded that they originated from a huge volcanic outburst on the Tibetan side, which was forbidden to travellers at that time. Subsequently Edward Suess (1885), on the basis of von Kraft's description, inferred that the phenomena were related to the great thrust. The fossils from the Ordovician and the Silurian were described by Reed (1912).

Auden (1935, 1937) introduced the term *Tethys Himalaya* for the fossiliferous sequence in the north of Central Crystalline axis, toeing the line of interpretation by Suess. An authoritative and monumental work on this region of the Central Himalaya was produced by the Swiss Expedition of 1936 led by late Professor Arnold Heim and Professor A. Gansser of ETH, Zürich. Heim (1937, 1938, 1956) published structural studies of the Central Himalaya. Their publications, a travelogue *Throne of Gods* (1938, 1939b) followed by a monograph *Central Himalaya* (1939a), mark the advent of systematic geological study of the regions. Their collection of fossils was identified by Jeannet and published much later in 1959. After their publication nothing new appeared until India achieved independence in 1947.

Post-independence work was initiated by the Geological Survey of India. Mehdi (1966) traversed 250 km passing through Dawe (30°27′: 80°29′) on the Darma Ganga and Lesser valley. Sarvashri Gopendra Kumar, S. H. Mehdi, and Gyan Prakash subsequently published their review of stratigraphy of parts of Uttar Pradesh Tethys Himalaya in 1972. The GSI organized expeditions for the furtherance of palaeontological investigations which resulted in the publications by Sastry *et al.* (1970), Sastry and Mamgain (1970, 1975, 1977), and Mamgain and Sastry (1975) giving an account of Cretaceous microflora and biostratigraphic data from Mesozoic and Palaeozoic sediments.

After the inception of Wadia Institute of Himalayan Geology in Delhi, a plan was drawn up to explore the geological knowledge of unknown and inaccessible regions of Himalayan terrain. With this idea, the multispecialization expeditions in geosciences started to be organized from its then base at Delhi University Campus. The result of the first expedition in the Kali valley was subsquently published by Valdiya and Gupta (1972), Valdiya *et al.* (1972), Powar (1972), and Banerjee (1974). Dave and Rawat (1968) attempted a geological report on the Girthi valley. Lithostratigraphy and trace fossils has been published by Kumar *et al.* (1977). Since 1973, after joining the scientific nucleus of WIHG, I started leading and participating in the expeditions of 1973, 1974, 1975, 1976, 1978, 1979 and 1980. As a prelude to the present monograph, my publications have been coming out from time to time covering the different aspects of geological problems from this inaccessible *terra incognita*. The regional geological, structural, tectonic, biostratigraphical, geochronological, mineralogical–petrological, and metallogenic problems have been partly covered and touched upon in the following publications. Shah and Sinha (1974a,b,c), Banerjee, Tandon, and Sinha (1975); Sinha and Jhingran (1977); Sinha (1977a,b,c,d); Sinha and Bagdasarian (1977); Ashgirei *et al.* (1976, 1977a,b, 1982); Fuchs and Sinha (1974, 1978); Sinha and Pal (1978); Mehrotra and Sinha (1978, 1981), Sinha and Srivastava (1978, 1979, 1986); Sharma *et al.* (1978); Sinha and Nanda (1979); Sinha and Raaben (1979); Sinha and R. Mehrotra (1979). Azmi and Sinha (1981); Bist and Sinha (1982); Negi *et al* (1982, 1983, 1984a,b); Sinha and Dmitrenko (1983); Khanna *et al.* (1985); Schwan *et al.* (1985); and Sinha (1972a,b; 1974a,b; 1975a,b; 1977a,b,c,d; 1978a,b; 1980a,b; 1981a,b; 1982; 1983; 1984; 1987).

CHAPTER 3

Outline of the Geology and Stratigraphy of the Central Himalayan Zones

To begin the discussion of the stratigraphic problems of higher part of Central Himalaya, it is necessary to recapitulate the salient features of the southern lesser zone. Besides a large number of contributions from different authors (Valdiya, 1983) over many decades, the author started contributing on the stratigraphic problem of the lesser Himalaya during his student and research career at the Friendship University, Moscow, and subsequently at the Wadia Institute of Himalayan Geology from 1973 (Sinha, 1972a,b; 1974a,b; 1975a,b; 1977a,b; 1978a,b,c; 1981, 1983). On the basis of the new data, fossil discoveries, and a detailed lithologic investigation of different structural–formational zones coupled with careful measurement of the sections, a new tectono-stratigraphic column emerged.

3.1 Southern Himalayan Zone

From south to north the southern part of Lesser Central Himalaya is divided into following tectonoformational zones: (1) Siwalik molasse zone; (2) Tertiary sedimentary zone; (3) terrigenous–carbonate Krol zone; (4) Parautochthonous flyschoidal Simla formation and terrigenous–carbonate Shali formation zone of Riphean age; (5) zone of early geosynclinal aspidic–phyllitic formation of Late Palaeozoic Chail thrust-sheet and the metamorphosed crystalline thrust-sheet of Jutogh (Sinha, 1980b, p. 102; Figure 7).

The foundation of geological studies in the southern zone of Himalaya started with the Simla area in 1864 by H. B. Medlicott. In his memoir (1864), he expressed his bafflement at the dipping of less metamorphosed rocks in all directions beneath the more metamorphosed rocks. The answer to this problem was given by Pilgrim and West (1928) in their historic memoir. In this work the nappe theory, earlier propounded in the Alps, was applied to the Himalayan region for the first time. This concept was further elaborated by West (1939) in his attempt to explain the Shali tectonic window. Subsequent confirmation of these views was provided by Auden (1934, 1970) whose studies embraced a large part of Garhwal and Kumaun of Central Himalaya. The recent works by Srikantia and R. P. Sharma (1976), Bhargava (1972, 1976, 1979, 1981), V. P. Sharma (1977), and Chaterjee and Swaminath (1976, 1977) attempt to tackle the problems in many different ways and a lot of reappraisal and rethinking have been done.

A large number of papers on different geological aspects of tectonics and stratigraphy have been published: such as Ray and Naha (1971), Bhattacharya and Niyogi (1971), Ravindra Kumar and Pande (1972), Rupke (1974), Valdiya (1970, 1973, 1980b), Fuchs (1967), Berthelsen (1968), Ashgirei *et al.* (1975, 1977), Prasad and Bhatia (1975), Tewari (1969), G. Kumar (1981), Fuchs and Sinha (1978), Sinha (1972a,b,c, 1974a,b, 1975a,b, 1976, 1977a,b, 1978a,b,c, 1980, 1981a,b, 1987), Ashgirei *et al.* (1975, 1976), Fuchs (1981), and many others. The tectonic interpretation has been attempted mainly on the basis of the inversion of metamorphism of the exposed rocks. Except for the dependable Tertiary fauna, no other formation has so far revealed

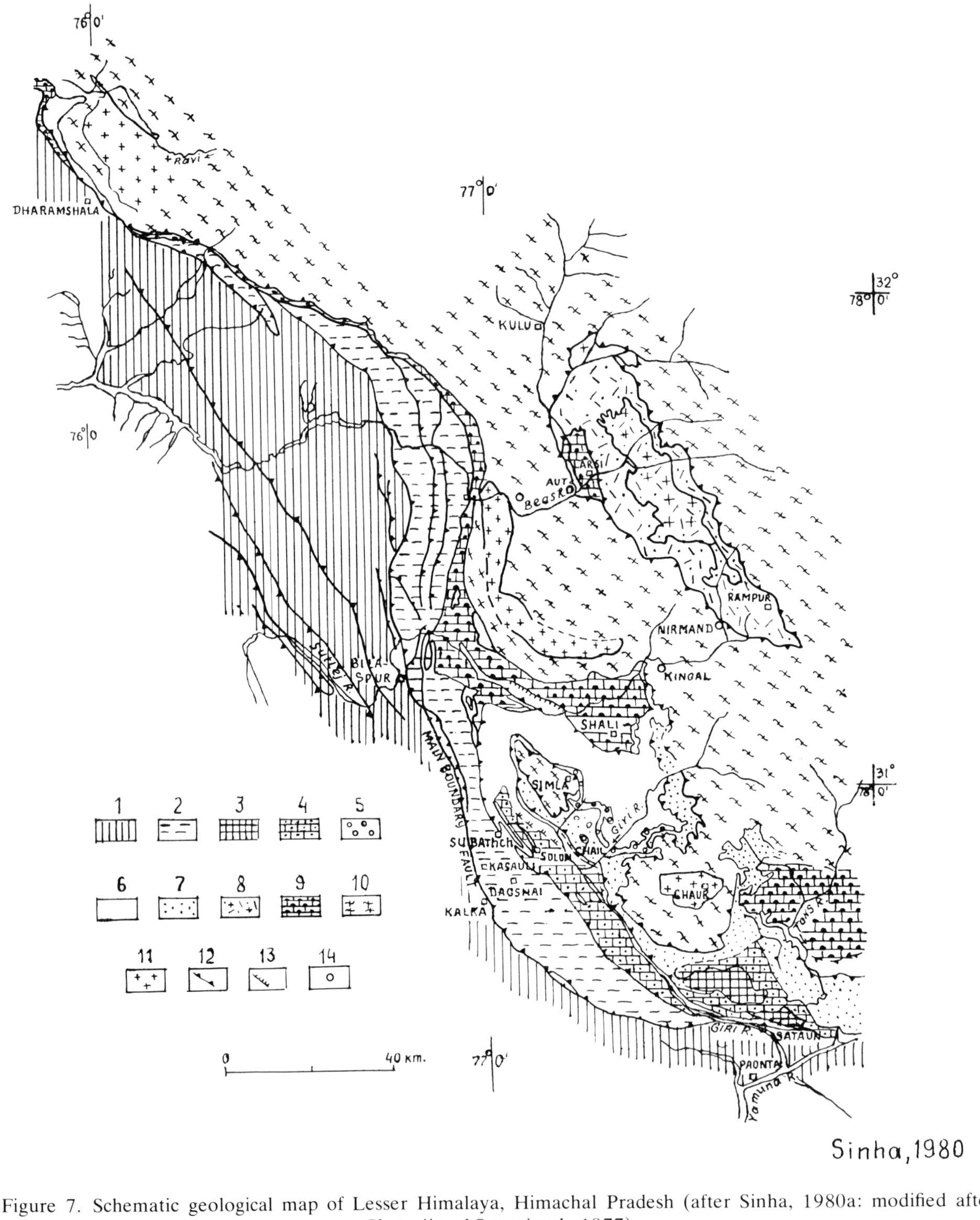

Figure 7. Schematic geological map of Lesser Himalaya, Himachal Pradesh (after Sinha, 1980a: modified after Chaterji and Swaminath, 1977).

Key: 1. Siwalik molasse; 2. Tertiary undifferentiated; 3. Tal; 4. Krol (Mesozoic); 5. Blaini (Permo-Carboniferous); 6. Parautochthon flyschoidal Simla (Upper Riphean Vendian); 7. Chail/Mandhali, Chandpur, Nagthat (Carboniferous–Devonian ?); 8. Porphyroids in quartzite/slate complex of the Rampur–Manikaran–Banjar formations; 9. Shali–Deoban limestone with basal quartzites and basics (Riphean); 10. Crystalline Jutogh and Chail phyllonite, chlorite schists, undifferentiated; 11. Intrusive granites; 12. Thrust; 13. Faults; 14. Fossil location from Chail unit

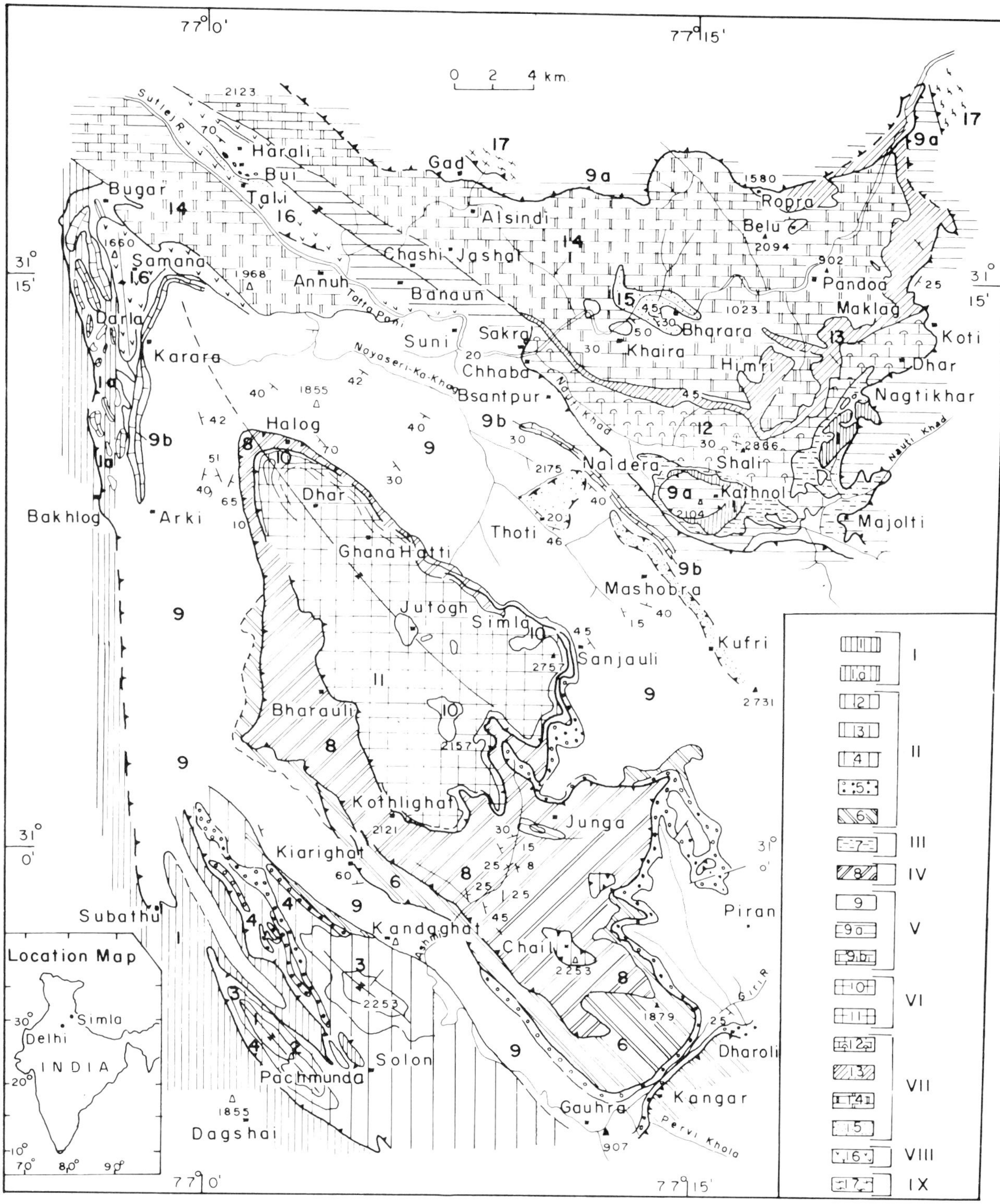

Figure 8. Geological map of Simla–Shali area, Himachal Pradesh, based on author's personal observation and the published work of Pilgrim and West (1928); West (1939); Srikantia and Bhargava (1967); and Srikantia and Sharma (1976). I. Tertiary (1. undifferentiated, 1a. Palaeocene); II. Krol thrust-sheet (2–3. Krol A, B, C, D; 4. Infra-krol; 5. Blaini; 6. Jaunsar); III. Madhan slates; IV. Chail; V. Simla flyschoidal type in general (9a. Basantpur; 9b. Naldera and Kakarhatti stromatolitic limestone); VI. Jutogh (10. graphitic schist and limestone; 11. Boileaugunge quartzite, mica schist); VII. Shali unit (12. Upper Shali stromatolitic limestone; 13. Shali slates; 14. Lower Shali stromatolitic limestone; 15. Khaira quartzite); VIII. Basic rocks suite; IX. Granito-gneiss

any megafossils to shed light on the age of the lithostratigraphic formations. In the absence of age data, the palaeogeography could not be reconstructed, neither could an accepted stratigraphic column be built. In recent years, the evidence of microfossils such as spore-pollen, Conodonts and the algal stromatolites (Prasad and Bhatia, 1975; Srivastava and Venkataraman, 1975; Sinha, 1975b, 1977b; Valdiya, 1969; Sah *et al.*, 1977; Sinha and Raaben, 1979; Azmi *et al.*, 1981, Ashgirei *et al.*, 1981; Sinha *et al.*, 1983) helps to place the lithostratigraphic formations in their proper positions in the stratigraphic column.

3.1.1 Geological setting

The geologic set-up of the area is shown in Figure 7—*Schematic geological map of Lesser Himalaya, H.P.* in Sinha (1980a, p. 100)—and its continuation to the southeast has been covered by Fuchs and Sinha (1974, 1978a, Plate 1). The detailed geology of Simla is depicted in Figures 8 and 9 and that of the southeast-adjoining Garhwal and Kumaun regions in Figure 10, modified after Fuchs and Sinha (1978). The tectonostratigraphic column (Figure 11) depicts the lithologic characteristics of different units. Stratigraphic units

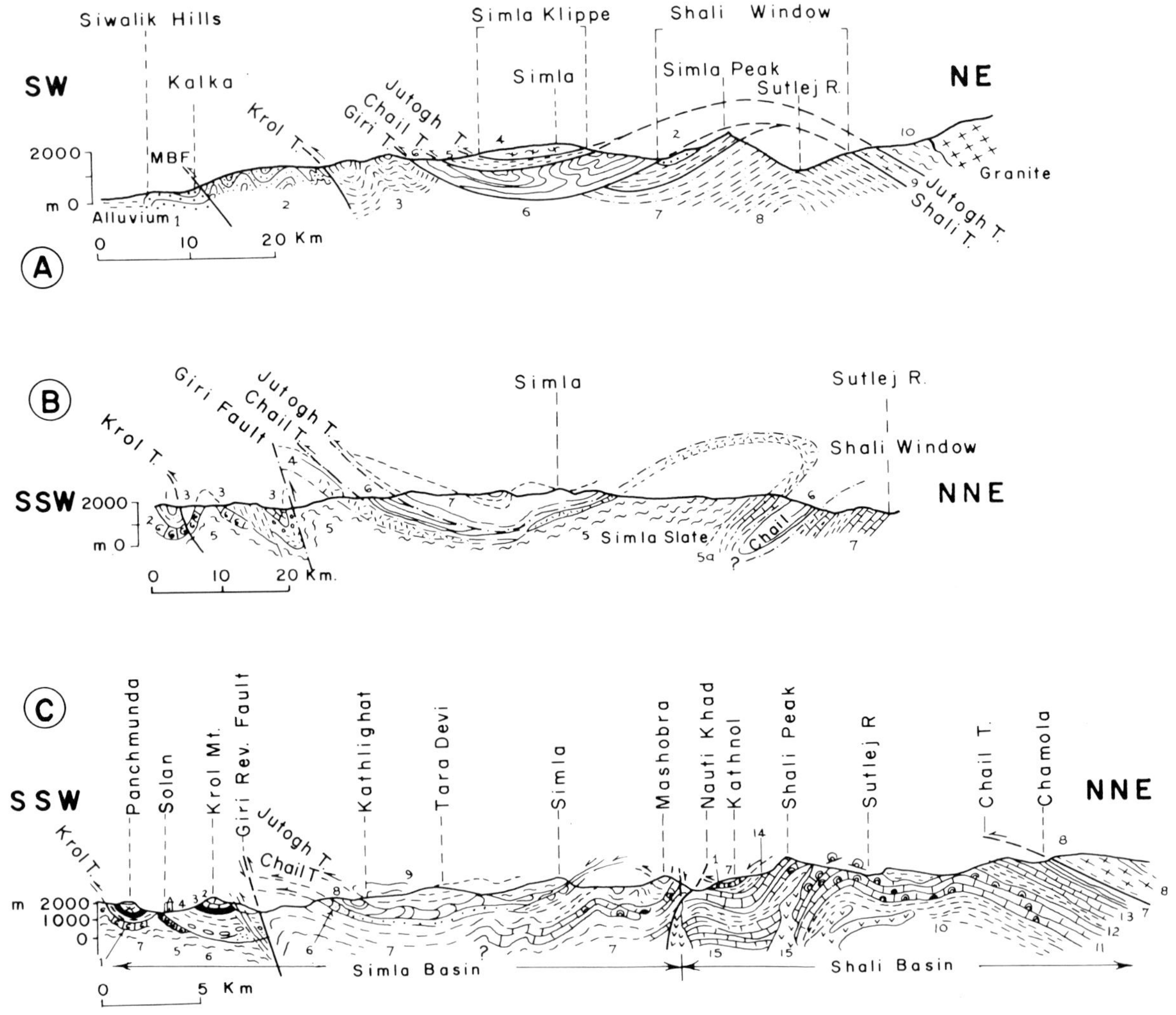

Figure 9. Cross-section along the Kalka–Simla section after (A) West (1939), (B) Berthelsen (1968), and (C) Sinha (1978a). (A) 1. Siwalik; 2. Tertiary; 3. Krol; 4. Jutogh; 5. Chail; 6. Simla; 7. Shali; 8. Khaira Quartzite; 9. Gneissose rock and migmatized rock; 10. Granites. (B) 1. Younger Tertiary; 2. Eocene with Nummulites; 3. Krol; 4. Jaunsar; 5. Simla slates; 5a. Naldera Limestone; 6. Chail; 7. Jutogh. (C) 1. Subathu with Nummulites; 2. Krol C.D.; 3. Krol B with Mesozoic coccolith; 4. Infra-Krol; 5. Blaini; 6. Jaunsar; 7. Simla; 8. Chail phyllite, metavolcanics, and gneiss; 9. Jutogh; 10. Khaira Quartzite; 11. Lower Shali limestone; 12. Shali slate; 13. Upper Shali limestone; 13. Madhan Slate of West (1939) 15. Basic volcanics

superimposed over one another are explained by thrust-sheet movements.

3.1.2 Autochthonous/parautochthonous units

This unit consists mainly of the Shali–Deoban–Pithoragarh–Tejam group of carbonate–terrigenous formation and the Simla flyschoidal formation of Riphean age with the capping of Palaeogene, unconformable Subathu and Kakara formations with nummulites and other fossils.

Shali Formation (Riphean)
(Deoban–Pithoragarh–Tejam Formation)

Pilgrim and West record a gradual transition from the grey slates (Madhan slates), through slates which contain thin bands or lenticles of limestone, often impure, to typical Shali limestone with its black chert bands. The limestone and slates, with some quartzites at the top and bottom, have been grouped together as the Shali Series which undoubtedly underlies the Madhan slates. The Shali Series has been subdivided by West (1939) and Srikantia and Sharma (1976). The Shali dolomitic limestones show profuse development of Riphean stromatolites (Valdiya, 1969; Sinha, 1977b; Sinha and Raaben, 1979; Kumar and Tewari, 1978). The notable forms reported from the Shali limestones from the Shali and Larji windows are: *Tungussia* Semikh., *Conophyton* ex gr. *Cylindricus* (Grabau), *Colonella* Komar, *Newlandia*, and *Irregularia*. Moundlike stromatolites, *Kusiella*, *Conophyton–Colonella* assemblage and presence of *Tungussia* and *Newlandia* indicate that the Shali limestones were deposited mainly in the Lower and Middle Riphean, but Upper

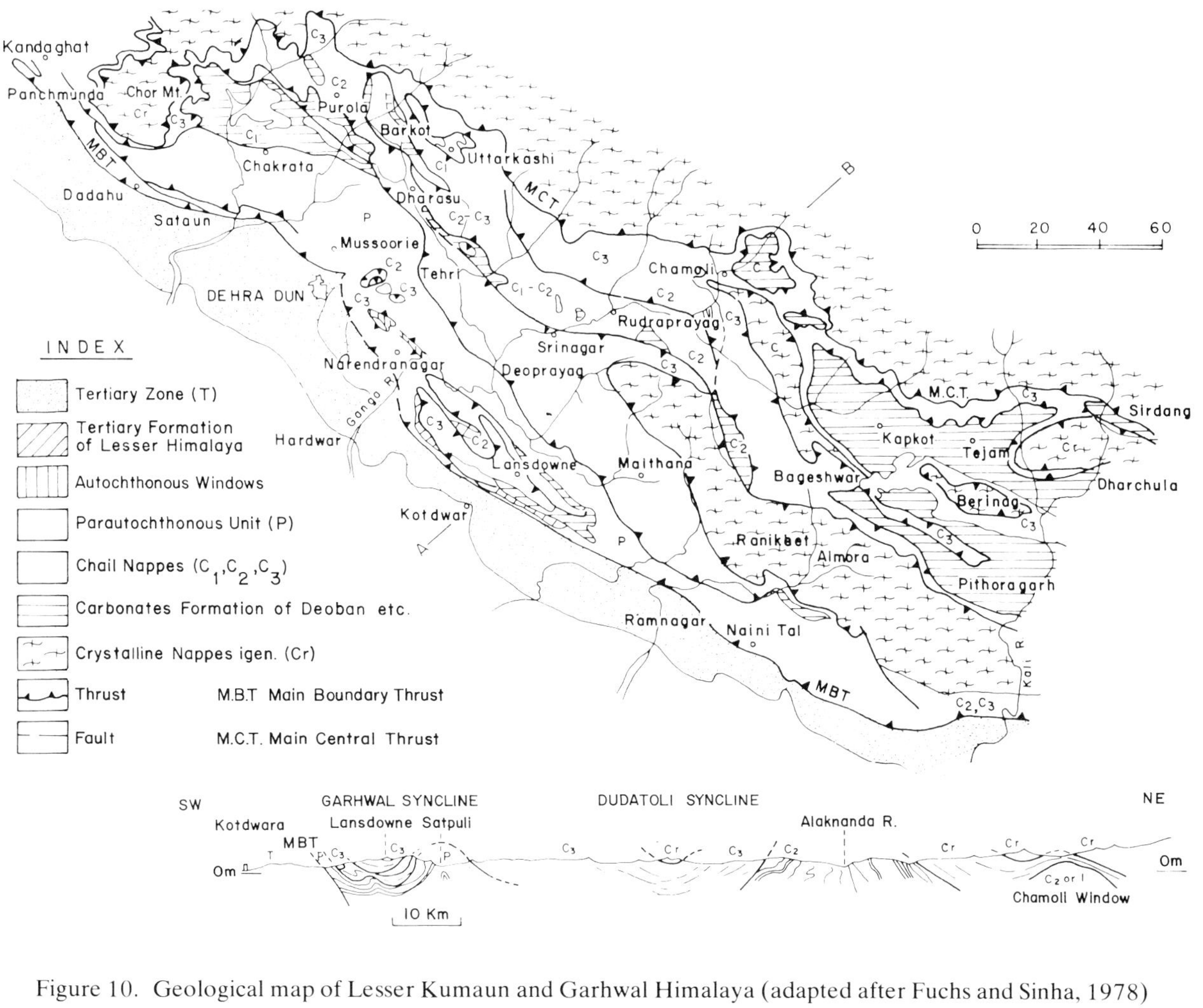

Figure 10. Geological map of Lesser Kumaun and Garhwal Himalaya (adapted after Fuchs and Sinha, 1978)

Riphean to lower Cambrian cannot be ruled out since the Deoban limestone has yielded *Ilicta deobanica* (Sinha and Raaben, 1979) which can be as young as Cambrian. The base of the Shali Formation is the Sundernagar Group (Srikantia, 1977), characterized by the abundance of clastic sediments consisting of white orthoquartzite and red quartzitic sandstones, slate, and basic magmatic rocks. These basics from Darlaghat and Yamuna Valley of Kuwa, dated by the K/Ar method, have given ages of 1190 ± 35 Ma (Sinha, 1977b).

Simla Formation (Riphean)

Well-bedded grey slates, without sandstone lenticles or carbonaceous matter, underlying the limestone and boulder beds of the Blaini, were first referred to as 'Simla Slates' or Infra-Blaini by Medlicott (1864). Pilgrim and West (1928) grouped the dark unaltered slates and micaceous sandstones, limestone with pseudoorganic structures, and also possibly the Deoban limestone under this Simla Series. Auden (1934) described the puckered leafy phyllites, massive quartzitic bands, finely ripple-marked shaly quartzite, and green nodular micaceous sandstones between Subathu and Arki, as the Simla Slates. According to him, one of the features of this formation is the intercalation of Kakarhatti Limestone. Ravindra Kumar and Brookfield (1978) suggest that the Simla Formation is a classic illustration of the collapse of a carbonate platform and consequent filling in with clastic material.

Lithopetrographically, the rocks of the Simla turbiditic flyschoidal formation are divisible into following four lithofacies groups (Sinha, 1976, 1978a):

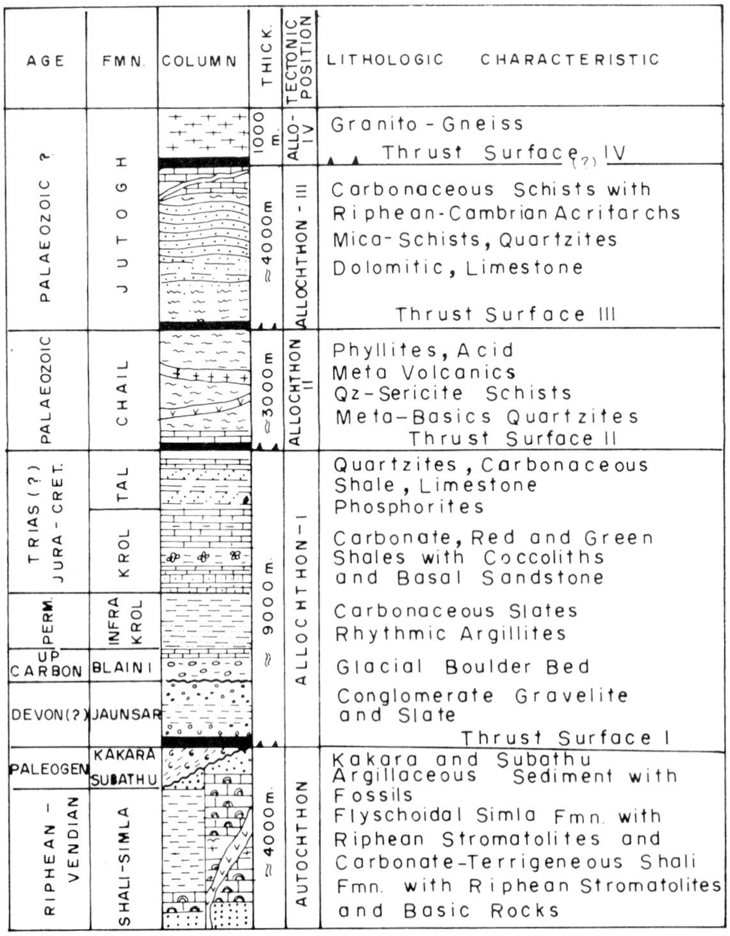

Figure 11. Tectonostratigraphic column of the Simla area

Group I: Terrigenous rocks—
(a) Arenaceous and silty–arenaceous
(b) Silty and argillaceous–silty
(c) Argillaceous and silty–argillaceous
Group II: Terrigenous–carbonate rocks
Group III: Carbonate rocks (Naldera–Kakarhatti limestones)
Group IV: Mixed rocks (Rhythmites).

The Naldera and Kakarhatti limestones have yielded Upper Riphean–Vendian stromatolites. The presence of stromatolites of Jurusania Group and a new form *Jurusania himalayica* Sinha date the lower part of Simla with greater accuracy, i.e. Upper Riphean–Vendian (Sinha, 1977a, 1984). The slaty rocks in our collection with Professor Ashgirei have yielded Precambrian acritarchs: *Kildinella hyperboreica* Jim, *K. sinica* Jim. (personal communication Dr (Mme) N. A. Volkova, Geological Institute of Academy of Sciences, Moscow, USSR). However, the identification is tentative but it confirms the presence of acritarch and its Precambrian age. Nautiyal (1985) records Middle Algonkian forms from the Naldera limestone.

The Kakara and Subathu Formations

In the autochthonous tectonic unit there are also two formations of Palaeogene age. The first one, the Kakara Formation, is composed of greyish-brown shale and greenish purple and yellow silty-shale. The fossils reported from this formation are: *Lockartia conditi*, *Ranikothalia caudri* (Srikantia and Bhargava, 1967), indicating Palaeocene age. Detached outcrops of Kakara are exposed in the southwestern peripheral part of the Simla Klippe.

The second unit is known as the Subathu Formation. The Eocene in the southern zone of Himalaya is represented by a formation named after the hill station of Subathu where it is well exposed. The Subathu beds are succeeded by the Dagshai and Kasauli formations. The Subathu comprises olive-green and purple, oily-looking gypseous shales with some subordinate lenticular bands of impure limestone and sandstones. The shales are characterized by minute, irregular, and ramifying veins of calcite. The limestone may be shelly or unfossiliferous, sheared and full of calcite veins. The sandstone is green and white in colour. *Nummulites* are seldom seen and well preserved fossils are rare, the shells being made up mostly of broken oysters.

The Subathu formation unconformably overlies both the Simla and Shali formations, and often marks the tectonic contact of the window under the thrust-sheets (see Figures 8 and 9; geological profiles through the southern Himalayan zones).

3.1.3 First lowermost thrust unit: Krol thrust sheet

This tectonic unit, occupying the lowermost position, is comprised of the following formations from bottom to top: (1) Chandpur–Nagthat (Jaunsar); (2) Blaini boulder bed (Mandhali); (3) Infra-Krol; (4) Krol; and (5) Tal. For thickness and lithology refer to Figure 11, tectonostratigraphic column.

Chandpur–Nagthat (Jaunsar) (Upper Palaeozoic)

The Chandpur–Nagthat phyllitic and quartzitic complex exposed in the Garhwal and Kumaun regions of Himalaya is known as the Jaunsar in the western region of Simla Hills (Auden, 1934; Fuchs and Sinha, 1978; Valdiya, 1980). There is no direct proof of the age of the Jaunsar but on the basis of our recent discovery of spores and pollens (Ashgirei, Umnova, and Sinha, 1981) the age of the Chandpur is taken to be Upper Palaeozoic. The Nagthat is mainly represented by quartzitic but the Chandpur is characterized by phyllitic and greenish spilitic-like formations.

The name 'Jaunsar Series' was first proposed by Oldham (1883). He classified this series into three divisions:

3. Traps and volcanic ashes (Upper),
2. Red quartzite and slates (Middle),
1. Grey slates and blue limestones (Lower).

Pilgrim and West (1928) restricted the name 'Jaunsar' to Oldham's middle division but extended the scope by including light yellow, lavender and purple phyllites, crushed conglomerates, light green–brown, brown and dark slates and massive white quartzites. According to Pilgrim and West (1928) 'Jaunsar' are older than the 'Simla Slate Series', and younger than the 'Chail Series'. Later, West (1939) considered the 'Jaunsar' to be younger than the 'Simla Slates', and stated that there was an apparent lithological gradation from 'Simla Slates' upwards to the

'Jaunsar' in a normal stratigraphic sequence.

Srikantia and Bhargava (1974) argue that the Jaunsar is part of the Simla Slate Series. But their arguments hold good only where the tectonic contact between the two is not apparent, such as near Kiarighat on the Kalka–Simla highway. However, here too the thick band of quartzite, exhibiting a definite style of thrust and drag-folding and flattening of deformation of the pebble conglomerate in the direction of thrust plane, provides unmistakable proof of thrusting. However, the conspicuous thrust plane over the Blaini in the eastern slope of Pervi Khala clearly suggests overthrusting of the Jaunsar over the Simla–Blaini complex. To take it as a part of the Simla Formation would be the denial of the existence of the Jaunsar Thrust. Furthermore, the quartzite always forms the lower part of Krol thrust-sheet in the Krol Mountain area, although in the Mashobra Syncline Pilgrim and West (1928) in their map show it as an integral part of the Simla Formation with the rhythmite below and at the top of it. Here movement over the base has to be taken into account since breccia, invaded by hydrothermal solutions, are present below the quartzite band at places. It is conspicuous especially in the right bank of Dugog Ka Nala flowing southward from the Naldera Golf links (Sinha, 1978a). As a matter of fact, Jaunsar is the undifferentiated horizon of the Chandpur and Nagthat of Auden (1934), which constitutes the lowermost part of the Krol thrust sheet. The Chandpur is correlated with the Chail on the basis of discovered palynomorphs (Ashgirei *et al.*, 1981; Sinha *et al.*, 1983) and its wider distribution in the southern zone of Himachal, Garhwal, and Kumaun (Fuchs and Sinha, 1978; Ashgirei and Sinha, 1982).

Blaini Boulder Bed (Permo-Carboniferous/Eocambrian?)

The Blaini Formation comprises of three lithounits, *viz.* the boulder beds, the intervening arenaceous and argillaceous horizon, and the pinkish dolomitic limestone. The boulders, cobbles and pebbles in the Blaini conglomerates are mainly of quartzite and less commonly of slates, chert, and limestone. The matrix of the boulder bed is mostly sandy, while at places it is argillaceous. The shale of the Blaini Formation is usually bleached due to oxidation and leaching. In a few sections the dolomite is seen to pass into the overlying Infra-Krol shales. In the Krol Belt there are more than one horizon of the Blaini Boulder Belt which are closely associated with the Infra-Krol Formation. This situation is often seen in the Kumaun and Garhwal region (Fuchs and Sinha, 1974).

Infra-Krol Formation (Permian)

The Infra-Krol rocks are described by Auden (1934) and Fuchs and Sinha (1974) from the Krol Belt of the Simla and Naini Tal areas. They consist of dark grey to black shales, interbedded with a silty component in a varve-like manner. White and yellow effervescences are characteristic on weathered surfaces of the sulphurous rocks. Silty layers are common and rocks are often thinly laminated. Blue carbonate layers are rather rare. The carbonaceous components are abundant in the lower horizon of both Simla and Naini Tal areas. Sometimes the quartzite layers are carbonaceous and contain disseminated pyrites (Pascoe, 1973–1975, vol. II, pp. 819–820). In the Naini Tal area it includes *Riri slates* features (Fuchs and Frank, 1970, p. 15) described from the Nepal region, and attains a thickness of approximately 1000 m in the southern limb of Naini Tal syncline. The Riri slates are light to dark grey, rarely green or slightly phyllitic slates and siltstones. They represent a less euxinic facies than the Infra-Krol rocks, but both are found alternating and intertonguing. In addition, there are also red and purple intercalations in the lower part, which indicate interfingering with the Blaini Formation as mentioned earlier. Sitholey *et al.* (1954), Lakhanpal *et al.* (1958) and Sah *et al.* (1968) discovered Gondwana microflora from the Naini Tal area. Our collection macerated by Dr Ilze Draxler in 1973 in the field and in the laboratory of the Austrian Geological Survey yielded only some unidentifiable cuticles.

Table 3.1. The Krol Formation as subdivided by Auden (1934)

Subdivision	Units	Lithology
Upper Krol Limestone	Krol E	Light coloured limestone
	Krol D	cherty limestone and shale
	Krol C	Dark blue limestone
Red Shale	Krol B	Red shale with limestone lenses
Lower Krol Limestone	Krol A	Limestone and shale
	Krol Sandstone	(Local)

Krol Formation (Mesozoic?)

Auden (1934) subdivided the Krol Formation as shown in Table 3.1. A reclassification and reconstruction of the lithostratigraphic column of the Krol rocks have been prepared by the author with Dr Fuchs (Fuchs and Sinha, 1974) in which the Krol is divided into A, B, C, D, E, and F members from the Naini Tal area.

The Krol sandstone is a local phase of the Infra-Krol formation. Around Solan it can be seen passing down into the Infra-Krol through a passage zone of about 30 m.

Krol A: The Lowermost member consists of dark thinly bedded limestones, dolomites, marls and shales. Oolites and algal detritus were observed in some of the carbonate rocks. The total thickness is approximately 25 m.

Krol B: It consists of red and green, laminated shales and slates with light cream dolomite layers (0.75 m). The thickness is approximately 80 m. Author recovered a rich assemblage of coccolith from this horizon (15 m) exposed in the northern limb of Panchmunda syncline of Solan area in the Simla Hills (Sinha, 1980b, Fig. 6, p. 107). The following forms were identified (Sinha, 1972b, 1975b) with the help of Dr (Mrs) Olga Dmitrenko: *Tergestiella margereli* (Noël) Rein; *Microrhabdulus orbitosus* Shumenko; *Micula staurophora* (Vekshina) Gardet, Stradner; *Zygolithus concinnus* Martini; *Tetralithus* cf. *gothicus Deflandre; Lucianorhabdus cayeuxi* Deflandre; *Lithraphidites carniolensis* Deflandre; *Discorhabdus* sp., *Deflandrius* sp. and ex gr. *Nannoconus*. Most of these forms which also occur in other parts of the world indicate an age of Upper Jurassic to Upper Cretaceous.

Krol C: It is composed of dark grey to blue–grey dolomite. These are laminated to massive and contain chert. Towards the top a few black shale layers of dm thickness appear and oolites become frequent. The approximate thickness is 300 m.

Krol D: These are green and red, fine to medium-grained micaceous sandstones (15–20 m). In the Naini Tal bazaar area we find red, green, and grey micaceous sandstones and sandy shales with a bed of light quartzite. Near the colleges on the ridge SW of Naini Tal, dark grey to green slates and sandstones with some light quartzites (approximately 100 m) are exposed. Azmi and Pancholi (1983) have recorded Early Cambrian (Tommotian) conodonts and other, shelly microfauna from limestone horizon of Mussoorie hills.

Krol E: It consists of thick bedded to massive, dark blue dolomite (300–400 m). Stromatolites are growing conspicuously with branching forms, but of long ranging Phanerozoic age (Personal communication Dr Maria Raaben, USSR Academy of Sciences, Moscow). Phosphorites have been noticed in association with stromatolites. P_2O_5 content varies from 17 to 30 per cent (Choudhury and Banerjee, 1966). In the southward rock face from the former Viceroy's Palace, the lowest beds, only a few metres thick, are conspicuously light grey and full of stromatolites and oolites (2 cm diameter).

Krol F: It is the youngest member of the Krol formation and builds up the western part of the ridge southwest of Naini Tal and the upper part of the rocky south face of the range. There is a transitional contact between Krol E and F with the development of more clastic beds. Thus, the dolomite sequence of Krol E passes into an alternation of dark dolomite, black to dark grey fine splintery shales, silty and arenaceous shales, grey sandstones, quartzitic sandstones, and light grey to white quartzites. There are also some layers of red to purple shale and sandstones with ripple marks. An approximate thickness of 350 m for Krol F may safely be assumed, however, considering that this series is exposed in the core of a syncline and some part has been removed by erosion.

Dykes (15 m) of fine to medium grained diabase are found to penetrate Krol F and E. K/Ar age determination gives 51 Ma (Sinha and Bagdasarian, 1977).

Tal Formation

Auden (1934) divided the Tal Formation in two units: upper and lower, while Bhargava (1976) proposes a three-fold classification on the basis of distinct lithologic assemblages. Gopendra Kumar (1981) also summarizes on the basis of recent works (see Table 3.2).

At the base of the Tal Formation in the Maldeota phosphorite mine, lower Palaeozoic conodonts have been recently reported (Azmi *et al.*, 1981). Earlier the age controversy has also been revived with the finding of so-called Permian fos-

Table 3.2. Tal Formation subdivisions (Kumar, 1981)

	Subdivision	Lithology
C	Upper Tal	Calc-arenite band at base, sandstone, orthoquartzite, arkosic sandstone, shale, algal limestone.
B	Middle Tal	Greywacke, subgreywacke, siltstone, and shale.
A	Lower Tal	Chert, black shale, siltstone, sporadic limestone, phosphatic shale and nodules, and phosphorite.

sils from the Tal horizon in the Lansdowne and Naini Tal areas (Valdiya, 1975, 1980b). Azmi (1983) on the basis of conodonts is contending a Cambro–Ordovician boundary age for the Tal phosphatic horizon, whereas Tripathi *et al.* (1984) report Early Cambrian brachiopods from the Upper Tal. Bhargava (1984) reports Cambrian (?) Trace fossils.

3.1.4 Second tectonic unit: Chail thrust-sheet

The name 'Chail Series' was given by Pilgrim and West (1928) after the Chail township situated southeast of the Simla ridge. But interestingly the exposed graphitic schists on the top of Chail hill belong to still higher tectonic unit, i.e. to the 'Jutogh Thrust-sheet', and according to their description the following rock types constitute the series: (1) Light grey and brown schistose slates and quartz-schist; (2) Slightly talcose flaggy quartzites and quartz-schists; (3) Talc-schists; (4) Slightly puckered grey phyllites; and (5) Grey slates with interbedded, banded, crushed limestone. In fact, different types of black, carbonaceous (graphitic), dark grey and greenish-grey shales converted into phyllites and black slates constitute the major part of the Chail complex.

They have been regionally metamorphosed to the grade of greenschist facies. In the Simla and Kulu areas the slate–phyllite component constitutes up to 60–70 per cent. Sericite, chlorite and quartz grains are dominated in the slate. The content of carbonaceous matter is not very high (locally 5 per cent). Silty slate and phyllites in many cases have thin alternating bands of metasandstones. Locally graded bedding is observed. Arenaceous phyllites contain quartz, sodic plagioclase (An_8–An_{12}) and feldspars of 0.05 to 0.1 mm size. The calcareous phyllite additionally contains minute crystals of calcite up to 20–30 per cent. The sandstones are a subordinate component in the Chail. They are oligomicitic greywacke, and tuffaceous sandstone occurring as thin layers in phyllites. Volcanosedimentary origin of the tuffaceous sandstone is without doubt. The rocks of this kind are characteristics of the eugeosynclinal regime (Ashgirei *et al.*, 1977). According to chemical analysis, the sodium content in the tuffaceous sandstones is conspicuously higher than that of potassium. This is a pointer to the presence of spilitic–albitophyric magmatic material in its lithological composition during deposition in an early eugeosynclinal environment. The age of the greenschist metamorphism according to K/Ar isotope analysis is upper Jurassic boundary, i.e. 124 Ma. Acid orthogneisses in the Chails gave surprisingly old Rb/Sr ages: 1840 ± 70 Ma (Frank *et al.*, 1976) and 1430 ± 150 Ma (Bhanot *et al.*, 1978).

The Chail thrust-sheet extends from the Mandi area in the northwest in Beas valley through the Sutlej valley in Himachal Pradesh to the Garhwal and Kumaun regions in the southeast of Uttar Pradesh (Fuchs and Sinha, 1978), and further to Nepal. Thus its strike extension is more than 300 km whereas 100 km in depth of Lesser Himalaya, and the total investigated area constitutes about 50,000 km^2. Average thickness of the Chail unit is 2000 m.

Micropalaeontological investigations of our collection of Chail slates from the four different localities, situated quite far apart in the lesser zone of Himalaya in Himachal Pradesh, have yielded about eighty different recoverable and identifiable well-preserved forms (Ashgirei *et al.*, 1981; Sinha *et al.*, 1983), identified by Dr (Mme) Nina Ivanovna Umnova, Moscow. The specimens were collected from the following localities of the Beas and Sutlej valleys: Pandoh (31°41.5′: 77°05′) outcrop 74; Larji (31°42′: 77°5′) outcrop 128; Kingal (31°18.5′: 77°28′) outcrop 137; and Nirmand (31°26′: 77°37′) outcrop 149.

A rich assemblage of some 80 and 55 pieces have been obtained from outcrops 137 and 74 respectively. Predominating forms are *Hymenozonotriletes pusillus* (Ibr.) Isch. (*Lycospora pusilla* (Ibr.) S., W., B), constitute 50 per cent of total recovered spores. Besides them, between one and four fragments have been recovered of *Trachytriletes lasium* (Waltz) Naum, *T. subintortus* Naum, *T. lobophorus* (Waltz) Byv., *Leiotriletes glaber* Naum., *L. kaevis* Naum., *Acanthotriletes parvispinus* (Lub.) Kedo, *A. sublitus Tusch*., *A.* sp., *Lophotriletes scrobiculatus* Naum., *L. macropunctatus* Namum *L.* aff. *inflatus* (Lab.) Naum., *Densoporites variabilis*

(Waltz) Pot. et Kr ? *D. irregularis* (Andr.) Byv., *D. intermedius* (Waltz) Byv., *D. aculeatus* Playf., *Cingulizonates bialatus* (Waltz) Sm. ct. Butt., *Euryzonotiletes literatus* (Waltz) Isch., *E.* sp., *Camarozonotriletes* sp.; *Trilobozotriletes inciso-trilobus* (Waltz), Naum, *T. concavus* Naum, var. okensis Tusch., *Diatomozonotriletes granulosus* Naum., *Perisaccus primigenius* Naum. and *Hymenozonotriletes rugosiusculus* Kedo.

At outcrop 128, there has been found four fragments of *Stenozonotriletes conformis* Naum., three fragments of *Hymetozonotriletes pusillus* (Ibr.) Isch. and a single fragment of *Stenozonotri-letes* aff. minor Naum., *St. luculentus* Naum., *Lophozonotriletes* sp., *Retusotriletes puhovae* Naum., *Trematozonotriletes punctatus* Naum., *Simozonotriletes minutus* Isch., *Peteino-sphaeridium* aff. *snigirevskaise* Stok. and Wil., *Baltisphaeridium* sp.

At outcrop 149 there have been recovered *Densosporites dentatus* (Waltz) Byv., *Diatomozonotriletes granulosus* Naum., *Euryzonotriletes salcatus* (Waltz) Isch., *Camarozonotriletes* sp., *Stenozonotriletes* sp., *Trematozonotriletes punctatus* Naum., *Euryzono-triletes sunluculentus* (Jusch. and Kedo).

The above-mentioned forms are characteristic of the Carboniferous sediments, whereas *Hymenozonotriletes pusillus* and forms of *Leiotri-letes*, *Trachytriletes*, *Lophotriletes*, *Acanthotriletes* have a wide diapazonic distribution in the entire Carboniferous, and *Densosporites crassipterus*, *Cinqulizonates bialatus*, *Stenozotriletes conformis* and *S. pumilus* are more characteristic for the deposits of early Carboniferous age. There have been recovered the species of *Euryzonotriletes*, *Trilobozonotriletes*, *Diatomozonotriletes*, *Simozonotriletes* and pollens *Perissaccus primigenius* known only from Visean (C_{1v}) and Namurian (C_{1n}) sediments. Amongst the re-covered forms there has been not a single form characteristic of the middle Carboniferous.

The above-mentioned forms have wide geo-graphic distribution and are characteristic of the lower Carboniferous sediments of the European part of USSR, Western Europe, and North America, occupying the lower Carboniferous Euroamerican floral province.

The age of the rocks of the Chail tectonic unit, on the basis of above-mentioned spores and pol-lens is, possibly, early Carboniferous, although the possibility of reworking or infiltration due to erosion of these forms from the physically above-lying Jutogh thrust-sheet could not be completely ruled out. But in both causes it is definite that in the realm of Northern Himalaya there has existed early Carboniferous eugeosynclinal regime (Sinha *et al.*, 1983).

3.1.5 Third tectonic unit: Jutogh thrust-sheet

The name Jutogh Formation has been assigned to a lithological association comprising schists, phyllites, slates, metaquartzites, limestones, and gneisses which were earlier described as the 'Jutogh Series' by Pilgrim and West (1928). They are subdivided in three lithounits. The metaquart-zites of the Boileaugunge member are generally massive, and with increase in the proportion of micaceous minerals become quartz schists. The quartz-schist is garnetiferous at many places. The carbonaceous member is a very conspicuous hori-zon. It is associated with graphitic schists and crystalline limestone or marble. The microschist member has a fairly extensive occurrence, and comprises mainly biotite schist, muscovite schist and chlorite schist. They are frequently garneti-ferous and contain in places green porphyroblasts of penninite. Within the mica schist there are thin bands of quartzite. The whole series are regionally metamorphosed to amphibolitic facies, and usually covered by overlying granite-gneisses. The average thickness of this tectonic unit is about 2000 m (Sinha, 1972b).

3.1.6 Fourth tectonic unit: Garhwal Kumaun crystalline thrust-sheet

This topmost thrust-sheet of Southern Himalaya is composed of crystalline rocks like gneisses, mica schists and garnetiferous gneisses. It is more widely exposed in the southeastern region as nappes and klippen (Fuchs and Sinha, 1978) from Chor Mountain to Almora township in the Kumaun. It is noteworthy that all the hill-stations, *viz.* Almora, Askot, Gwaldham, Lansdowne etc., are situated on the crystalline rocks of this thrust-sheet. The detail description of these rocks would be taken up in the next subheading under 'Vaikrita Group'.

3.2 Higher and Tethyan Zone

Main Central Thrust (MCT) divides the Lesser Himalayan zone from the Higher Himalayan zone in the Central Himalaya which consists of two major structural formational zones of crystalline rocks of 'Vaikrita Group' and sedimentary thick sequence of Tethyan Himalaya. These two prominent Higher Himalayan zones are separated by the Main Central Axial (MAZ) zone. Each of these structural formational zones (SFZ) has their independent geological history of development.

3.2.1 Geological setting and stratigraphic approach

The Tethys of Suess (1885) was introduced by Auden (1935, 1937) to interpret the fossiliferous sequence lying in the north of Central Crystallines. This northern facies of sediment were called the 'Tethys Himalayan facies' or 'Tethyan facies'.

The 'Tethyan' facies comprises the fossiliferous Precambrian – Palaeozoic – Mesozoic – Tertiary sequence resting on a basement of the Central Crystallines, which in turn overrides the sediments and metasediment complex of the Lesser Himalaya to the south along the main Central Thrust. It is no wonder why the first and pioneer observer of Central Himalaya, Captain Strachey (1857), was fascinated by this region, collecting the marine fossils of Silurian age. A great leap forward in establishing the stratigraphic column of this region was made by the end of last century by the Austro-Germans of the Geological Survey of India, *viz.* Diener, Griesbach and v. Kraft. Hayden's (1904) work in the adjoining Spiti zone has also contributed in the stratigraphic correlation of this region.

Fig. 12
Fig. 13
Fig. 15

The notable classic contribution to this region of Central Himalaya came after the Swiss Expedition in 1936, led by late Professor Arnold Heim, and the Doyen of Himalayan geology Professor Augusto Gansser. Their monograph published in 1939 remained until today the only authentic literature with the modern concept of the first half of the present century in geological sciences. A gap of forty years is not sufficient to revise substantially the biostratigraphic concept with the available new techniques, tools, and the new index fossils in microform studied under high magnification with binocular and electron microscope. My intention is to refer to the discovery of scolecodont, dinoflagellates, coccoliths, holothurian sclerites, conodonts and new findings of foraminifera etc. from the different horizons of the Tethyan column. These valuable findings have revolutionized stratigraphic research.

Apart from the difference in lithology and presence of prolific fauna, the 'Tethyan' rocks display an entirely different structural set-up as compared to the rocks of the Lesser Himalayan domain. This is manifested in the general absence of isoclinal and tight foldings and prevalence of Jura type, disharmonic, gravity, and simpler open folding patterns. In the regional perspective of the Tethyan zone of the entire Higher Himalayan terrain, the Tethyan basin of Garhwal and Kumaun (Figures 13 and 14) is neither continuous with the Spiti basin in the northwest nor the Nepal Basin in the southeast. The depositional basin in question is being closed by the axially uplifted crystalline rocks on the either side. As far as the classification or rock definition and delimitation of the stratigraphic unit is concerned, there has crept in some confusion because of quantities of new data and the synonymous use of lithostratigraphic and biostratigraphic terminologies. The code of Stratigraphic Nomenclature (1871) has been followed to rectify and simplify these discrepancies. The biostratigraphic assemblage zones have been superimposed on the formal lithostratigraphic units. The first such attempt was made in our earlier publication (Shah and Sinha, 1974a,b,c), and with the additional data produced by the author in collaboration with Dr Nanda, a supplemented and up-to-date biostratigraphic column has emerged (Figures 15 and 16, after Shah and Sinha, 1974c; Sinha and Nanda, 1979; Sinha, 1980b,c). The stratigraphy is prepared in summarized fashion in Table 3.6 giving the details of biostratigraphic zonation.

3.2.2 Higher Central Himalayan zone (Figures 12 and 13)

Vaikrita Group

It was Griesbach (1891) who introduced the Sanskrit word to describe the highly deformed crystalline complex of rocks of Central Himalaya. In Sanskrit, *Vaikrita* means the deformed, and the metamorphosed. Subsequent workers have called this unit variously the Central Crystalline or basement complex. Metamorphic rocks of this group show grading from low to high grade catazone sillimanite–kyanite schist with an appreciable thickness of hundreds of metres of quartzite, sometimes occurring with kyanite–sillimanite as well as garnets in the schistose and gneissose part of the argillaceous sediments. The other components of the crystalline complex are various grades of gneisses and schists including calc-silicate schistose rocks and granitic gneisses of

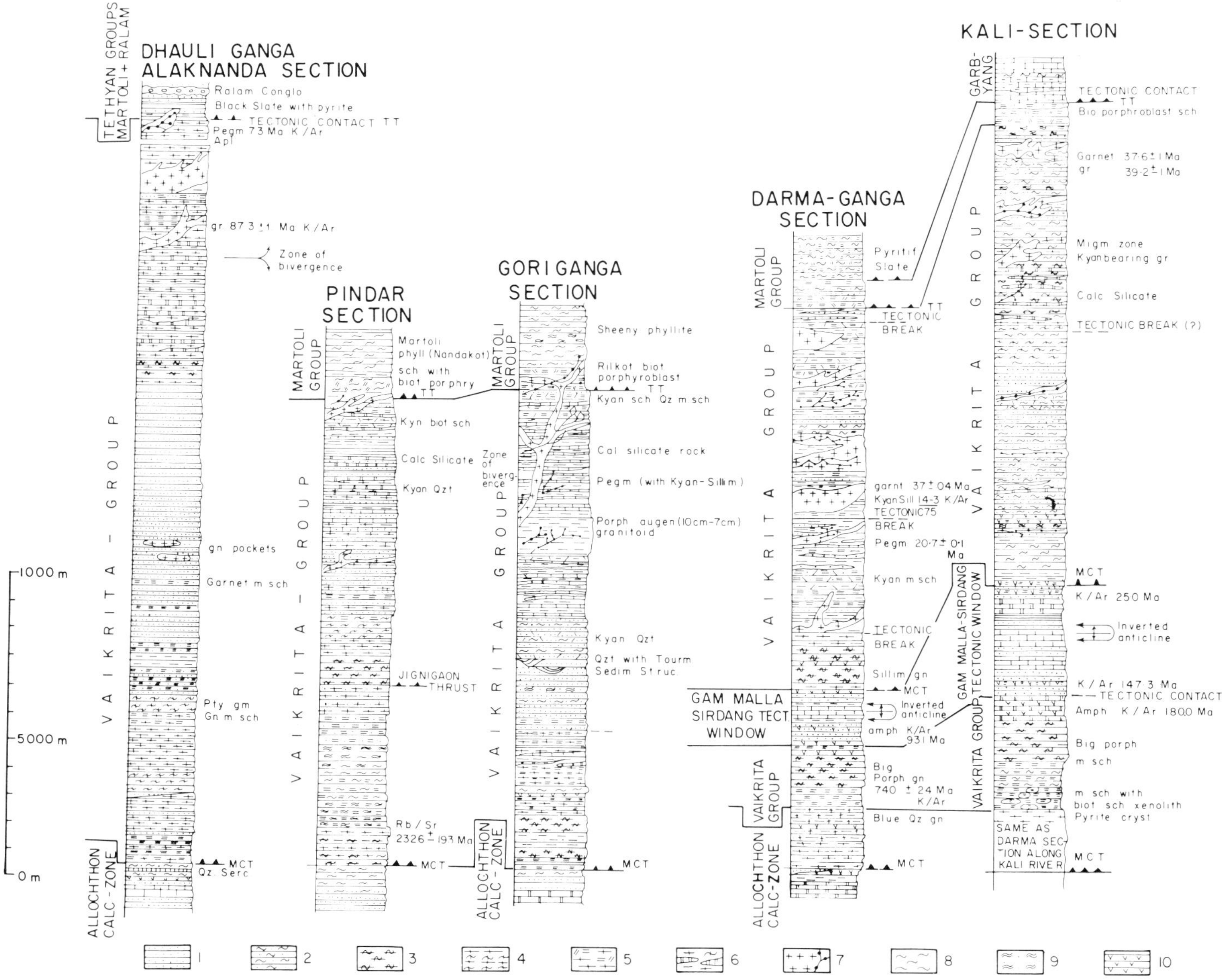

Figure 14. Tectonostratigraphic lithologies of the Central Crystallines along the profiles. 1. quartzite; 2. quartz sericite-schists; 3. porphyritic gneiss; 4. gneiss; 5. schists with biotitic porphyroblasts and schists in general; 6. gneiss and schists with calc-silicate rocks; 7. tourmaline granite and pegmatite; 8. phyllite/argillite and sedimentary rocks; 9. schists with quartz components; 10. amphibolite and other basic rocks

different phases (Figure 14). The earlier phase is sillimanite–kyanite bearing, whereas the last phase is of aplitic–pegmatitic type. This younger phase of granites is mostly confined to the Tethyan sediments tectonized zone. K/Ar date of aplitic-granite exposed at Malari (Figure 12) in the Dhauli Ganga section gives 73 Ma (Sinha and Bagdasarian, 1977). Along certain zones there is intense development of migmatites. This deformed and metamorphosed group of rocks have suffered different grades of geodynamic forces at different places of the entire mapped area (Photo Panorama No. 1). In the northwest corner it has an immense exposed thickness of more than 20 km whereas following along the arcuate structure in the eastern part in Goriganga and Kali gorges it reduced to nearly 2000 m. The *Annapurna* Gneissic complex (Bodenhausen and Egeler, 1971) is the continuation of this zone further east in the Nepal.

Pan 1

Tectonically, the crystalline complex show one very prominent earlier phase of recumbent isoclinal folding. This phase is conspicuously seen also in the quartzitic horizons. It is to be noted that such a phase is completely absent in the Tethyan zone sediments.

Chapter 5 incorporates the detail study of different crystalline sections, and the litho and structural aspects are depicted in the sectional sketches itself.

The tectonolithological column of Vaikrita in the Gori Ganga valley between Girgaon and Lilam has been portrayed in Figures 17 and 18.

The entire Vaikrita complex is intensely

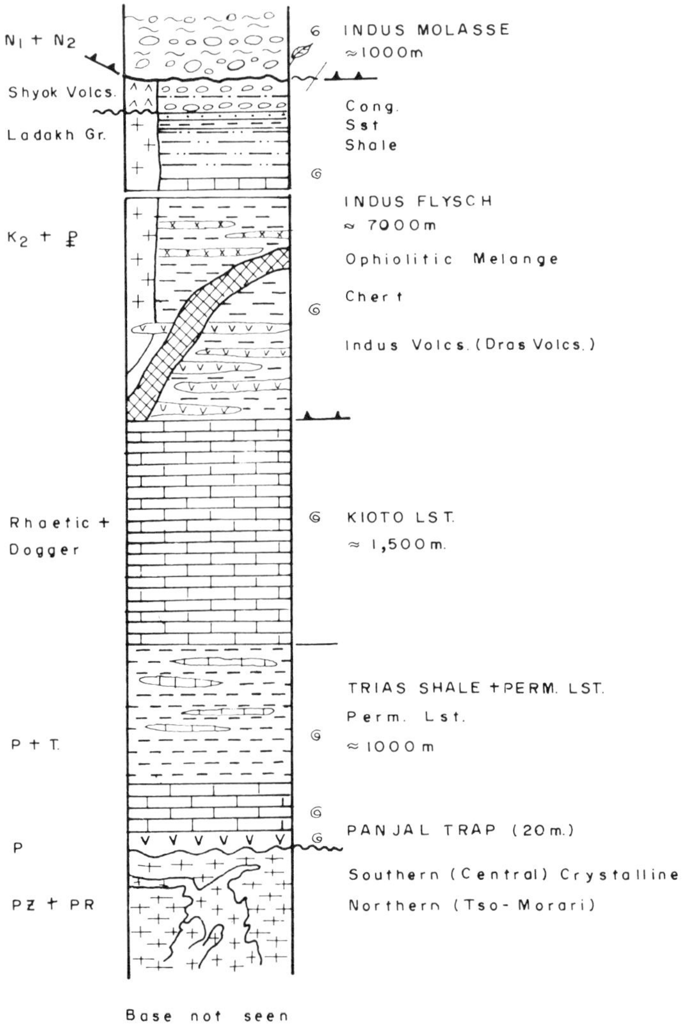

Figure 16. A stratigraphic column of the Ladakh–Spiti region in the northwest of mapped area reproduced for comparative view (Sinha, 1982)

deformed and metamorphosed due to remobilization and it would be a Herculean task to reestablish the parent rock composition and the original thickness, since the whole psammitic series has suffered tight isoclinal folding. In the entire south to north section we come across a thick quartzite sequence which has preserved the sedimentary structures like cross and graded bedding.

The role of the central crystalline in the palaeogeologic reconstruction of Himalaya has been crucial and remained an enigma for the workers without any authigenic fossils record and radiometric dating. In the literature this Vaikrita complex has often been called a 'Precambrian basement for the Tethys zone'. Powell and Conaghan (1973a) and Picket *et al.* (1975), giving a synthesis of Vaikrita of Rohtang, summarized the different schools of thoughts and rightly pointed out that they can be divided into three groups. Firstly, those who regard the Vaikrita complex as a remobilized ancient geanticline which separated a northern Tethyan zone from southern unfossiliferous basin (Fuchs, 1967). Secondly, the school which advocated the Vaikrita complex to be a slab of remobilized and upthrust basement beheaded and elevated during the Tertiary orogeny (Auden, 1935; Odell, 1943; Gansser, 1964, 1974b; Bordet *et al.*, 1972). And thirdly, those who think the Vaikrita complex to be the collision suture between the Indian subcontinent and Asia (Valdiya and Gupta, 1972; Valdiya, 1980a, 1983). All these workers with their assumed models proposed that Vaikrita was derived from the Precambrian basement which had been remobilized and remetamorphosed during the Cainozoic.

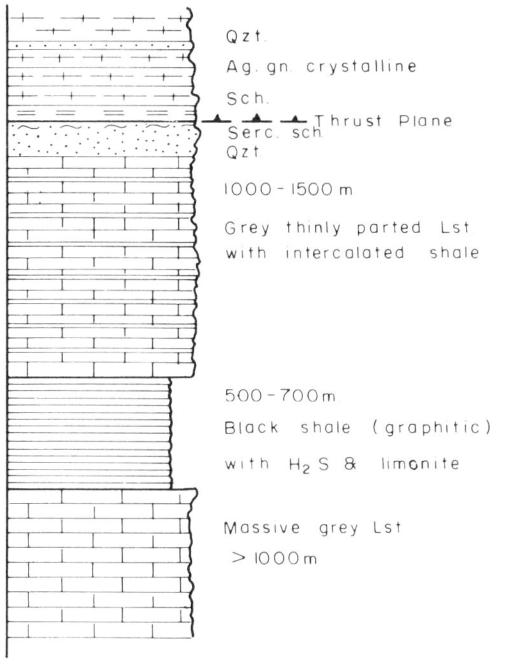

Figure 17. Tectonostratigraphic column at the calc-zone base of the Main Central Thrust with physically overlying Vaikrita rocks along the Tejam–Girgaon (Jakula Nadi) traverse

But now with the finding of fossils representing the Permian conodonts *viz. Neogondella bisselli* (Clark and Behenken) along with foraminifera *Hyperammina* sp., *Thuramminoides* sp., *Hemidiscus* sp., *Lituotuba* sp., *Ammobaculites* sp., and Ostrocod *Bathocypris* sp. in the Thanglang La Formation constituting the upper part of the Tso–Morari Vaikrita crystalline of Ladakh region (Virdi *et al.*, 1978), is apparently clear that there is no reason to consider them as Precambrian in the conventional way (Figure 16). This group of the Vaikrita complex, with its digitation in two directions from Leo Pargial peak (6974 m) seen in the right wall of Satluj gorge, represents the central crystallines of Himachal, Garhwal, and Kumaun including the Rohtang Pass area. Picket *et al.* (1975) describe the discovery by Powell and Conaghan (1973a) of cephalopods and corals in highly-deformed metasediments from within the Central Gneiss of Lahaul, *viz.* nautiloid *Cenoceras* and macrocephalitid ammonites indicating a Middle Jurassic age, as well as foraminiferal material and scleractinian corals, *Thamnasteria* sp. A., *Thamnasteria* cf. sp. A., *Montlivaltia* ? sp. A. *Montlivaltia* ? sp. B. etc.

This fauna clearly and most authentically advocates that high grade metamorphism is not Precambrian. We have to take note of Fuchs' (1979, p. 531) statement: 'From all my traverses and also from the observations of many other workers I came to the conclusion that now here in the Himalayas apparently there is proof of a transgressive contact of the sedimentary series with the crystallines. A sharp boundary cannot be drawn, and by decrease in metamorphic grade the crystallines pass into slightly altered sedimentary facies'. He further adds: 'There is no crystalline basement existing beneath the sedimentaries of Tibetan zone in the SW. From the descriptions of Hayden (1904), Berthelsen (1951) and others I also doubt the existence of a "Precambrian basement" underlying the Tibetan sediments in the NE.'

Well, with these lines from an extensively travelled Himalayan and Alpine geologist, I will attempt to show that all along the northern contact in the Central Himalaya of Kumaun and Garhwal there is no transitional passage to the sedimentary zone as assumed by Heim and Gansser (1939a, p. 201). The Vaikrita complex is abutted by the Martoli–Garbyang sediments with a marked tectonic contact. The intervening layer of porphyroblastic Pindari–Rilkot–Baling and Budhi schist zone (250–350 m) represents contact thermal metamorphic zone. This contact is of typical deep-seated nature, dividing two structural facial zones of the Vaikrita and Tethyan zones, and marked by multiphase magmatic episodes along them (Figures 13 and 14). In the later Himalayan phases of the orogenic cycles, it has assumed the nature of thrust, but as a matter of fact it assumes a vertical zone of fracture in depth and passes down to the mantle.

With all the facts of fossil data, metamorphic episodes, tectonic lineament, and phases of deformation, there is now ample evidence to dispute the Archean age of the Vaikrita group. Let

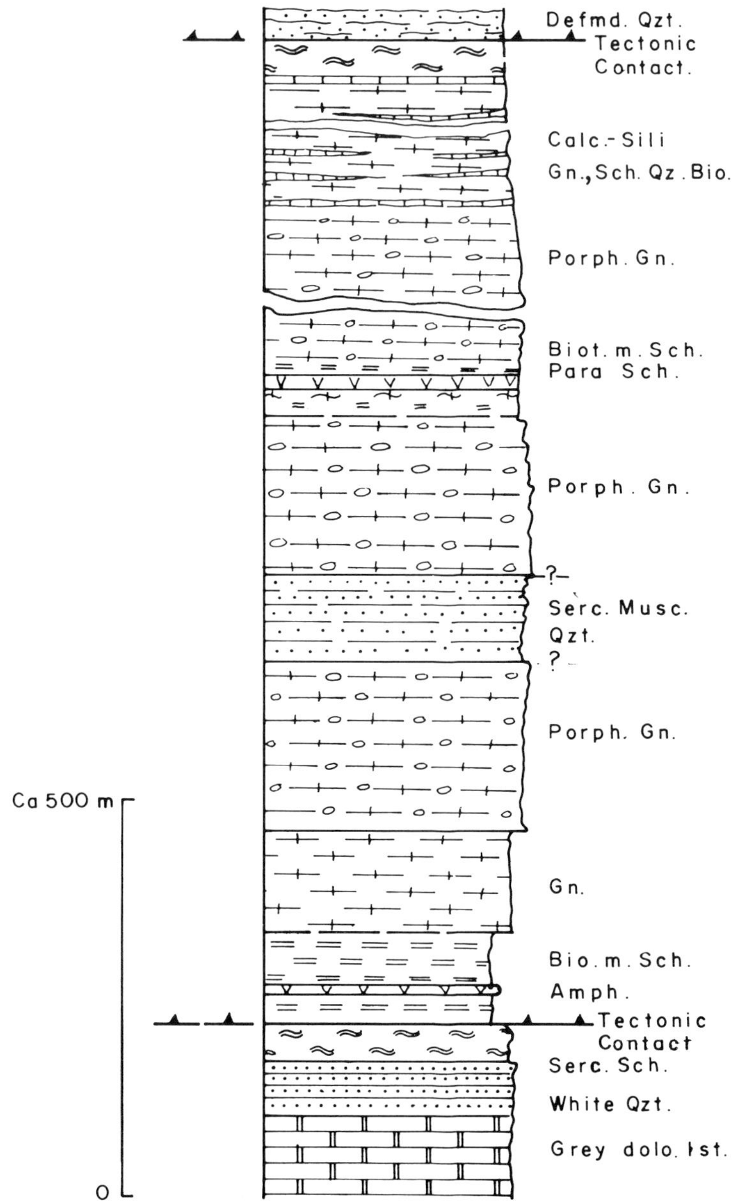

Figure 18. Tectonolitho column of traverse between Girgaon and Lilam in Gori Ganga valley (upward continuation of Figure 17)

us keep this question open until there are more authentic data to draw a final verdict on the issue.

3.2.3 Tethyan zone

Martoli Group (*ca*. 3250 m)

Griesbach (1891, p. 42) observed that the Vaikrita complex, which rests conformably on the granitic gneiss of the southern range along its northern slope, passes gradually into Haimanta ('snow-covered' in Sanskrit) system. He gave a classification as follows:

Silurian: Series of quartz shales and slates; shales and silty phyllites with great thickness of quartzites.

Haimantas: Quartzite, generally purple, with great thickness of conglomerates.

Vaikritas and older gneiss.

Griesbach describes (1891, p. 41): 'In nearly all sections which I have hitherto examined between the Kali river and Spiti, the schists seem to pass gradually into the overlying slates, phyllites, and quartzites of the Haimantas. The latter, therefore, belong structurally to the schistose strata below, and take part with them in all the complicated flexures and minor folds which have affected the Himalayas generally. This series of schists I distinguish north of the Central Himalayas as the Vaikrita System'.

With the passage of time Griesbach's classification was modified by Heim and Gansser (1939a), who remarked (1939a, p. 202): 'Indeed, in the maps and sections, Griesbach has drawn an almost straight line or fault, which does not exist, as the lower boundary passing arbitrarily through the middle part of the great phyllitic quartzite series, for which we propose the name of *Martoli* Series. This important village Martoli on the Gori Ganga below Milam lies in the Central part of this huge sedimentary series, which, in Griesbach's map, is included in the "Metamorphic and Granite".' Thus the Martoli nomenclature was introduced in the Tethyan sedimentary sequence. The Martoli has been given a group status in this monograph in view of modern stratigraphic code (1971) (Figure 19).

In the entire mapped area of Central Greater Himalaya (Figure 13), the Martoli Group of sediments tectonically overlie the Vaikrita crystalline complex and in turn they are unconformably overlain by Ralam conglomerates and quartzites. The sandwiched biotite porphyritic horizon, *viz*. Pindari, Rilkot and Budhi schists, represents the locally metamorphosed Martoli sediments at the base. The intense tectonic and magmatic activities, and dynamothermal phenomena, caused the schistosity of sediments with the development of random growth of porphyritic biotite.

Looking at the geologic map (Figures 12 and 13), it would be clear that the distribution of Martoli Group is not uniform from the Dhauli valley in the northwest to Kali gorge in the extreme southeast. The maximum thickness (*ca*. 6000 m) is attained in the central part of the arcuate disposition of basin in the region of Nanda Devi High (7820 m) and Gori Ganga gorge (Figures 13, 15 and 19). Notably, the highest and other higher peaks of the Great Himalayan range, *viz*. Nanda Devi (7820 m), Api (7140 m), Nanda Kot (6861 m) etc. are all formed by the Martoli Group of sedimentary rocks.

At both ends of the arcuate structure the tectonic contact is very conspicuous at Malari (Figure 12; the section in Figure 54; Shah and Sinha, 1974c) and at Budhi. The thickness at Malari is considerably reduced due to shearing off and a wide zone of mylonite involving the rocks of the Vaikrita Group. The Martoli Group exposed to the north of this tectonic zone comprises phyllites and dark grey to greenish sandy slates showing a distinct graded bedding. Towards the top the slates become cherty and also contain some greywacke. Ripple marks are visible in the quartzitic lithology.

In the Gori Ganga valley, the Rilkot biotite porphyroblastic schists pass into less metamorphosed biotite–quartz–sericite schists which near Burphu is transitionally overlain by greenish silty and quartzitic rocks showing beautifully preserved graded and cross-bedding synsedimentational structures. This zone of siltstone and phyllite is followed by a thick (*ca*. 200 m) sequence of pyritiferous quartzite. The uppermost part of the Martoli Group sequence near village Milam and the traverse along the Gonkha Gad is represented by silty argillite, calc-phyllite, and siltstone with conspicuous sedimentary structures of cross and graded bedding.

Considering the huge thickness and basal position of the Martoli Group and the unconformable Ralam Conglomerates at the top, it is considered that a division on the line of modern stratigraphic nomenclature ought to be proposed, dividing the entire group into four distinct formations with appreciable thickness and distinct lithological identity. These four formations, *viz.* Rilkot, Burphu, Bilju, and Milam, are best exposed only in the Gori Ganga section from Rilkot to Milam, and their systematic study could be done only in this part of Central Himalaya. The other sections across the Martoli Group either do not have a complete development of the constituent member formations or some member formations are drastically obliterated by tectonic faulting or the juxtaposition of thrusting slices overriding from the north.

In the Martoli Group at the base of all the sections, *viz.* Dhauli, Gori, and Darma gorge, is characterized by carbonaceous slates or argillite typically dotted with sedimentary pyrite crystals (Figures 20, 71 and 72). This sort of lithology, in association with a high percentage of carbonaceous material and sedimentary pyrites, are typical earlier geosynclinal black slate formations (aspidic in the genetic sense; unlike flysch in the general sense of the term as proposed by Heim and Gansser (1939a, p. 202)). Of course, the upper part, where we get a massive pile of silty argillite and quartzitic horizons, could be flyschoidal in nature but not a flysch type. The author has personal experience of working in the black slate pyritiferous zone of the northern slope of the Caucasus and the occurrence of such horizons in association with phyllite in geosynclinal troughs has always been regarded as earlier geosynclinal facies. However, the presence of shallower sedimentary environments could be expected as the fluctuation or pulsational movements in the mobile belt in the Precambrian Tethyan realm.

Considering the age of Garbyang Formation, it could be inferred that the Martoli Group goes back to Precambrian era. Suitable samples from different horizons of the Martoli Group were macerated by the USSR, Academy of Sciences laboratories of the Institute of Precambrian Geology, Leningrad, under the supervision of the late Dr B. V. Timofeev for microfossils, but they appeared to be barren without yielding any positive results.

The top of this enormously thick, low to almost unmetamorphosed sequence is overlain by marked angular unconformity in the Gonkha Gad valley near the Milam glacier region and in Darma valley at the confluence of the Lesser with the Darma river. The overlying Ralam Conglomerate horizon dips at low angle of 35°–40° due northeast parallel with the plane of unconformity, whereas the Martoli Group rocks stands erect with high angles of dip and show irregular deformation patterns.

Refer to Figures 13, 14, 16, 19 and 20 for descriptions pertaining to the lithostratigraphic description of the Martoli Group.

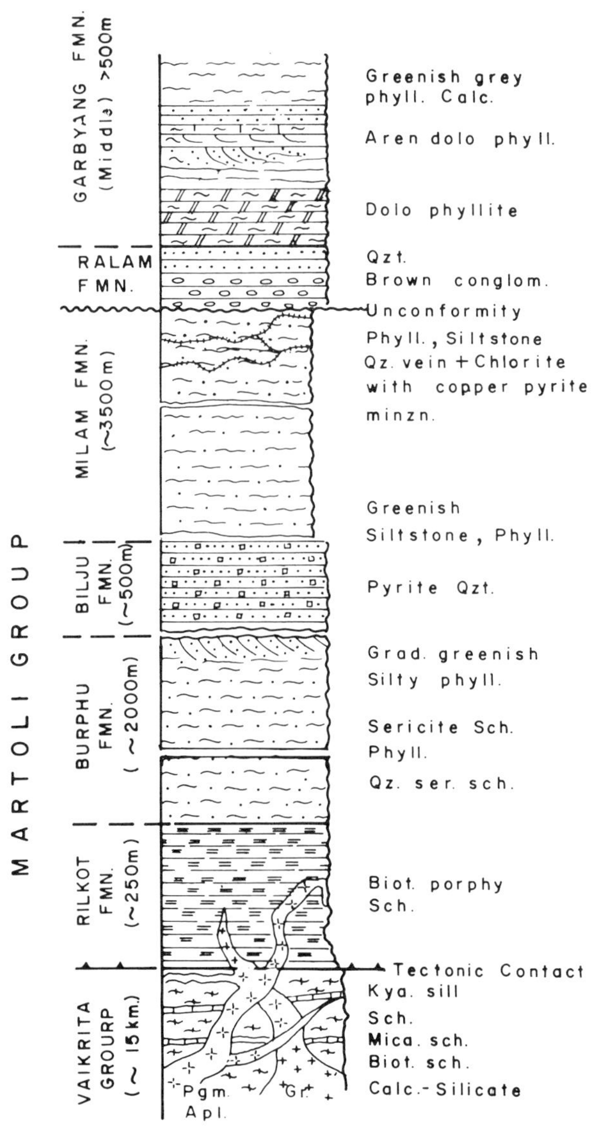

Figure 19. Stratigraphic column of the basal Tethyan sequence with subdivision of the Martoli Group

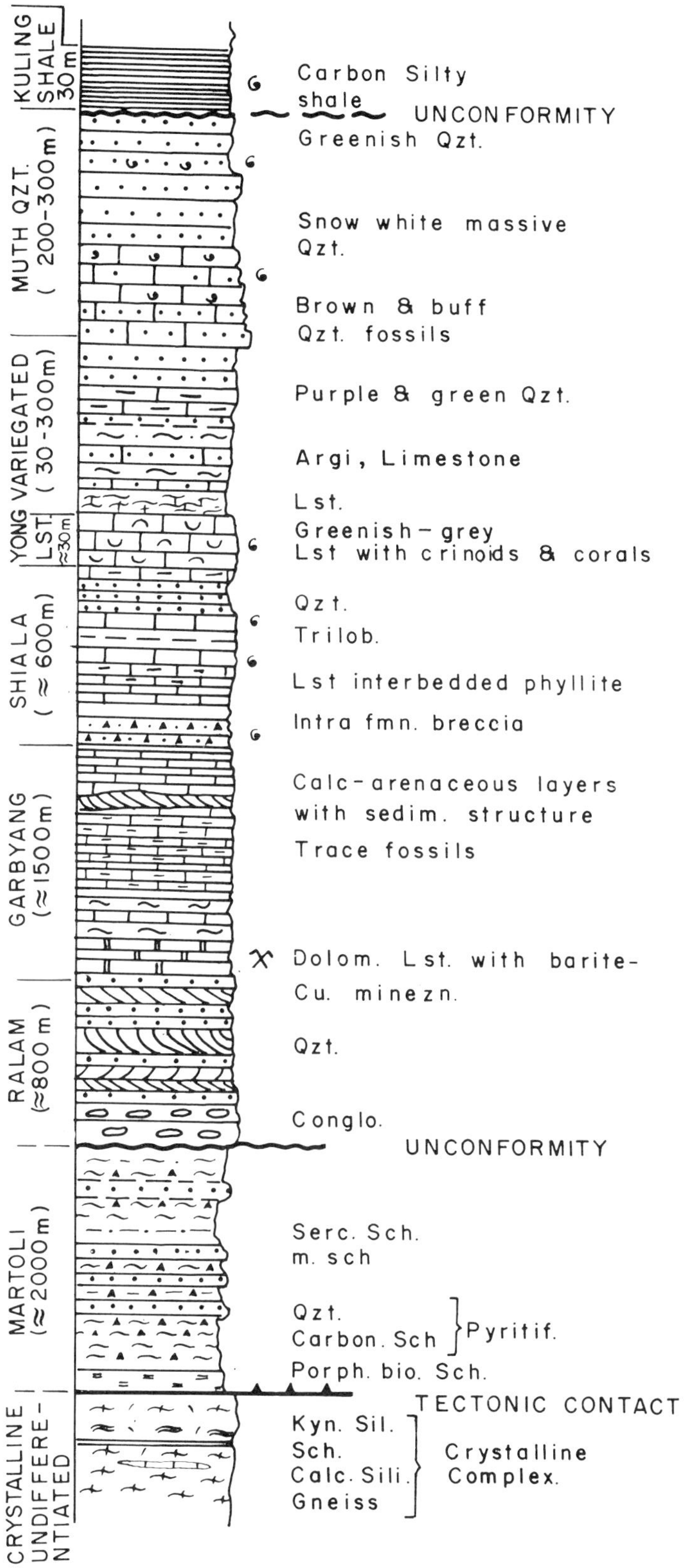

Figure 20. A generalized stratigraphic column of the Tethyan sedimentary sequence in the Darma valley

A. The *Rilkot Formation* (*ca*. 250 m) in the Goriganga valley is represented by the schists with biotite porphyroblasts. This passage bed is sandwiched between the high grade catazone kyanite–sillimanite bearing rocks of the Central Crystalline Vaikrita group and the huge geosynclinal Precambrian sedimentary pile of the Martoli Group. In hand specimen, rectangular biotitic porphyroblasts are dotted randomly in the ground mass across the foliation plane, and in the transverse section phyllitic groundmass is easily recognizable. Thus it proves that the parent sedimentary Martoli group is metamorphosed to give rise to biotite porphyroblasts. In thin section the matrix is observed to be composed of quartz, sericite with epidote, tourmaline, and magnetite. Gansser (in Heim and Gansser, 1939a, p. 88) described the rock under the microscope and states that the large biotite porphyroblasts with sieve-like quartz inclusions are pleochroic in pale yellow to dark olive-green. The quartz inclusions were primarily parallel to the schistosity but occasionally they are at right angles. The space of mineral alteration has been filled by the quartz. Often the biotite is xenomorphic with relict reticulation. The origin of the biotite porphyroblasts could be attributed to the deep-seated metamorphism of early deposited argillaceous sediments at the deep seated fault contact zone of the Vaikrita and Martoli groups.

This formation is equivalent to the Pindari and Budhi formations of the Pindari Glacier and the Kali river valley. In Dhauli, due to truncation by a fault, this formation is missing. The younger granite and pegmatites are found to intrude them in the Pindari, Gori, and Kali valleys (Figures 14 and 15).

B. *Burphu Formation* (*ca*. 2000 m, in the Gori Ganga valley)
This horizon, exposed between Rilkot village as far as Burphu village in the Gori Ganga valley and having the Martoli village in between, is characterized by quartz–biotite–sericite schists, sericite schists and phyllites, silty phyllites, and graded greenish quartzite with intense synsedimentary structures. The lower part of the formation is more carbonaceous schists with the presence of pyrite crystals, reminding us of the aspidic euxinic slate formation of the Jurassic slates in the geosynclinal regime of the Caucasus.

C. *Bilju Formation* (*ca*. 500 m, in Gori Ganga valley)
This is a pyritiferous, dirty white quartzite exposed near Bilju village. This horizon is observed in the Darma valley near Dugtu village, and in the Girthi valley about 4 km from Malari on the way to Barmatiya.

D. *Milam Formation* (*ca*. 500 m, in Gori Ganga valley)
This thick uppermost part of the Martoli Formation is exposed in between Bilju and Milam in the Gori Ganga valley and in Gonkha Gad as far as Lakhuri Bhel. It is represented typically by greenish siltstone and calc-phyllite. The arenaceous component is medium to unsorted fine grained with beautifully developed graded and cross-bedding structures. The calc-phyllite and siltstone in the Gonkha Gad valley near the unconformable contact with Ralam Conglomerate is marked by hydrothermal activity and quartz veins with mineralization of copper, pyrite, and barite with quartz and calcite.

This formation is unconformably overlain with angular unconformity by conglomerates of Ralam.

Ralam Formation (*ca*. 800 m)

The Ralam series, proposed by Heim and Gansser (1939, p. 202), is exposed in the spur of the Ralam pass (5500 m) between the gorges of Lissar and Gori. The characteristic lithology is represented by basal conglomerate and quartzite. This formation is exposed in the Dhauli gorge, the Girthi gorge, the Gori–Gonkha Gad gorge, and the Darma gorge. Due to tectonic truncation, this horizon was found to be completely missing in the Kali gorge along the slope of Chhiyalekh climb. The brick-red conglomerate has quartzitic clasts of hand to smaller size, unsorted, with a hard quartzitic matrix with bands of quartzites, where clasts are absent, or with a decrease in the number and size of clasts. The rock imperceptibly grades into grey to purple and green, very hard, massive quartzite in the upper horizon (Figures 12, 13, and 15). The best-exposed section for studying this horizon is in the Dhauli–Girthi and Gonkha Gad section near Samgaon, where a repeated thrust of Ralam conglomerate is exposed.

The clasts are almost invariably of quartzite. The upper part of Ralam assumes a dolomitic lithology and transitionally crosses over to the Garbyang horizon.

Since it lies conformably below the Garbyang dolomitic rocks and with a marked unconformity over the Martoli Group, the age of the Ralam

Formation is regarded as Precambrian–Cambrian. The marked unconformity may represent the hiatus of *Baikalian* orogeny.

Garbyang Formation (*ca*. 1500 m)

This stratigraphic formation, first studied by Heim and Gansser (1939, p. 203) in the Kali river valley, was named after a flourishing Bhotia village, 'Garbyang' functioning as the centre of Indo-Tibetan trade. It is situated on the fluviolacustrine varve type and fine clay sediments of the Pleistocene period. The geologic aspects of this area have been covered in the sections describing the traverse along the Kali River. In the earlier works of Griesbach (1891, 1893), this was considered to be the upper part of Haimanta system. The best exposure to measure and prepare a litholog is available along the newly-cut road in the Girthiganga valley between the 8 km camp point and the confluence of Girthi with Kiogad on the way to Sumna. This section was carefully measured for a litholog and studied with the samples collected during expeditions in 1973 and 1974. During the 1974 expedition, the Garbyang zone was especially studied since the rich barite–polymetallic mineralization has been discovered to be confined in this zone (Shah and Sinha, 1974b; Sinha, 1977a).

The characteristic lithological column of Garbyang, divisible in the two members, is best exposed in the Girthi Ganga gorge. Although the Garbyang section exposed in the Gonkha Gad (Gori valley) and the Kali valley does not present a complete sequence hence this significant horizon could not be divided in accordance with the modern stratigraphic nomenclature since the work of Heim and Gansser in 1936. The 1973 expedition of WIHG provided us an opportunity to examine the complete exposed sequence of Garbyang and raised its stratigraphic status to formation with the division of three A, B, and C members. The lowermost Member A conformably overlying the quartzite of Ralam assumes a thickness of *ca*. 250 m and the characteristic lithology is shaly dolomitic limestone, dark blue calcareous phyllite and sericite schist. Its total thickness would be *ca*. 250 m in the Girthi gorge exposure between 8 km milestone point and 16 km milestone point (Barmatiya from Malari). This Member A is conformably overlain by massive grey dolomitic limestone with marly bands and intercalated shaly horizon to an insignificant amount in a huge thickness of 200 m. This 'Member B' of Garbyang in its lower part is confining barite–pyrite mineralization. This massive zone is also exposed in the Darma valley before the Rama camp, in the Gonkha Gad (Gori valley) as well as in the type locality of Garbyang. The uppermost member C of the Garbyang Group (Figure 15), having a thickness of *ca*. 550 m, is well exposed in the Girthi Ganga near Dobala at the confluence of the Girthi Ganga and the Kiogad. The boundaries between members A, B, and C could be ascertained by the presence of crinoid and oolitic limestone with pyrite crystals over the massive horizon, characterized by the occurrence of unidentifiable remnants of trilobite (glabellar pieces and possibly the fragments of genae) and some indeterminate lingulid brachiopods. Within the sandstone shaly unit, about 3 km southwest of Sumna are rocks showing a lingoid type of ripple mark and a horizon of flat gastropod *Eccliopteris kushanensis* Grabau.

To recapitulate the division between the Garbyang A, B, and C members, it should be noted that lower Garbyang consists of alternating bands of calcilutite, calc-siltstone and brown dolomitic limestone in the lower part and crinoidal and oolitic limestone in the middle, the two being separated by a band of brown gypsiferous marl about one metre in thickness. The upper part consists of cross-bedded calcareous sandstone and alternating bands of sandstone and shale with the shaly members predominating towards the top. It is to be noted that crinoids are the first fossils to be observed in the entire sequence. As described earlier, in the succeeding graded-bedded shales a few fragments of ill-preserved trilobite have been observed.

There has been attempt to split the Garbyang in different ways (Valdiya and Gupta, 1972; Kumar *et al*., 1977). But it seems more justifiable that only three members, A, B, and C, should be considered for the entire Garbyang sequence exposed in the Central Himalayan region of Garhwal–Kumaun. Although the Garbyang has been considered to be Cambrian, the finding of Ordovician forms does not warrant the change of name of a purely lithostratigraphic unit (Code sec. 4.03; Code of Stratigraphic Nomenclature of India, 1971; GSI page 1–2). This section studied in a different valley shows that there is no *lithological break* anywhere in the Garbyang formation. The *Eccliopteris kushanensis* Grabau of the Middle Ordovician has been also recorded from the Dhaulagiri limestone in West Nepal. Fuchs and Frank (1981) also assigned a Cambrian

(?)–Ordovician age to the Dhauladhar limestone. Bodenhausen *et al.* (1971) have suggested that the Garbyang sequence may be older than Cambrian referring to the Nilgiri (Dhaulagiri) limestone of West Nepal. The Garbyang sequence in the Dharma valley, especially in the upper part, is profusely dotted with the trace fossils (Figure 110). In the Gori Ganga valley, the development of the Garbyang suite of rocks is mostly confined to its middle part, which has been reportedly brought by the repeated thrusting along with the Ralam conglomerates at the base.

It would not be proper to apply the term flysch facies to either the Garbyang or the Martoli, as suggested by Heim and Gansser (1939a). This sedimentary facies classification was acceptable then but it would not be relevant now. The sedimentological aspect of the Garbyang formation has been dealt with in detail by Kumar *et al.* (1977), to which readers interested in the sedimentological environment are referred. Descriptions of trace fossils from the Cambrian succession in various parts of the world reveal a rapid and sudden differentiation in the animal phyla in existence in this period. Recently a Cambrian conodont has been discovered from the Indian subcontinent in a similar set up in the Spiti region of Himalaya. The dolomitic samples were collected from the Parahio river 1.7 km upstream at Thidim (32°02′: 55°50′) (Bhatt and Kumar, 1980). The Middle Cambrian reported trilobites are: *Agraulos* sp., *Opsidiscus* sp., *Oryetocephalus* sp., *Emmrichella* sp., *Liostracina* sp., besides abundant *Ptyohoparia* spp. and brachiopods. In our sequence of Garbyang, this type of dolomitic limestone is characteristic of the Middle Garbyang sequence and as such it should fit with our Middle Garbyang or 'B' member. Occasionally, as in the case of the Garbyang Formation of Kumaun Himalaya, the fauna is fairly meagre until the close of the Cambrian. The succeeding Ordovician Shiala Formation, however, shows a greater variety of animal phyla. Both body and trace fossils are useful in starting the study of these phyla. Some trace fossils are sufficiently distinctive to allow reasonable inferences to be made regarding the phylum of the animal. The studies on modern animals and functional morphology of fossils together indicate that most biogenic structures in Palaeozoic sediments were produced by the phyla Annelida, Arthropoda, and Mollusca.

As described earlier, the Garbyang is more or less devoid of identifiable fossils except gastropods, *viz. Eccliopteris kushanensis*, and doubtful fragments of trilobites and Lingulids. The sedimentary structure in the Garbyang Formation includes small and large scale cross stratification, ripple cross lamination, climbing ripple lamination, parallel lamination, ripple lode cast, piled load casts, convolute lamination, rainprint wrinkle marks, and current and wave ripples. The lower Garbyang arenaceous dolomite sequence with shale interlayers have the trace fossil Ichnogenus: *Aulichnites* Fenton and Fenton, 1937 (Banerjee *et al.*, 1975). It occurs as bilobate epichinial ridges showing very low relief over the rippled brown arenaceous dolomitic limestone traversing the ripple crests and troughs. The animal has trailed on the surface without making a significant dent on the calcareous substratum. Thin intervening marly layers are common, which have aided the preservation of the trailing activities on the sediment interface. The epichinion consists of two equal lobes marked by a pronounced median depression. The surface is unornamented and smooth. The trail width varies from 0.5 to 1 cm and extends for almost 30–40 cm. In some outcrops the ridges are seen extending up to a metre, riding over the crests of the ripples. These traces are similar to *Roualtia de Tromelin* Hantzschel but the name is now invalid. *Conchostracan cyzicild* trails with smooth bilobate furrows separated by a lower median ridge, as described from the Ordovician of Czechoslovakia, are closely comparable. These can also be confused with *Scolicia de Quaterfrages* 1849 but lack the elevated median ridge.

These trails appear to have been produced by gastropod-like animals on very shallow tidal flat type of environment with only a few centimetres of water over the rippled surface.

In summary it should be said that the characteristic trace fossils in the Garbyang Formation of the Central Himalaya are as follows (Tandon and Bhatia, 1978): *Cruziana*, *Gorylhorte*, *Isopodichnus*, *Lennea*, *Phycodes*, *Planolites*, *Rusophycus*, *Scolicia*, *Teichichus*, *Lavicyclus*, *Aulichnites*, and *Zoophycus*.

Shiala Formation (*ca.* 600 m)

The 'Shiala Series' is assigned to the Ordovician and was first described by Heim and Gansser (1939a). It was named after the Shiala Pass (Figure 13). It has been described by them that the major part of the outcrop of this formation is covered by glaciers but even then the exposed part could convincingly help establish the Ordovician horizon in the Central Himalaya. Keeping in mind the stratigraphic position of the Shiala Series, we started our observation with the measured litholog and established the biostratigraphic zonation in the gorge of the Kiogad, a few kilometres before the Sumna campsite. It is obviously visible that there is a perfect conformity between the Garbyang Formation and Shiala Formation. The main distinction in lithology between the two formations is the presence of a biostromal band in the latter. This should be taken as a distinctive and important criterion to mark the boundary between the Garbyang and the Shiala Formations. In this section, the Shiala Formation comprises a sequence of green shale with silty bands towards the base. With the gradual increase in the arenaceous component, the Shiala Formation becomes essentially quartzitic towards the top. Throughout the formation, more towards bottom, two bands of biostromal limestone (Figures 20 and 21) consisting of fragments of arenaceous, calcareous algae, coral and shells of brachiopods etc. are met with. They vary in thickness from a few centimetres to about 60 cm, and are seldom persistent. Very often they occur as podshaped bodies within the predominantly shaly rocks. As presented in the column (Figure 15), the imperceptible transition zone between the Shiala and the Garbyang formations could be characterized by the lithology of green shale with arenite followed by the *Orthis* (*Dalmanella*) *testudinaria* zone. As a matter of fact, the occurrence of *Orthis* should be taken as the first and the best exposed fossils in the Tethyan column of Central Himalaya. This zone is again overlain by biostromal limestone-shale which becomes sandy in its upper part giving a hard quartzite marked by the preservation of innumerable fossils remnant of *Rafinesquina aranea*. This sandstone horizon with minor shale is *ca*. 200 m in thickness. Walking towards Sumna in the Kiogad valley, the upper part of the *Rafinesquina aranea* zone is followed by the *Monotrypa* zone and the *Rafinesquina alternata* zone. Finally, as the *Calostylis* zone appears, we cross over to a new biostratigraphic horizon discovered by us and named the *Yong Limestone Formation* (Figure 15) (Shah and Sinha, 1974a; Khanna *et al*., 1985).

Heim and Gansser (1939a) have also mentioned the presence of crinoid breccia in the Shiala lithology. They write: 'Summarizing our description of the Shiala pass, we may state that above the typical Garbyang division follows a series of about 400 and 500 m of variegated shales with intercalations of sandy limestone crinoid breccia of brown weathering. The Ordovician trilobites and the brachiopods mentioned on page 121 were collected *in situ* from the calcareous sandstone of about 60 m thickness on top of the division which we have proposed to call the "Shiala Series". Perhaps Capt. Strachey, without mentioning the localities, also had meant in 1848 the presence of "Lower Silurian" fossils, which should not be called Ordovician zones.

The Shiala Formation is rich in fauna and can be divided into four biostratigraphic zones. These four zones have been marked in addition to the horizon which contains indeterminate fossils. They are named after the characteristic fossil taxa, from the base, as: *Orthis* (*Dalmanella*) *testudinaria* zone, *Rafinesquina aranea* zone, *Monotrypa* zone and *Rafinesquina alternata* zone. The fossils found in each of them are given in Table 3.5 while the position in the stratigraphy column could be had from Figures 15, 21, 22, 23 and 24.

The faunal assemblages clearly indicate an age ranging from Middle to Upper Ordovician.

The exposures of the Shiala Formation studied in the valleys of the Goriganga, Gonkha Gad, Darmaganga (Figure 20), and in the Gunji Kalapani sections in the Kali valley (Figure 22) have given more detail picture of the formation.

The Shiala Formation is exposed in the right escarpment and deep-cut gorge of Gonkha Gad between Shamgaon Shillong and Dung camp. This remains inaccessible and only the part of the tectonically-complicated exposure ahead of the Shilong camp site could be seen on the left bank along the track. However, while trekking from the Shuntpani camp site via the Dung old campsite over a temporary bridge crossing the Gonkha Gad on the way to the Unta Dhura pass, we came across a big chunk of fossiliferous quartzitic sandstone. This big block (13 cm × 20 cm) with typical fossils of *Refinesquina* zone was picked up opposite Dung camp, the place called Balthun camping ground situated exactly at the right bank of Gonkha Gad at the confluence of Bamras Gad and Lesser Gad. The exposure did not become visible on the left bank of the Gad due to the very thick cover of scree material. The following fossils could be primarily identified: *Refinesquina alternata*, *R. aranea*, *Leptaena*, *Puecelles*, Strophomena, etc.

The exposure in the Darma valley about 2 km

ahead of Rama camp upstream was carefully studied and an attempt was made to give a systematic collection of fossils, attaching it with the different lithologies. It would be worth mentioning that the lowermost part of the Shiala Formation is characterized by intraformational breccia followed by limestone and phyllite, dolomitic limestone, sandy dark grey limestone, and greenish shaly phyllite, followed by trilobite-bearing horizon, 300 m thick and characterized by limestone. And finally at the top of Shiala there is an exposed 200 m thick quartzite horizon with a few metres of green phyllite before transitionally passing over to reef limestone of the Yong Formation

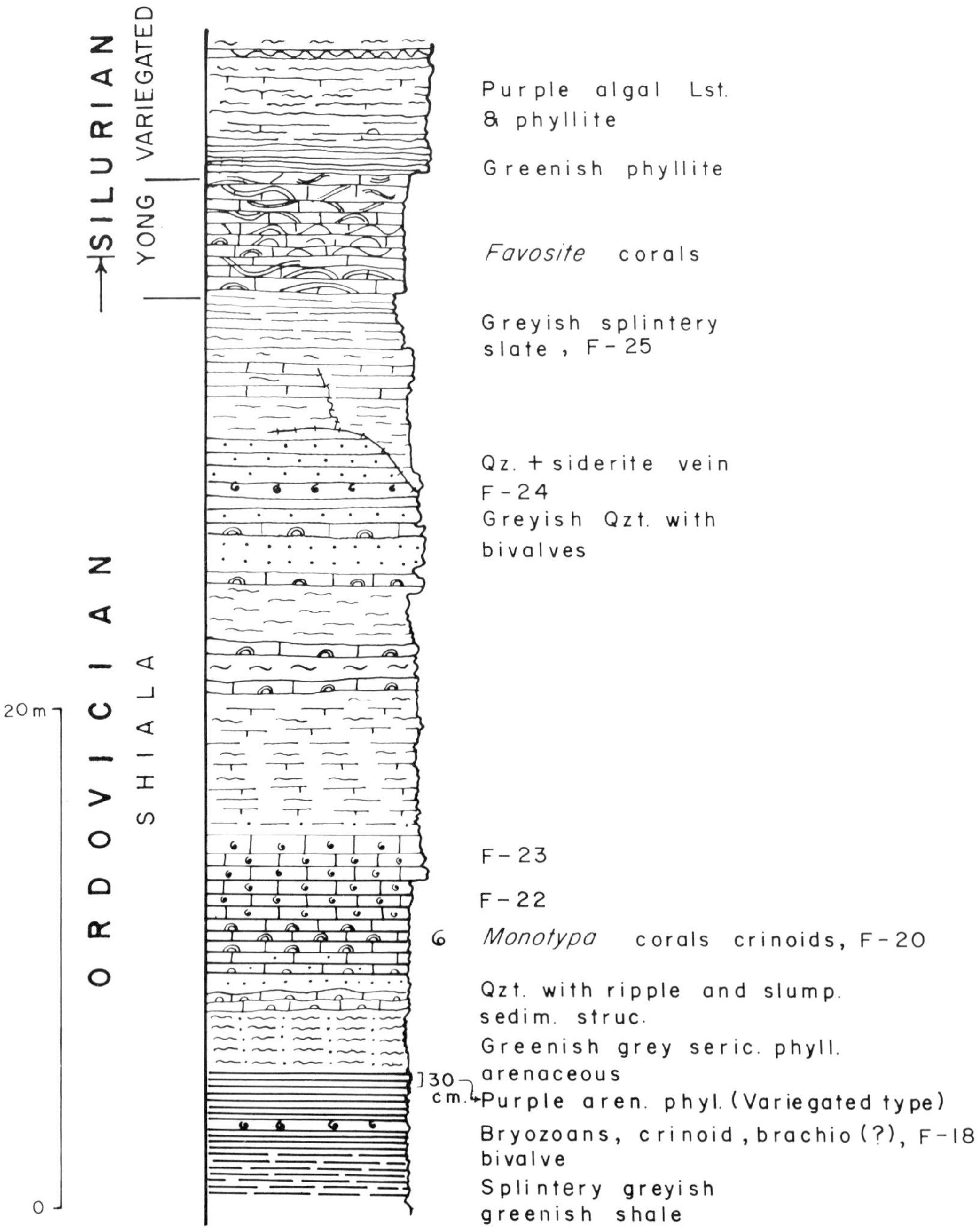

Figure 21. A measured biostratigraphic column of Shiala–Yong–Variegated formations in the Kali Valley between Gunji and Kalapani

that appears with a typical type of coral fossil.

The measured column of Shiala studied at the Gunji locality in Gunji–Kalapani sections could be clearly recapitulated in Figure 21. Here it has been found that this splintery grey and greenish shale contains the horizon of bryozoans, crinoids and brachiopods. It is followed by 30 cm of purple arenaceous phyllite of variegated type in ascending order. It is noted that greenish grey sericite phyllite of arenaceous type (5 cm) is overlain by an algal limestone band followed by quartzite with ripple marks and slump sedimentary structure, and then an algal fossiliferous limestone (*ca.* 10 m) characterized by the *Monotrypa*, corals, crinoids. Further 12 m of thickness is of typical calc-phyllite overlain by biostromal limestone phyllite and green quartzite sandstone with bivalves. The last 20 m are quartzitic sandstone intruded by a quartz sericite vein in the splintery slate. The Yong Formation with *Favosite* coral appears over this splintery horizon of slates. Thus the major horizon on the millimetre graph paper depicts the lithological unit of Shiala. Some of the samples from this horizon collected from different local Kiogad, Darma and Kali gorges have recently been reported to bear scoelecodont and similar organic chitinous organic remains.

The description of Shiala in Kali valley (Valdiya and Gupta, 1972) is being summarized for a comprehensive distribution for recapitulation in the eastern part of basin. The incomplete and poorly preserved pygidium of a trilobite resembles *Calymene nivalis* described by Reed (1912) from the Chorhoti Pass, Kumaun. The broken specimen of *Trochonema* (?) has closer affinities with the *Trochonema humifusa* reported from Malla Johar, Kumaun. *Orthis (Dinorthis) straitocostata* has been reported from the Shiala Pass (Heim and Gansser, 1939a) and the Upper Rinkin in Kinnaur in Himachal Pradesh, whereas *Orthis* (*Dalmanella*) *testudinaria* is known to occur in the Ordovician rocks of the Pin Valley, Spiti (Reed, 1912). *Strophomena chamaerops* has been described from the Hangrand Pass (Kinnaur) and *Rafinesquina alternata* from the Pethathali Nala in Kumaun. Two specimens of *Leptaena* closely resemble *Leptaena trachealis* from the Hangrand Pass in Kinnaur. *Rafinesquina loxorthis*, *Sowerbyella rugosa* and *S. clarkvillensis* are now known to the Himalaya, but they have been described from the Upper Ordovician rocks of Ohio Valley in the USA. The occurrence of *Monotrypa*, similar in form to that found near Gunji has been described from the Upper Ordovician Gauran Beds of Kashmir. One specimen of coral looks like *Holophragma* which is known to occur in the Upper Ordovician rocks of North America. *Pasceollus mellifluus*, which occurs in profusion near Gunji, has been described by Reed (1912) from the Ordovician rocks near Chorhoti Pass and Lissar Valley in Kumaun and near Muth in the Pin Valley of Spiti. In lithology and stratigraphic position, the Shiala 'Series' is broadly comparable with the North Face Quartzite of the Kali Gandaki Valley in West Nepal (Bodenhausen *et al.*, 1971).

Valdiya and Gupta (1972) have identified a separate stage called the Gunji stage within this formation, based on the fact that it contains Upper Ordovician fossils. It is to be emphasized that the Shiala Formation is, however, a distinct lithostratigraphic unit and was meant to be so as is also clear from the nomenclature proposed by Gansser (1964), enlarging his original work with A. Heim of 1939.

Trace fossils: The Shiala as such is profusely dotted with the trace fossils and some of them are Ichnogenus: *Planotias* Nicholson, 1837 (Banerjee *et al.*, 1975). *Phycodes* pieces (Figure 4 of Banerjee *et al.*, 1975), Ichnogenus: *Chonodrites* Strenberg, 1833 Ichnogenus: *Loranzinia (?)* Gabelli 1900, *Rusophycus* Hall, 1852, *Scolicia de Quatrefages* 1849, from Kali and Darma valley (near Rama in Dharma valley) and Bedang in Dharma Valley, and near Gunji in Kali valley have been described. *Skolites* Heldeman, 1890 (near Sumna in Birthi Ganga valley) have been described by trace fossils confined in the Shiala Formation are: *Chondrites*, *Lorenzinia*, *Planolites*, *Skalolithous*. Kumar *et al.* (1977) have also identified the following trace fossils in the Kiogad gorge: Ichnogenus *Dimorphichnus* Seilacher, 1955, Ichnogenus *Fuousopsis* Vassoevitch, 1932, Ichnogenus *Helminthopsis* Heer, 1977, Ichnogenus ? *tomaculum* Groom, 1902, Ichnogenus ? *Planolites* Nicholson, 1873, Ichnogenus Form.

Yong Limestone Formation (Figures 12, 13, 15, 20, 21)

This unit was first identified as an independent formation during our 1973 expedition (Shah and Sinha, 1974a), since then it was logged in all the sections in Garhwal–Kumaun Tethyan zone. This is the horizon characterized by biostromal debris. The biota consists of calcareous branching algal, stromatoporids, crinoids, corals, and bryozoans

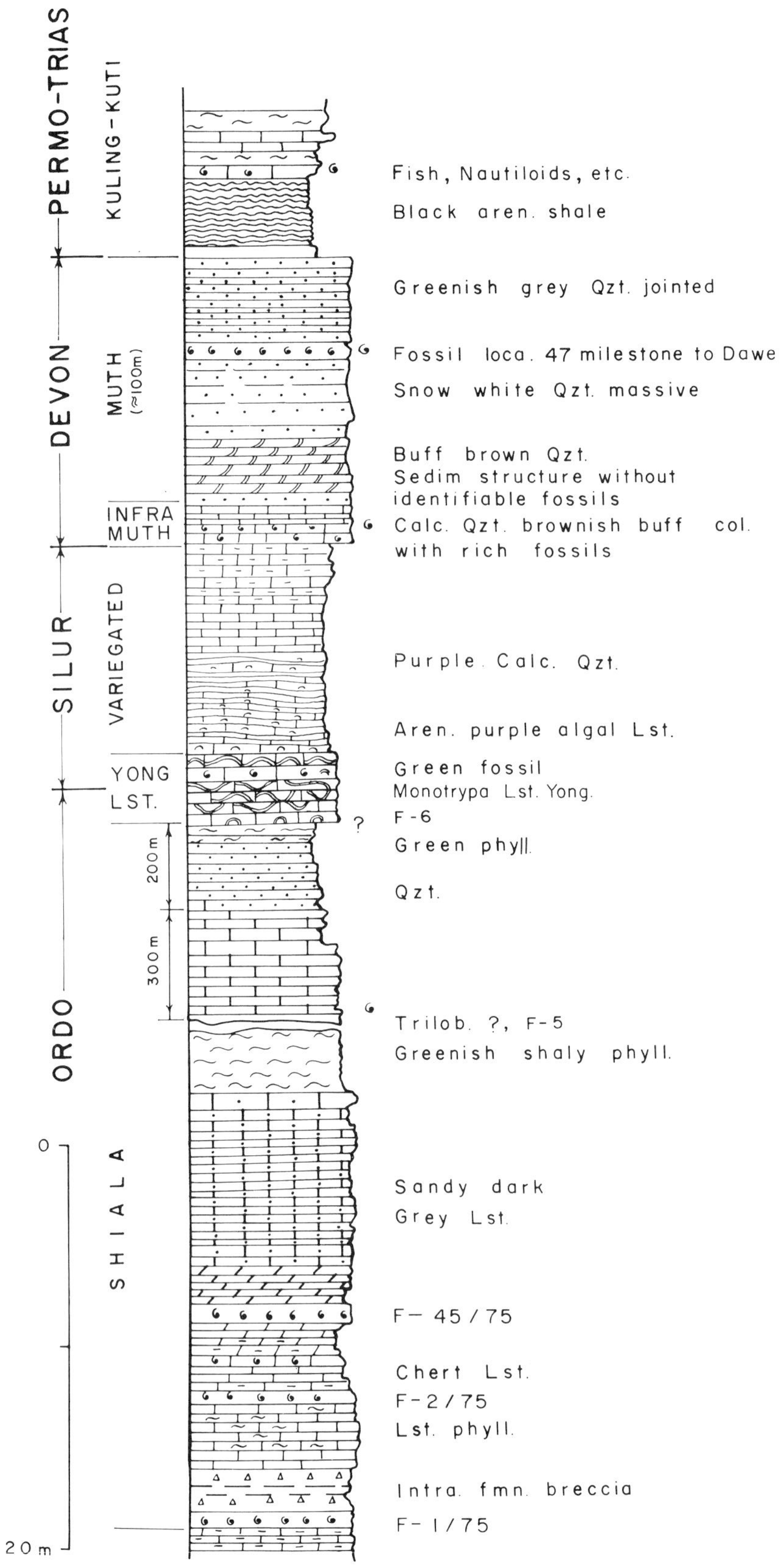

Figure 22. Measured stratigraphic column from Ordovician to Permo-Trias in the Kuthi and Kali river sections

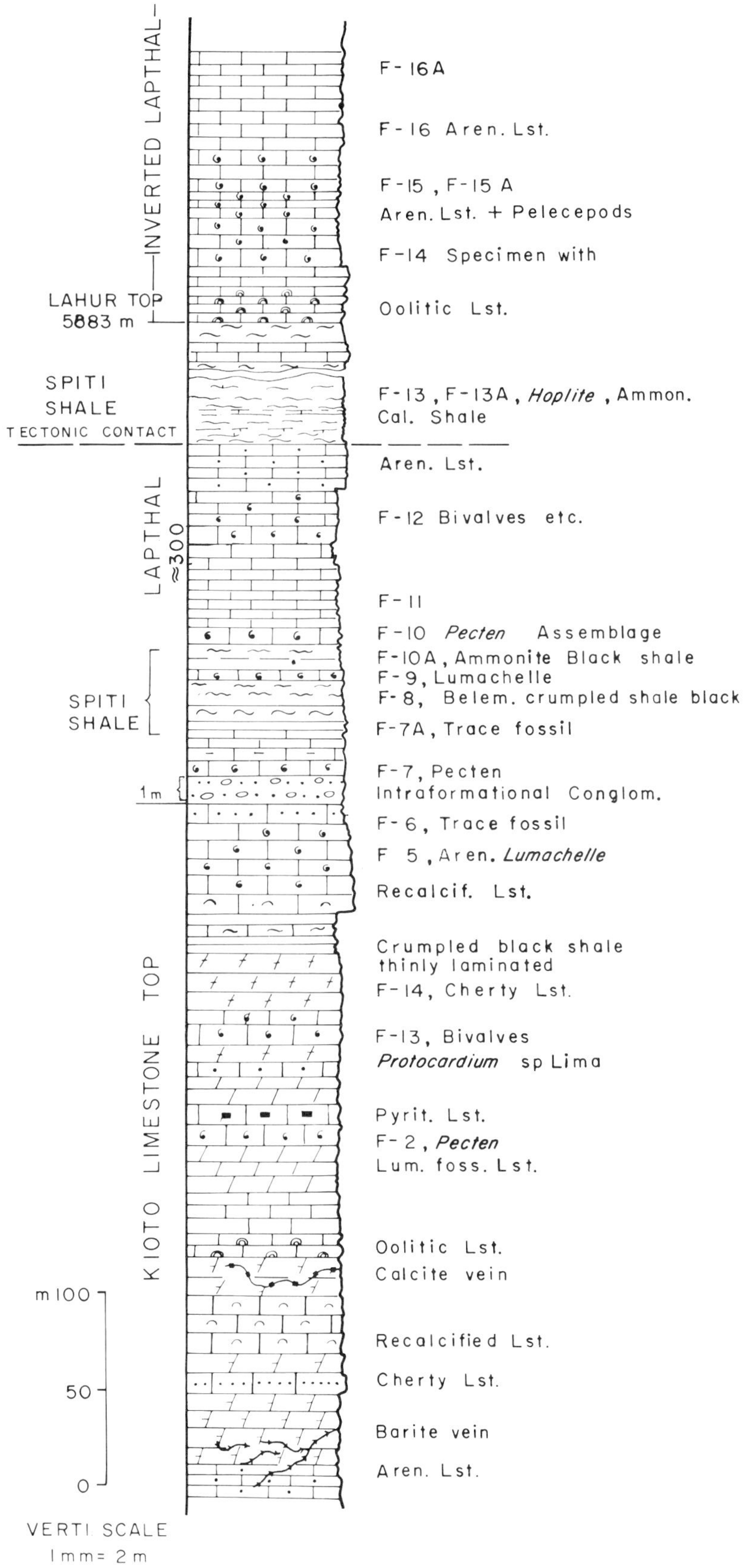

Figure 23. Tectonostratigraphic column of Khingru Dhura Pass climb involving the western limb of Lahur anticline

etc. In the Yong Gad ravine it presents a rather domed character with imperceptible bedding features. This horizon's thickness varies in fifties of metres.

One of the most interesting rock unit discovered during our 1973 expedition is an algal biohermal limestone. It is the first of its kind found in the Himalaya above the Shiala Formation (Shah and Sinha, 1974a; Khanna *et al.*, 1985). It is greyish-green in colour and continuously met with from Rimkhim to Yong Gad. It is characterized by a helmet-like disposition by which it can be spotted readily even from a distance. It comprises a series of reefal structure with quaquaversal dips in all directions of domal shape and interconnected with one another by bedded reefal debris. The biota that comprise the reefs are calcareous branching algae, stromatoporids, crinoids, corals, and bryozoans. The reefal cores lack bedding, though occasional partings of marl or calcareous mudstone are found. The reefs are discordant with the overlying 'Variegated' Formation although the latter lies concordantly over the bedded limestone intervening between the reefs. The limestone extends beyond Yong Gad in the Kiogad valley also and occurs exposed for about a kilometre to the east of Sumna, but here it is biostromal in nature and no reefal structures are present.

The formation has been named after the deserted village of Yong in the Yong valley where one of the reefs is exposed on the mule track leading to Rimkhim. This reef is about 750 m across and has an exposed thickness 45 m. Towards the margin it merges into a bedded limestone of biostromal nature but, due to the dome structure of the reef, the dip of the overlying rocks is disturbed. This phenomenon is noticed throughout the area and as a result of which the corresponding outcrops show almost a bedded

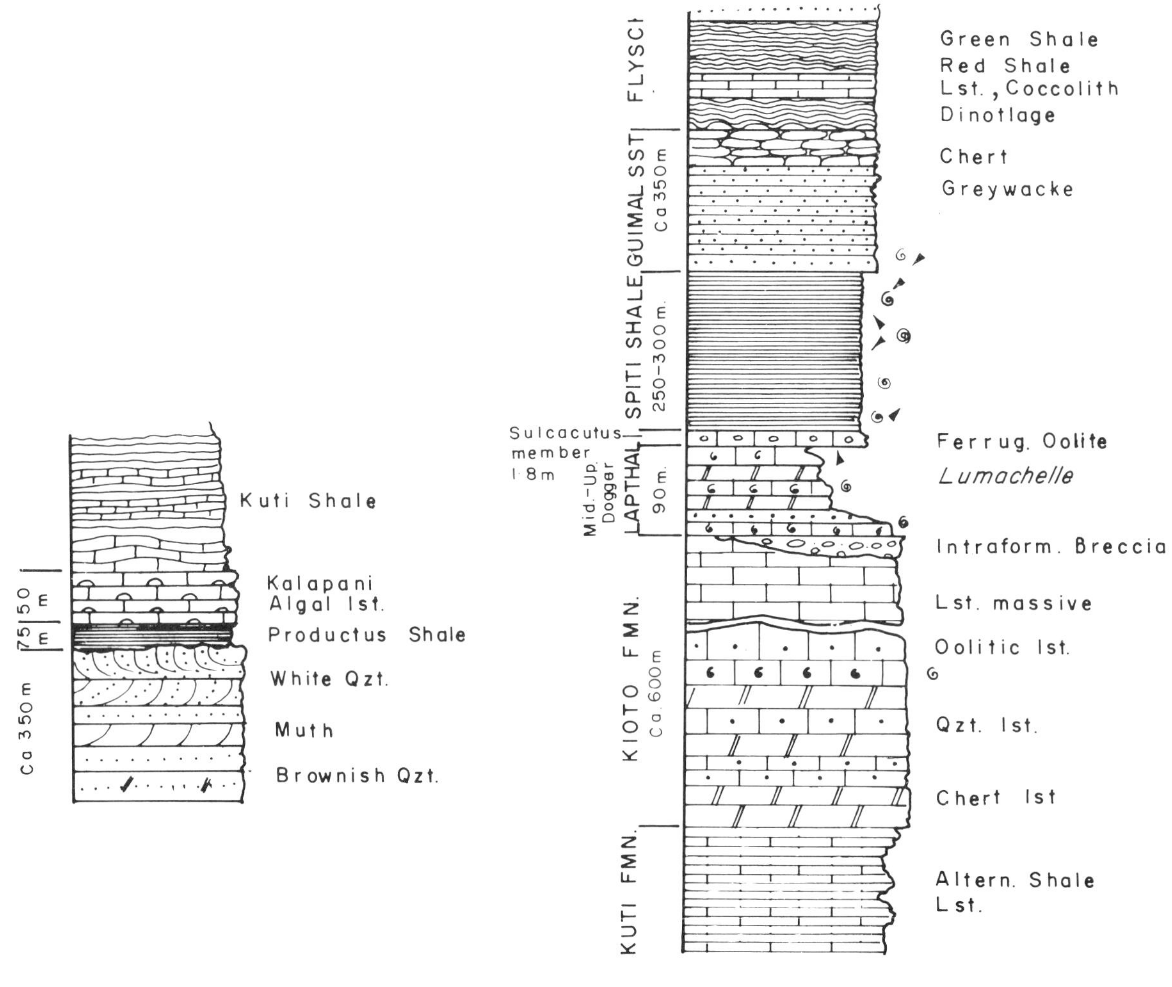

Figure 24. Detailed biostratigraphic column along Dung–Unta Dhura–Belchi Dhura Pass traverse

appearance. Dave and Rawat (1968) have mentioned the presence of a green and nodular limestone in the formation which they describe as the Somna Series. This limestone appears to be the same formation. Their lithological sequence however, indicates the green limestone at the base and nodular limestone at the top separated by pink shale and limestone, etc. Actually the green limestone and nodular limestone are part of the same formation, the Yong limestone, and the pink shale and limestone belong to the overlying 'Variegated' Formation. They are, however, repeated at Yong because of faulting as is indicated in the map (Figure 12).

Heim and Gansser (1939a, p. 204) have mentioned the occurrence of 'a dense gray, somewhat nodular limestone' on top of the Managshang pass and at Dung which they regard as 'Special facies of the variegated Silurian sequence'. Reference to this has also been made by Kumar *et al*. (1977, p. 92). This description and the stratigraphic position corresponds to the Yong Limestone. If it is so, it would indicate that this formation has a widespread distribution in the Kumaun Himalaya. As a matter of fact it was found in all the traverses, author could take in the Tethyan basin.

Although the Yong Limestone is fossiliferous, the preservation of fossils is extremely poor. Massive stromatoporids are recrystallized and have lost most of their structure. Same is true of the corals. *Calostylis* probably referable to the species *C. dravidiana* Reed has been identified with some degree of certainty and on the basis of that the assemblage zone has been named. This form has essentially Lower Silurian affinities. The Upper Ordovician–Lower Silurian age is indicated, although a little extension below into Upper Ordovician cannot be ruled out. The Yong Formation exposed in the Gonkha Gad valley has been tectonically complicated by a thrust zone which could be traced from the Kali, Kuti and Dharma sections but the typical character of reefal limestone is so characteristic that it would be easily identified as a separate horizon over the Shiala and below purple variegated calc-slate. Besides that fossils assemblage described in Table 3.6 is itself a marking horizon to identify this formation. Initial microfaunal studies have revealed that reasonably rich palynological assemblage, consisting of 11 chitinozoan and nine scolecodont forms occur. These forms tend to suggest an upper Ordovician–Lower Silurian, probably Caradocian to Llandoverian age, to the Yong Limestone and the Ordovician–Silurian boundary lies in the lower half of this stratigraphic horizon (Khanna *et al*., 1985).

For the purpose of mapping sometimes this horizon has been taken together with the Variegated Silurian especially in upper reaches of Darma and Kali valleys (Figure 21).

Variegated Formation (33–400 m) (Figures 2, 3, 5, 20, 21)

The name of this horizon is derived from the 'Variegated Silurian' proposed by Heim and Gansser (1939a, p. 204) in the Kumaun Himalaya. They recall (p. 204) that this formation was called 'Red Crinoidal Limestone' by Griesbach (1891), and was regarded as carboniferous in one place, and as basal Silurian (quartz shales of Kali) in another. The mistakes were rectified by Hayden (1904, p. 29), who ascertained that the Spiti representative of the 'red crinoid limestone' is certainly not younger than upper Silurian. This resulted in course of correlation of Spiti sequence with that of Kumaun–Garhwal. Heim and Gansser (1939a, p. 204) could not establish the coral limestone, which in Spiti underlies the red crinoidal limestone, and also noticed by Griesbach (1891) in Kali valley, although it was established that there is no gap between Shiala and Variegated Silurian. Nevertheless, the coral limestone horizon of Hayden to my mind is the reefal biostromal zone of Yong formation.

The Shiala pass exposure of Variegated Formation was estimated by Heim and Gansser (1939a): (a) 200–300 m red to green and grey shales with layer of white limestone, characterized by smaller blood red shale and rusty weathered dolomitic in the upper part; and (b) 500–600 m of shaly lenticular limestone with crinoid fragments. Somewhat nodular limestone of about 100 thickness has also been observed by them and regarded as special facies of Variegated Silurian sequence. But as a matter of fact now it has been established that this nodular limestone facies is nothing but reefal Yong horizon. Similar sequence was also observed by them at the top of Manshang pass at the Indo-Tibetan water divide and at Dung. It is interesting to recall that this seems to be already known by Griesbach who then regarded it as Ordovician. Near Gunji in the Kali valley (Figure 21) the Variegated section was measured and plotted on graph paper column with the Yong Formation. It was found that they overlie the *Favosite* coral-bearing Yong limestone of 15 m thickness overlying grey splintery shale of Shiala. While making observation Heim and Gansser (1939a) marked near Gunji in Kali valley the violate siliceous and sandy shales reduced less than 50 m. It seems to overlie directly the Garbyang series with a discontinuity. In left escarpment on the Kali gorge the violate shales are reduced to 100 m or less, but obviously this part of section is involved in tectonic complication as shown in detail sketch section referred in Chapter 5, section 10 (Figures 47 and 50). Such repetition is obvious also on the southeast extension along the strike being repeated by Nihal thrust over the younger Permo-Triassic strata.

In general the lower part of the Variegated formation is represented by purplish algal limestone and phyllite, sometime arenaceous followed by upper part mainly composed of purplish and green calcareous quartzite. Their thickness was found varying from section to section and in some places it is reduced to 30 m whereas in Dharma valley the thickness goes up to 400 m. The red coloured calc-phyllite in Kali valley shows a perfect bedding plain cleavages pattern of secondary calcite giving a look of tension gash. This formation is separated distinctly on account of colour contrast with the underlying green Yong Limestone and the overlying white Muth quartzite.

The name 'Variegated Formation' is used here to perpetuate the nomenclature adopted by earlier worker, which has become popularly known and understood by the geoscientists of Himalaya in India and abroad.

Though the formation is poorly fossiliferous one characteristic assemblage zone was recorded from Kiogad valley section named as *Strophonella* Zone. Valdiya and Gupta (1972) recorded in their intermediate horizon the various forms: *Orthis* (*Dalmanella*) *bouchardi* and *Orthis* (*Dalmanella*) *basalis* var. *muthensis*. The upper limestone members have yielded *Favosites spitiensis* and poorly preserved *Streptelsma* (?) and broken blades of polygnathids.

Trace fossils recorded by Tandon and Bhatia (1978, p. 383) is represented by *Chondrites* Sternberg, 1833 in their plate III, Fig. c. The form collected from the Darma and Kali valleys are having plant-like structures on the shale–silt interface. They range in age from Cambrian to Tertiary and considered to represent asymmetrically ramifying horizontal burrows. Ichnogenus *Volkichnium* Pfeiffer, 1965, are radiating burrows parallel to bedding plane. Burrow system are slightly irregular overlapping and branched. Presence of scratches suggests that the burrows have been produced by some arthropod.

Pan 2

Muth Formation (400–100 m, on average *ca*. 400 m)

The name Muth series has been introduced by Griesbach and since then it has been accepted equivalent to the Muth quartzite of Spiti region. In this subchapter, the Muth series is depicted as the Muth Formation according to the stratigraphic code of nomenclature.

The Variegated Formation is overlain by the Muth quartzite with an angular unconformity which is clearly visible in the Yong valley section. Heim and Gansser (1939a, p. 102) also mentioned this angular unconformity in one of the sections and regarded it as lithological rather than tectonic. The stratigraphic situation reveals that deposition at the bottom of the deep sea possibly might have been interrupted for a short period. It may be mentioned, however, that neither the 'Variegated Formation' nor Muth quartzite represents a deep sea deposit. In the section studied presently with a special reference to Yong and Kiogad valleys, the angular unconformity is sharp and clearly defined.

In the Kiogad valley the lower part of the Muth quartzite comprises about 30 m of chocolate brown quartzite and dolomitic limestone succeeded by about 100 m white sugary orthoquartzite. The lower unit, however, pinching and repeated to the northwest is completely absent in the Yong valley where the sugary quartzite directly rest on the Variegated Formation. Although lithologically the lower unit resembles the Variegated Formation in colour, it is conformably overlain by white quartzite.

In the upper reaches of Gonkhagad the Muth quartzite forms a cliff and its resistant wall helps carve an amphitheatre-like glacial geomorphic feature. For example, the Haldua Pinglua amphitheatre (5915 m) and its eastern extension is characterized by partly resistant Muth Quartzite. Further to the northeast, a huge faulting anticlinal structure with the Muth quartzite is exposed in its core as we proceed towards Untadhura pass near Pari Tal lake.

In the Darma valley, the tectonically complicated and repeatedly thrusted and folded, Muth quartzite exhibits escarpment of almost vertical wall from Bidang along the left bank of Darma valley towards Dawe pass (Photo Panorama No. 2).

In the Kali valley the disposition of Muth quartzite series assumes a larger thickness reaching up to 1000 m as an average in the Kali and Kuti valleys. It is mainly composed of brownish weathered quartzite with dolomite layers and invariably capped with massive sugary quartzite.

The thickness of Muth Formation is obvious. The facies difference is observed between NW to SE part of the Central Himalayan Tethyan basin. This is obvious from the thickness of Muth quartzite of 200 m in the Yong Gad valley to nearly 1000 m in the Kuti–Kali valleys. Moreover the tectonic complications and back folds developed in the later region has got mostly the Muth quartzite invariably in their core. White top quartzite not only varies in thickness of 200 m it is also replaced by dolomitic in the area of Kuti and Tinkar Lipu. It has also been observed that the difference in thickness might have not been caused primarily due to palaeogeologic but also by later denudation of the surface marking the unconformity. In this part of Central Himalaya it has been called 'Hercynian' gap, by earlier workers. The sedimentary study by Kumar *et al.* (1977) suggests that the environment of deposition represents the barshoal sand of a tidal sea palaeocurrent, characteristically polymodal. Main current direction is north easterly. The white sugary horizon is showing all signs of sedimentary feature, e.g. the cross bedding, mega ripples, wave ripples, etc.

The lower chocolate quartzite and limestone is fossiliferous and contains fragmentary brachiopod shells. *Pentamerus* was, however, identified from the recorded fauna. The sugary orthoquartzite is generally unfossiliferous but from some transported blocks *Schellwienella williami*, *Leptaena rhoenloidali*, *Atryapa reticularis* and bryozoans are found. Position of this horizon is not possible to fix in the column (Figure 15) since the quartzite form vertical cliffs. This was given the name of *Schellwienella* lithozone on the basis of collected fossils identified by Shah in 1974 from fallen block. Crinoidal limestone from Lipu Lekh indicating special facies has been recorded by Heim and Gansser (1939). It is expected to be Devonian.

Chondrites and *Pholeous* trace fossils are described from this horizon by Tandon and Bhatia (1978). *Pholeus* Fiege, 1944, collected from the Muth Formation at the Kali section near Gunji, appears in semi-relief on the bedding plane, 0.7 width, 3–7 cm length, larger compact nearly cylindrical dwelling burrows, longitudinal axis horizontal walls distinct, corrugated margins clearer.

In the Western Nepal mapped by Fuchs (1977, p. 196), the Muth Formation is dominated by dolomitic component while the quartzite horizons are rather interbedded. The dolomites are generally cross bedded, occasionally also rather massive in the southern line of Dolpo synclinorium, about 300 m below the top 800 to 1200 m thick Muth Formation. There is a zone about 150 to 200 m thick, rich in fossils. It is being further reported by Fuchs that the fossils are confined to a zone of marls and dolomite interbedded with brown and grey quartzite, black, silt, micaceous shale and siltstone. The fossil zone has yielded algal, corals, bryozoans, brachiopods, highly spired gastropods, and crinoids.

Heim and Gansser (1939a, p. 204) also mentioned about two lithological units in 'Muth Series'. Valdiya and Gupta (1972, p. 13) record white quartzite as the lower unit and call the higher and great Silurian rusty green brown dolomitic and fine grained quartzite that conformably follow 'Kali Series'. This part of the Muth quartzite formation is mainly developed in the Kuti and Kali Valley area. Considering the overall description of Heim and Gansser (1939a) and Fuchs (1977) from eastern area of Nepal, the whole unit should be grouped as one unit.

Fenestella Shale (? m)

This horizon was first recorded by Valdiya and Gupta (1972, p. 15) in the gorge of Kali river. Half the way between Gunji and Kalapani some carbonaceous silty shale of crumpling types bearing the *Fenestella* was collected by us (with Dr S. P. Jain, Panjab University, Chandigarh) during 1975 expedition. Despite of our best efforts *in situ* fossils could not be collected as such because this horizon should be exposed above the massive Muth formation in inaccessible escarpments. Thus presently I have not been able to map these zone *in situ* but the description by Valdiya and Gupta (1972) is incorporated for the benefit of readers. The shale are characterized by Middle Carboniferous bryozoan fauna: *Fenestella plebei*, *Protoretepora* cf. *ampla*, *Spirifer middlemissi* and *Productus* (= *Waagenoconcha* cf. *scabriculo*). However, the specimen of *Fenestella* could not be collected further upstream along track towards Lipu lekh just climbing up along eastern slope of Teragad. This was the place where Valdiya and Gupta (1972) collected *Fenestella*. To my mind the characteristic fossils and the lithology of this Teragad zone is rather closer to the Kuti shales with alternate limestone bands and micaceous silty, carbonaceous shales, rather than to the *Fenestella* shale. Valdiya and Gupta's (1972) profile should be corrected to map this part as Triassic since there is the abundance of Triassic fossils. It would be interesting to recall that, while mapping the adjacent part of Nepal in eastern continuation of the area in question, Fuchs (1977, p. 198) recalls that Carboniferous bedding is also missing in Dolpo west of Tarakphole. Probably this is related to the shallowing of the basin in western direction. This facies change tendency noted earlier in the Devonian becomes more and more pronounced. Particularly in the Devonian strata with fish fossils. In the Kali valley region, it is dolomitic (western Dolpo) mixed dolomitic quartzitic (western Dolpo, Kumaun) and pure quartzite Spiti facies. The fact that the Carboniferous is missing in parts of Kumaun and eastern Dolpo region might have been caused due to the erosional activity.

Kuling Shale (20–100 m)

The original name has been derived from the locality in the Spiti basin which is called 'Guling' in the new map series. This persistent horizon of euxenic facies of black silty micaceous shale of *ca*. 20 to 100 m thickness with well preserved fossils exhibits contrasting geology by its situation over white sugar competent quartzite of the Muth formation (Figures 14, 15a,b, 20, and 22).

In the Kiogad and Rim–Khim Gad valleys a 70 m thick band of black friable Carboniferous shale is continuously present resting on the Muth quartzite. Within these shale thin lenticles of limestone are present. This formation is rich in nodules towards upper part, which on breaking yield fossils. In this section the *Fenestella* shale horizon could not be established. The first fossil horizon immediately over the Muth quartzite is the *Paramarginifera himalayensis* zone, rich in middle to upper Permian fossils. This would indicate that the Carboniferous lower Permian is completely missing from the section. Sastry and Mamgain (1971) described it to partly 'overthrusting' but no thrusting was seen anywhere in the area and the Kuling shale clearly rest over Muth quartzite without any disturbance.

Cyclolobus Oldhami assemblage zone was recorded in the Barahoti region. Crossing the Unta Dhura Pass in the Gori Ganga valley, about 50 m thick, steeply-dipping Kuling shale horizon, resting over Muth quartzite underlain by Kalapani limestone (Figure 83), is exposed in the eastern flank of Pari Tal lake. The fossils collected from these horizons by the author and Dr A. C. Nanda are *Paramarginifera himalayensis*, and *Waagenoconcha purdonie* in the Darma valley section along eastern slope of Bedang camp site, Permian and Triassic intercalated that their identity has been recognized as Permo-Triassic (Figure 15).

The exposure of Kuling shale has been recorded and sketched in the section along the Kali and Kuti river valleys and back folding in Kuling shale functions as a prominent marker horizon to decipher the Nihal thrust (Figure 47) in the Kuti valley.

The trace fossils reported from the Kuling shale are *Cochlichnus* (Tandon and Bhatia, 1978). The geological characteristics of the formation point to a reducing euxenic lagoonal sea environment. In the adjoining area of Nepal this horizon is recognized as Thini Chu formation, consisting of white–grey green and rather dark quartzite by Fuchs (1977). The thickness increases from Thini Chu (30–70 m) to 470 m in the Kali Gandaki section (Egeler *et al*., 1964).

The Triassic of Central Tethyan Zone

The Triassic of the Tethyan zone of Himalaya as a whole has been worked out in much greater detail in comparison to other geochronostratigraphic horizons. The basic reason for such special attention was attributed to the resemblance of fossils assemblage from the Alps. The pioneer work was done by the group of Austro-German geologists who were engaged by the Geological Survey of India as early as the end of the last century. The fundamental contributions have been made by Griesbach (1891), von Kraft (1904), and Diener (1912). Amongst them the contribution by Diener were the most valuable, starting in 1895 and culminating in 1912 with the GSI memoir volume 36, part III. von Kraft made special studies of the Triassic in Kumaun in 1900. His stratigraphical and tectonics results have been summarized in the interesting paper (GSI Memoir volume 32, part III). Soon after finishing his mission he died suddenly on 22 September, 1901, in Calcutta. His materials, note books, sketches and fossil collection made with F. H. Smith were passed on to Professor Carl Diener of Vienna University for completion of the publication of the unfinished valuable research. Professor Diener was assigned this job by the then director of GSI, Dr C. L. Griesbach. The early work on the development of lower Trias in the Painkhanda region of the higher Garhwal Himalaya has been known through work on the classical Shal Shal Cliff near Rim Khim (Griesbach, 1891; Diener, 1903). This sequence is summarized by Diener (1912) as follows, in descending order.

9. Hard, splintery, nodular limestone (Niti limestone, 60 feet).
8. Earthy, grey limestone, shaly near base, with *Rhynchonella* Griesbach and *Sibirites* Prahlada: 3 ft.
7. Thin-bedded, grey limestone with regular partings of shale, containing *Plemingites rohilla* and *Pseudomontis himaica*: 25 ft.
6. Grey limestone, divided into two bands by a shaly parting; no determinable fossils: 5 ft.
5. Dark concretionary limestone, containing the main layer of *Meekoceras markhami* and *M. varaha*: 8 ins.
4. Dark-blue shales, separated by a band of grey limestone. Unfossiliferous: 18 ft.
3. Dark blue limestone, with *Otoceres woodwardi* and *Ophiceras tibeticum*: 5 ins.
2. Dark hard clay, with limestone concretions, containing very few fossils (*Episageceras dalailamae*, *Proptychites scheibleri*): $1\frac{1}{2}$ ft.
1. Dark-blue limestone, main layer of *Otoceras woodwardi* and *Ophiceras sakuntala*. Near top rich in *Pseudomontis griesbachi*: 1 ft.

Heim and Gansser (1939a), keeping in mind the early recognized nomenclature, proposed the following main lithological divisions in ascending order:

1. Chocolate series (= chocolate limestone, Griesbach).
2. Kalapani limestone (after the famous place in Kali River).
3. Kuti shales (after the Kuti village situated in Kuti Yankti).
4. Kioto limestone (name from Spiti adopted from a sketch of the geography and geology of Himalaya by Burrard and Hayden and revised by Burrard and Heron (1934).

The author after making a thorough lithological and palaeontological investigation could also establish that the above-mentioned subdivisions of Triassic remains undisputable, and henceforth the description of the individual horizon would be taken up.

Chocolate Formation (30–50 m)

The horizon of the Chocolate Formation is not developed uniformly throughout the different observed sections from east to west. Their development is more pronounced in the upper reaches of Gori valley near Unta Dhura pass and exposure in Kuti and Kali valleys, previously known as Byans. Heim and Gansser (1939a) described this as formation with typical chocolate colour due to the brown weathering. They have observed that it is more shale than limestone. Their analyses together with a microscopic test have shown that dark blue fine grained dolomitic limestone contains 50 per cent or more of insoluble (silica with clay), and 50 per cent or less of carbonate (Ca, Mg, Fe). They have observed that black shaly parts deserve even less to be called limestone. The Chocolate series of Heim and Gansser (1939a) has been called, henceforth, the Chocolate formation in accordance with the code of stratigraphic nomenclature. At the confluence of the Shalshal and the Yong Gad this horizon was spotted with the underlying Kuling shale and overlying Kalapani limestone. Kumar *et al.* (1977) referred this horizon to the lower Triassic on the basis of conodonts found in them. Dr R. J. Azmi, WIHG (Azmi and Sinha, 1981) reported some typical lower Triassic conodonts from my collected samples at the contact horizon between the Kuling shale and the Kalapani limestone exposed in the Kiogad on the way to Rewlibagar. But it is to be noted that there is no Chocolate horizon available here. The total thickness of about 5 m consists mainly of a greenish band and limestone with alternating greenish-black shales. The fossils, reported by Dr Azmi from a sample of Kalapani Limestone from the Unta Dhura Pass in the Malla Johar region, are: *Neospathodus triangularis*, and forms transitional to *Neogondolella jubata–N. regalae*, indicating latest Scythian age.

The horizon exposed in the Unta Dhura pass area is not approachable for study but it looks like the Kalapani algal limestone horizon of greenish limestone directly succeeding the *Productus* shale.

However, the development of this formation obviously is best in the Kuti Yanti near the 'Castle hill'. I reproduce here the observation by Heim and Gansser (1939a) in the region of Kuti and

Kali valleys formerly known as Byans having a thickness of 32 to 50 m.

(a) 'Castle hill' horizon, 3 to 5 m of streaky or nodular impure ferruginous limestone with shaly layers forming a light brownish projecting bend on the cliffs, and rich in imperfectly preserved ammonites of the *Ophiceras* horizon. The following fossils were collected by Heim and Gansser from 2 to 5 m of nodular ferruginous limestone with clay ironstone layers, 3 km NWW of Kuti (identification by A. Jeannet):

Pseudocageceras aff. *drinense* Arthaber
Ophiceras cf. *demissum* Oppel
Anakashmirites nivalis Diener
Meekoceras hodgsoni Diener
Vishnuites sp. (cf. *pralambha* Diener)
? *Koninckites* sp. (aff. *Yudishthira* Diener)

(b) Middle division (30 to 50 m) a repetition of impure limestone, clay ironstone and shale with rare and badly preserved fossils.
(c) Upper division (3 to 6 m) usually well defined, of chiefly shaly layers, forming a black band below the wall of the Kalapani limestone.

The Chocolate series is Lower Triassic or Scythian.

Kalapani Limestone (*ca*. 50 m)

The Kuling shale in the northwestern part of the area is directly overlain by the Kalapani limestone, a bluish massive limestone towards the lower part and nodular limestone above. It appears to be of the nature of algal biostromes. The maximum thickness is available in the Kiogad valley and in this valley it does not exceed 20 m. The Kalapani limestone is highly fossiliferous in certain sections. One such locality is in the Yong valley on the pony trail to Rim Khim, about ½ km ahead of the confluence of the Shalshal Gad and the Yong Gad. Here, apart from large numbers of pelecypods, ammonites of typical Carnic affinity are found. These include among others, species of *Gymnites*, *Arcestes*, *Ptichites*, *Griesbachites*, *Holobia,* etc. This has been given the name *Ptychites* zone and seems to correspond the *Halobia* and *Tropites* beds of Spiti as also indicated by Heim and Gansser (1939a). Some conodonts have been reported by Azmi and Sinha (1981). They are *Neogondolella excelsa* and *Himalayalla dropla* indicating Lower Ladinian age from sample collected near Sumna in Kiogad valley.

Sahni and Chhabra (1974) reported middle Triassic conodonts recovered from the samples of Kalapani limestone 5 km east to Sumna in the Kiogad valley. The samples bear close affinities with European forms and consist of eight species which are abundant in the Alpine region. The conodont assemblage consists of 45 taxa of which eight are newly described. The conodont species from Malla Johar are from Upper Anisian to Lower Ladinian age.

The samples of Kalapani limestone analysed by Sahni and Chhabra were collected from two localities between Malla Johar and the Niti pass. The conodonts from the Kalapani limestone from Malla Johar are probably younger than the Niti.

The Kalapani limestone in the area of Unta Dhura pass assumes a thickness *ca*. 50 m and lower part having algal characteristic (Figure 24). The Kalapani limestone as a whole represents the three important periods, Anisic, Ladinic, and Carnic, which have already been recognized by Diener (1912). Its equivalent in the adjoining western Nepal is represented by the Mukut limestone of Anisian to Carnian age (Fuchs, 1977, plate 5). In the Kalapani–Pankhagad area, ammonites have been found at certain horizons throughout the limestones. Heim and Gansser (1939a) described that, lithologically, the rocks in the Kalapani–Kuti area are characterized by its orange colour patches of ankerite and ankeritic calcite in tiny rhombohedra. Throughout the Tethyan basin in Central Himalaya, while mapping, I found that the Kalapani limestone is persistent between the carbonaceous Kuling shale and the limestone bands of the Kuti horizon. Diener (1912) identified abundant fossils forming the *Tropites* limestone at the top of Kalapani horizon which were collected by Smith and von Kraft during 1899 to 1900 from Teragad, Lilinthi, Nihal, and Kuti.

There are on the whole 168 species known up to the present time from the *Tropites* limestone. Not less than 155 species are ammonites. Among them 102 are peculiar to this horizon, whereas 53 are identical to species either from the Alpine Hallstatt limestone or from the *Halorites* beds of the Bambanag section. But a very large percentage of the species peculiar to the *Tropites* limestone are closely allied with Alpine types. Thus the relationship with Upper Triassic fauna of the Mediterranean region are very strongly marked.

A. von Kraft was the first author to notice the strange assemblage of Carnic and Noric in one single bed of limestone with a thickness of only one metre (approximately 3 ft). From the sandy limestone of the Kalapani formation, corresponding in its stratigraphic position to the *Tropites* limestone of Diener, were gathered numerous cephalopods.

We thus have here an almost exclusive cephalopod horizon, the *Arcestidae* (Carnic) being preponderant. But there are also the Noric *Paratibetites*. In the author's idea it is a condensed product of Carnic and Noric age similar to the *Tropites* beds, although no true *Tropites* are preent at Tinkar Lipu.

We recall here the passage from page 208 of Heim and Gansser (1939a): 'In consequence of the above observation the Kalapani limestone is to be recorded as comprising not only the Muschelkalk but three Triassic stages: Anisic, Ladinic and Carnic. Diener (1912, p. 1–9) has shown the Carnic stage towards N to W from Byans (Kali section). The Pankhanala (267 m) and Spiti have a thickness of 517 m, while in Kumaun it is supposed to be represented by the upper part the Kalapani limestone of 30 m at the most. In this sense the Kalapani limestone even as whole is a somewhat condensed deposit'.

Agarwal and Kumar (1981) reports an ostracod assemblage from the Kalapani limestone. In all 13 taxa, belonging to 11 families of the Ostracoda which are all long ranging, are as follows: *Brinsteiniana galba*, *Bythoceratina* cf. *B. mastayerae*, ? *Monsmiralibia* sp., ? *Palaeomonamiralbia* sp., Gen. cf. sp. indet cf. *Kloedenellitina* sp., ? *Primatia* sp., *Thaumatocyrpus* sp., *Polycope* cf. *P. Orbiocularis*, *Cytherella* cf. *C. ovata*, *Hungarella* cf. *H. contractula*, *Monoceratina* sp., *Bairdia* sp., *Krithe* sp., and *Bairdia* sp. indet. have been recorded. This assemblage is associated with the foraminifera, conodonts, micro-molasses, microfish remains, and indeterminate fragments of skeletal remains. All the recorded ostracods are long ranging forms. However, on the basis of the earlier recorded conodont *Gladiogondolella tethydis*, Anisic to Ladinic age could be assigned to the assemblage.

Kuti Formation (*ca*. 500 m)

The name of the formation is derived from Kuti village (3750 m) situated in the bank of the Kuti Yankti river (Figure 13). The typical alternating horizon with shale and limestone lying over the Kalapani limestone formation belonging to Kuti shale is exposed throughout Tethyan zone of Central Himalaya (Figures 13 and 15). This typical lithological complex is found to be conformably overlying the Kalapani limestone. Its thickness varies from 300 to 500 m (Figures 17 and 26). This horizon is being given the status of formation in conformity with the mode of stratigraphic nomenclature since it represents a consistent, well demarcated, quite thick horizon throughout the Tethyan zone. Geomorphologically, the Kuti formation is exposed along the escarpment wall of the steep gorge in the Kiogad valley on the track through the campsite to the Rewlibagar top (Figures 12 and 13, Photo Panorama no. 3) and rooms the difficult passes of Unta Dhura (5456 m) Pan 3
and Jainty (5600 m). Its geomorphic feature are controlled by the competent horizons of the Kalapani limestone and the Muth below it and hard massive limestone of Kioto overlying it. The upper part of Kuti formation is characterized by regular layers of great limestone. Sometimes folded disharmonically in such a manner that it forms a vertical wall and a pillar-like feature. In the Shalshal Gad and Kiogad valleys, including the Niti pass area, the Kuti formation is represented by at least 400 to 500 m suite of disharmonically-folded alternating bands of shale and limestone. The shale is dominated by black micaceous compounds. The individual limestone bands do not exceed a metre thickness. This formation has been found to continue northeastward up to Rim Khim where it is cut out by a transverse fault (Figure 12). In spite of its thickness, it has been found to be specially fossiliferous in these northwestern parts. Some long-ranging pelecypods were collected by the authors with Dr S. K. Shah during 1973 expedition of the Wadia Institute.

In the region of the Unta Dhura pass near Lauka, situated on the northern slope of the trail down to Topidunga, fossils were collected by the author with Dr A. C. Nanda, who has identified them as *Juvavites* (?) *angulatus*, *Lilangina* cf. *novites*. The bryozoans were collected for the first time by us from this horizon. The Cephalopod collection of Heim and Gansser from the Kuti shales has been described by Professor Jeannet (1959).

Numerous tiny gastropods, pelecypods, and brachiopods are reported to be Noric fauna, characterized by the numerous *Parajuvavites* and *Steinmannites*. From the overlying shales, in the talus, was found *Nonotissalinaria* Bronn, also a Noric species.

In the adjacent region of Western Nepal, Fuchs (1977) described the shale horizon and named it the Tarap shales equivalent to Carnic–Noric age. In Eastern Dolpo, the Mukut limestone of Tarap shales are fossiliferous. There the carbonate becomes sandy and silty and pass into flyschoid Tarap shales. Upper portion of the Tarap shales consists of shaly siltstone and impure sandstone (150–200 m). Singh *et al*. (1980) record the presence of a Triassic foraminiferal assemblage from the lower part of Kuti Shale formation.

Kioto Formation (*ca*. 700 m)

The Kuti shale is conformably overlain by a thick suite of limestone, occasionally having an arenaceous composition. In some places recalcification of limestone is rather obvious. The exposure of the Kioto limestone formation forms a continuous ridge from Rewlibagar to Rim Khim separating the Lapthal–Shalshal plateau to the northeast from the deeply-dissected rugged to the southwest. The lower part of this formation is arenaceous and occasionally cross bedded while part is essentially calcareous with numerous bands of oolites, coquina, etc. Kumar *et al*. (1972) separate the lower part into the so-called passage zone. The creation of the separate unit, however, is not justified since it was also mentioned by Heim and Gansser (1939) that it is an integral part of the overlying limestone with which it gradually and transitionally merges. The lower part of this formation, in the escarpment of Rewlibagar climbing from Kiogad valley, contains characteristic pelecypods and *Megalodon* in addition to number of other fossils like *Pecten*, *Spiriferina*, and *Megalodon*, reported earlier from the Spiti region. In this section this is the only characteristic biozone of Kioto formation. Apart from 'Problematica' (Heim and Gansser, page 39, plate 16, 1939a) and a number of bands of Coquina of pelecypods deformed to such an extent that no determinable form is present. A measured column was prepared in the gorge of Girthiganga along the trail from the Topidunga base camp (Photo Panorama No. 4) as to Kingru Dhura Pass, exposed in the right bank of the gorge (Figures 3, 24 and 25). As depicted in the diagram the basal part of the Kioto limestone is arenaceous and sometimes quartzitic (Figure 39). The arenaceous limestone occupies the bottom most part of the column. It is criss-crossed by the secondary hydrothermal activities with barite and calcite veins of a thickness not exceeding 5 to 10 cm. The upper part of the first 50 m is dolomitic, then a band of about 10 m of cherty limestone is situated within the dolomitic limestone. A zone of about 25 m of recalcified limestone overlying dolomitic horizon is followed by 15 m dolomitic limestone with secondary calcite veins. A merely oolitic limestone horizon of 10 m thickness is situated between dolomitic horizon below and grey massive limestone 10–15 m thickness on the top. The upper part of the dolomite limestone is Lumachelle. Fossiliferous limestone with the fossils *Pecten* (F_2). Further 8 m is pyritiferous limestone which is followed by 20 m of massive dolomitic zone with arenaceous component on the upper part followed by *Protocardium* bearing horizon (F_3) and other fossil remains. The remaining 100 m are the horizon with the cherty limestone with *Lumachelle* bed and trace fossils. Over these arenaceous members fossil (F_6) bearing horizon a typical intraformational conglomerate of 1 m thickness is observed over the top and below Lapthal formation. It would be important to emphasize that the tectonostratigraphic column (Figure 23) was measured and plotted in a graph paper since it represented a good area for observation and measurement of the exposed rocks with fossiliferous strata, but on the other hand it should be noted that this section represents a repeatedly folded limestone of the western flank of the Lahur anticline and that is why it could be observed that the Spiti shale and the Lapthal formation in the inverted limb repeated twice. The intraformational conglomerate is quite typical in character with its 1 to 2 m thickness and is constituted by clasts of arenaceous dolomite. The cement is calcareous argillite.

Pan 4

A barite vein with a thickness of 10 to 30 m has been spectrally analysed and the geochemical aspect is described in the next chapter. Such intraformational conglomerates are not observed along the Rewlibagar climb in the Kiogad gorge. While comparing the Kioto limestone in his stratigraphical column in different parts of Himalaya, Gupta (1977) described that stratigraphical position of Kioto limestone succession is defined between Middle Upper Noric and Lower Dogger (Table 3.3). He has compared the columns of Kashmir, Ladakh, Spiti, Kumaun and Nepal.

The age of the Kioto limestone: The grey massive type was regarded as Rhaetic by Hayden

(1904, p. 73). Diener (1912, p. 128) discovered fossils in the ferruginous oolites and erroneously ascribed the '*Sulcacutas*' beds to the Middle Jurassic, having overlooked the discontinuity at the base of the oolite. In general, the Rhaetic age of the Kioto formation has been maintained so far. But in the report from the adjacent region in the Western Nepal, Dr G. Fuchs assumes a Rhaetic to Dogger age for the Kioto limestone quartzitic bed in the Kioto limestone sequence. He also mentioned that the obvious fact observed by Gupta (1977), that the uppermost Noric is also present in the succession, cannot be excluded.

Lapthal Formation (91 m, Middle to Upper Dogger)

The name of this formation was given by Heim and Gansser (1939a) from the locality of the Lapthal campsite situated on the Spiti shales between Chidamu and Sangchamalla. The exact thickness exposed in the Kiogad gorge had been carefully measured and it was found to be 91 m (Figure 25). On detail documentation it has been found that the main lithological component is calcareous but it is mixed up with arenaceous (Figure 23). These formations are characteristically folded, as seen from Chidamu through the Khingru Pass Lahur Anticline (see Photo Panoramas 3 and 4). In the section from Rewlibagar to the Lapthal camp site, this formation has a total thickness of about 80 to 90 m. About 10 m from the base of the Rewalibagar campsite (Photo Panorama No. 3, Fig. 15), a horizon bearing pelecypods was found and named as *Pecten* and *Astarte* assemblage zone.

The major section in the Kiogad gorge (Figure 25) up to 50 m represents the Lapthal formation from the type area. The topmost part of the Kioto limestone consists of black cherty massive lime-

Table 3.3. Stratigraphic position of the Kioto limestone exposed in Central Himalaya adapted after Gupta (1977, p. 195)

		Kumaun	
		Northwestern	Northeastern
	Liassic–Dogger	Sulcacutus stage containing Middle and Upper Dogger fossils *Lumachelle* Formation	Lapthal Beds containing Middle and Upper Dogger fossils
	Liassic–Dogger	Massive unfossiliferous dolomitic limestone *Lithiotis* limestone	Massive unfossiliferous limestone and dolomite *Lithiotis* limestone Limestone containing *Conchodon* Massive limestone containing poorly-preserved *Megalodon*
UPPER TRIASSIC	Upper Triassic	Massive limestone and dolomite with *Megalodon ladakhensis*, *Diceracardium himalayanse*, *Spirigera noetlingi*, *Lima*, *Entolium subdemissium*, *Spiriferina* cf. *haueri*	
UPPER TRIASSIC	Rhaetic	Bioclastic grey arenaceous limestone near Chhota Hoti containing *Neospathodus hernsteini* and *Stensiona*	Quartzite, limestone and shale
	Noric	Quartzite and limestone with *Spirigera (Athyris) maniensis*, *Aulacothyris joharensis*, Spirigera, cf. *noetlingi*, *S. dieneri*, *Pecten interruptus*	
	Noric	Brown limestone containing *Sagenite* (or *Anodontophora*) *griesbachi*	Limestone and shale containing *Monotis salinaria*

stone and oolites at the contact point, transitionally superseded by cherty limestone with calcite and secondary calcite quartz veins (Figures 24 and 25). Grey sandy limestone covers the rest 5 m to 10 exposed thickness from the bottom being followed by arenaceous limestone of greenish colour with shales and in this horizon the typical *Lumachelle* start appearing. The colour becomes brownish and composition is more sandy. Up to further 30 m the brown sandy limestone continues and at the point 45 m measurement from the top a horizon of 4 m thickness is characterized by black sandy limestone appears with large 7 to 10 cm belemenite. The rest part of the sequence as shown in Figure 24 is composed of cherty limestone, thinly laminated calc silty shale with intermittently occurring at least seven to eight bands of *Lumachelle*-bearing horizon. The last 10 m at the top contains belemnites, brachiopods etc. *Lumachelle* is so characteristic of this horizon that it assumes the name 'lumachelle formation' in Nepal (Fuchs, 1977, p. 209). Some of the typical fossils identified from this horizon are: *Rhynchonella*, *Isognomon* sp. The Lapthal bands contain Middle and Upper Dogger fossils. The *Lumachelle* formation appears to be a rather consistent horizon in the part of Central Himalaya including Nepal and it has been established as Upper Dogger in age by Fuchs (1967, 1977) and also Dogger according to Bordet *et al.* (1972). With the above-mentioned data, a Middle to Upper Dogger age for the Lapthal formation is accepted.

The characteristic Lapthal formation type is exposed in the western limb of the Lahur anticline. Repeatedly folded in the inverted limb has been document in the Figure 41. Since the repetition is so common and hence in the tectonostratigraphic column Figure 24 and its upper part over the Lahur top (5833 m) remain difficult to identify with respect to their corresponding stratigraphic zonation. I must recall that this is one of the most fascinating sections studied by von Kraft (1904) and Heim and Gansser (1939a), and myself with Dr A. C. Nanda during 1976 field work. Many interesting biostratigraphic features could be documented here, while climbing the difficult passes of Kingrudhura (5244 m) between the Lahur (5833 m) and Kingrudhura (5848 m) peaks.

Sulcacutus the Ferruginous Oolite (3 m Callovian)

This significant rather thin index bed was discovered in 1895 by Diener in Chirchun area and is easily referred to because of its abundant *Belemnites* as the Sulcacutus bed. In the southeastern part of Kumaun, it was found in 1899 by F. H. Smith in the Byans (Diener, 1912, p. 119) as a 'constant and well marked horizon on the top of the limestone series'. Heim and Gansser during their mapping in 1936 found them on the Chagu pass east of Kuti. The fossils collected from the Lapthal area fix the age of this horizon as Callovian (Figure 15). However, it has been found that the Ferruginous oolitic is not very consistent in the exposed zone between the Lapthal and Spiti shales. Heim and Gansser observed that Ferruginous oolite is absent in the Lapthal valley. The Lapthal beds lie over the Kioto limestone without any sharp discontinuity. The Middle Jurassic is represented in Central Himalaya by a sequence only 3 m thick, in a great contrast to the Tibetan region, where Hayden (1907) has establish a great sequence of fossiliferous limestones of Middle Jurassic age.

Although the Sulcacutus member is a persistent unit, it cannot be taken as separate formation since it is not mappable. It is preferable to keep the name of the Sulcacutus member in the lithostratigraphic sequence. From lithostratigraphic point of view, the author prefers to treat it as a member of the Spiti shale formation recognizing a chronostratigraphic break over it.

Spiti Shale (250–300 m)

This persistent horizon of crumpled black shale provides a picture of flat topography between the competent calcareous Kioto–Lapthal horizons and the sandstone cliffs of Giumal. The Spiti shale is highly fossiliferous but a systematic collection of fossils is not possible from the upper shale because it weathers easily and is consequently always soil covered. Furthermore, though the formation is full of fossils, their exact position within the shale is not determinable, for these are found within the concretions which are rolled into ravines or mixed with the soil. This fossiliferous horizon was exploited for a 100 years by orthodox religious Tibetan people and Hindus, treating the ammonites as '*Shaligram*, the symbol of Lord Shiva'. The excellent palaeontological script by V. Uhlig (1903–1910) remains as standard work on the ammonites in this formation. The thickness of soft black carbonaceous clay shale varies greatly, apparently by stratigraphical and technical causes. Some of the fossils collected by us includes already reported forms: *Oerisphinctes*, *Hoplites*, *Nacrocephalites*, *Himalites*, *Belemnopsis*, *Nucula* (Figure 26). The upper part of the Spiti shale has been designated the *Perisphinctes–Hoplites* zone. The author's collection was identified by Dr Jaikrishna (Banaras Hindu University, India) and the following ammonites (Figure 27) have been classified: F18 (2) *Boehmiceras*–Upper Tithonian; F18 (4) *Aulecosphinctoides*–Lower Tithonian; FLS (18)–*Virgatosphinctes/Aulecosphinctoides*–Tithonian (Lower and Middle): F18 (5)–*Aulecosphinctoides*–Lower Tithonian; F18 (15)–*Kossmatia*–(Middle Tithonian). These forms are previous known from India (Spiti Shale) through the work of Uhlig (1903–1910) and others. *Boehmiceras griesgorive* has been separated from the group of *Blandfordiceras boehmi–middlemissi–celebrent* group of species.

The excellent collection of fossils made by Heim and Gansser in 1936, and identified by Jeannet, is given below for the benefit of the stratigrapher correlating this area (p. 210, Heim and Gansser, 1939).

Upper Jurassic microfauna from the Spiti shale including foraminifera and ostracods from the exposure at the Lapthal camping ground, have yielded the following forms (Singh and Kumar, 1978). Foraminifera: *Lingulina mallajoharensis* Singh and Kumar, *Bentalina pseudocommunis* Franke, *Marginulina batrakiensis* (Myatliuk), *Marginulina* sp., *Lentinculina variana* (Bornemann), *Lenticulina muesteri* (Roemer), *Lenticulina* sp., *Saracenaria reesidei* Fox var. *shalshalensis* Singh and Kumar, *Veginulina constricta* (Terquem and Berthalin), *Pseudonodosaria* sp., *Involtina* sp., *Eoqutalina* sp., ostracods: *Bythocypris* sp., *Monoceratina* sp.

The microfaunal assemblages correspond well with the Callovian–Oxfordian microfaunal assemblage. Heim and Gansser have presumed a disconformity between the underlying Ferruginous Oolites (Callovian) and the Spiti shale (Portlandian). The present microfaunal assemblage indicates that the lower age of the Spiti shales may be Oxfordian and the presumed break in sedimentation after Callovian, may not at all be existing. Thus it may be suggested that there is a continuous sedimentation from the Ferruginous Oolites to the Giumal Sandstone.

Apart from these, the palynological assemblage also contains a few specimens of pteridophytic spores and acritarchs at particular levels. A perusal of the Upper Jurassic dinoflagellate assemblages described from different parts of the world reveals that the Spiti shale dinocysts are comparable to those of Australia and New Guinea.

The faunal studies suggest that *ca*. 250 m thick Spiti shale sequence near Lapthal represents a rather continuous succession ranging in age from Oxfordian to Lower Hauterivian. However, Isocrinics reported by Singh *et al*. (1980) representing *Isocrinus* and *Nielsenicrinus* from the upper part of the Spiti shale range in age from Cretaceous to Palaeocene, have revealed a rich assemblage of dinoflagellates and acritarchs along with some spores and pollen grains. Five distinct microassemblages are identified and the suggested age is Kimmeridgian–Tithonian.

Guimal Sandstone (400 m) (Lower Cretaceous to Upper Neocomian to Gault)

The term Guimal has been derived from a locality in the Spiti Basin. Its average thickness is *ca*. 400 m. This formation is represented by geomorphologically resistant cliffs composed of glauconitic sandy shale, sandstone, and chert. In the upper part they are very prominent in the Sangchatala Shalshal belt of the mapped area. The Spiti shale passes conformably into this thick bedded glauconitic sandstone. Formerly these rocks were called 'Lower Flysch' by von Kraft (1904), and thus the name has been retained. In the Central Himalayan region, Heim and Gansser (1939) used this term for the similar rocks of Spiti. The author has also retained this name, since it is quite a popular name in the geological literature and conforms to the code of stratigraphic nomenclature. Towards the upper part near Sanghatali these rocks contain thick bands of radiolarian chert which were reported first time by Shah and Sinha (1974c). Sinha and Nanda (1979) (Figures 13 and 15) recorded *Pseudomontis*, *Inoceramus*, and *Gryphea*. In the Giumal sandstone are a radiolarian assemblage dominated by *Cryptothorecis* (Jain *et al*., 1978), dinoflagellates, multisegmented *Theoperid* and *Nassellaria*, and single-shelled spherical spumellarians (? *Cenosphaera*). A few forms referable to speudoaoulaphacid and/or spongodiscid spumellarians also constitute a significant part of this assemblage. Glauconitization is conspicuous. Nassellarian genera identified are: *Sethocapsa* opp. *Lithocampe* sp. *L. elegantissima*, *Eycyrtis* sp. *Zhamboidellum* or (? *Cryptamphorella*), *Holocryptocapsa*, and *Williriedellum*. The

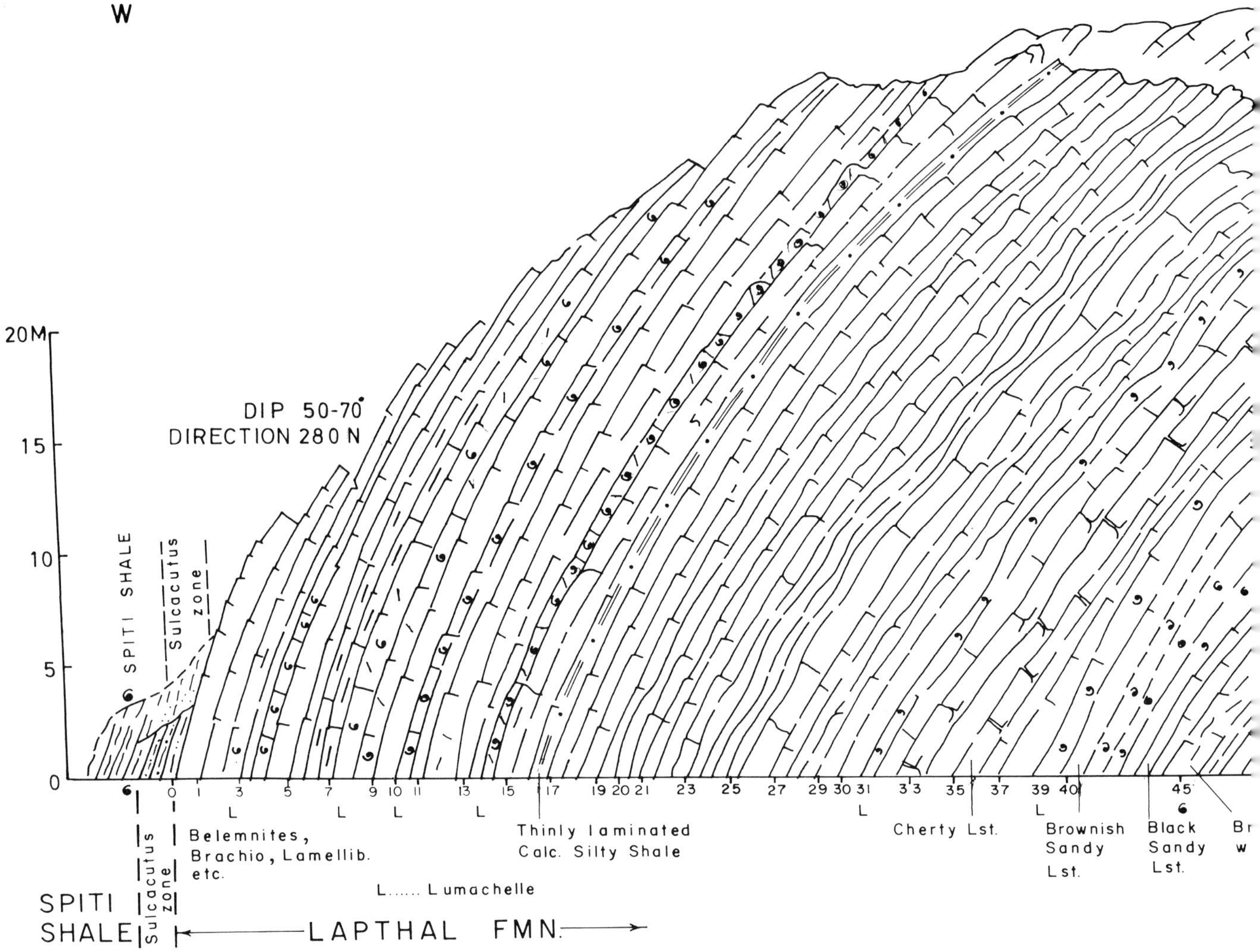
W
20M
15
10
5
0
DIP 50-70°
DIRECTION 280 N
SPITI SHALE
Sulcacutus zone
0 1 3 5 7 9 10 11 13 15 17 19 20 21 23 25 27 29 30 31 33 35 37 39 40 45
L
Belemnites,
Brachio, Lamellib.
etc.
L...... Lumachelle
Thinly laminated
Calc. Silty Shale
Cherty Lst.
Brownish
Sandy
Lst.
Black
Sandy
Lst.
SPITI
SHALE
Sulcacutus
zone
LAPTHAL FMN.

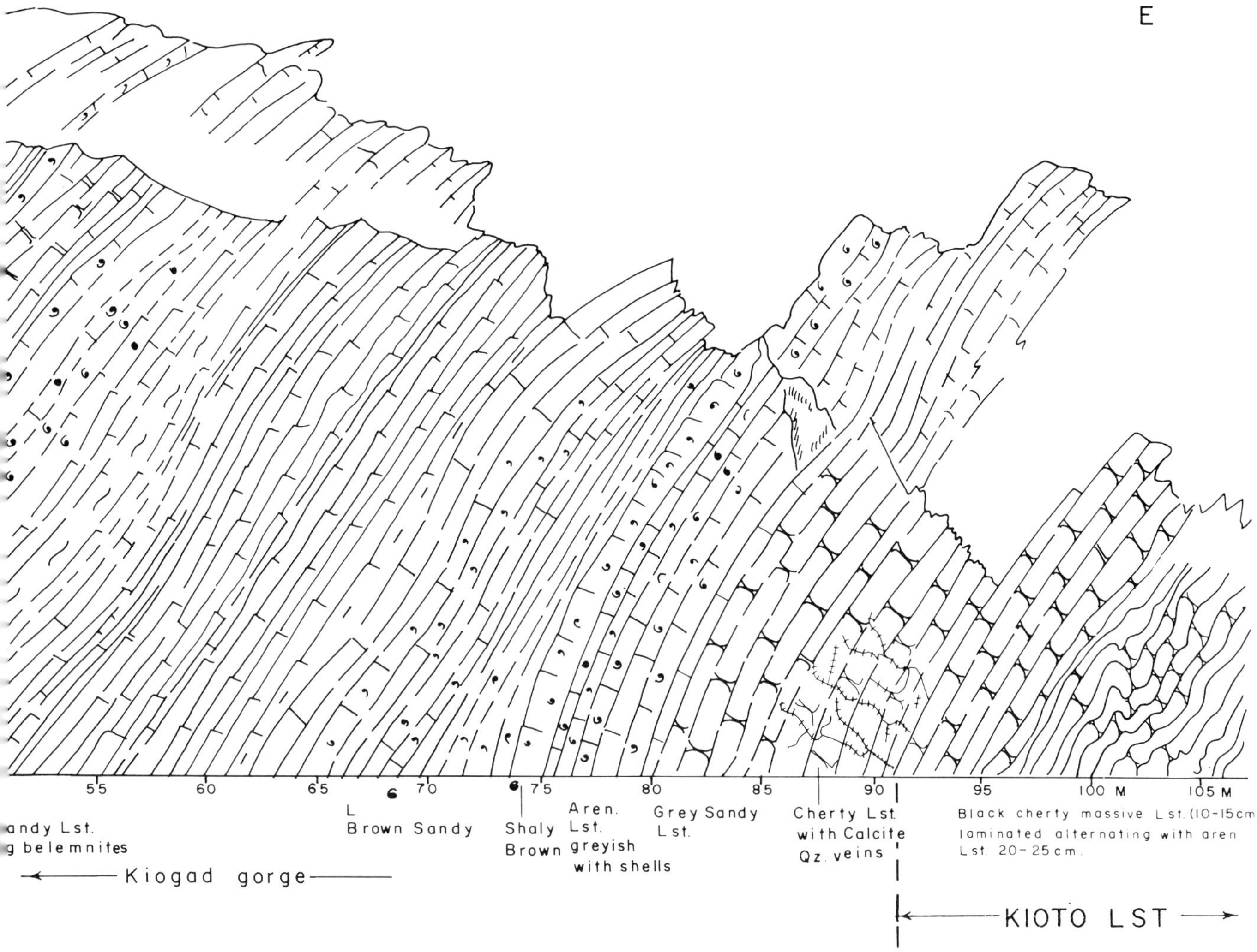

Figure 25. Tape measured exposed Section of Lapthal formation in the Kiogad stream gorge near old Lapthal campsite

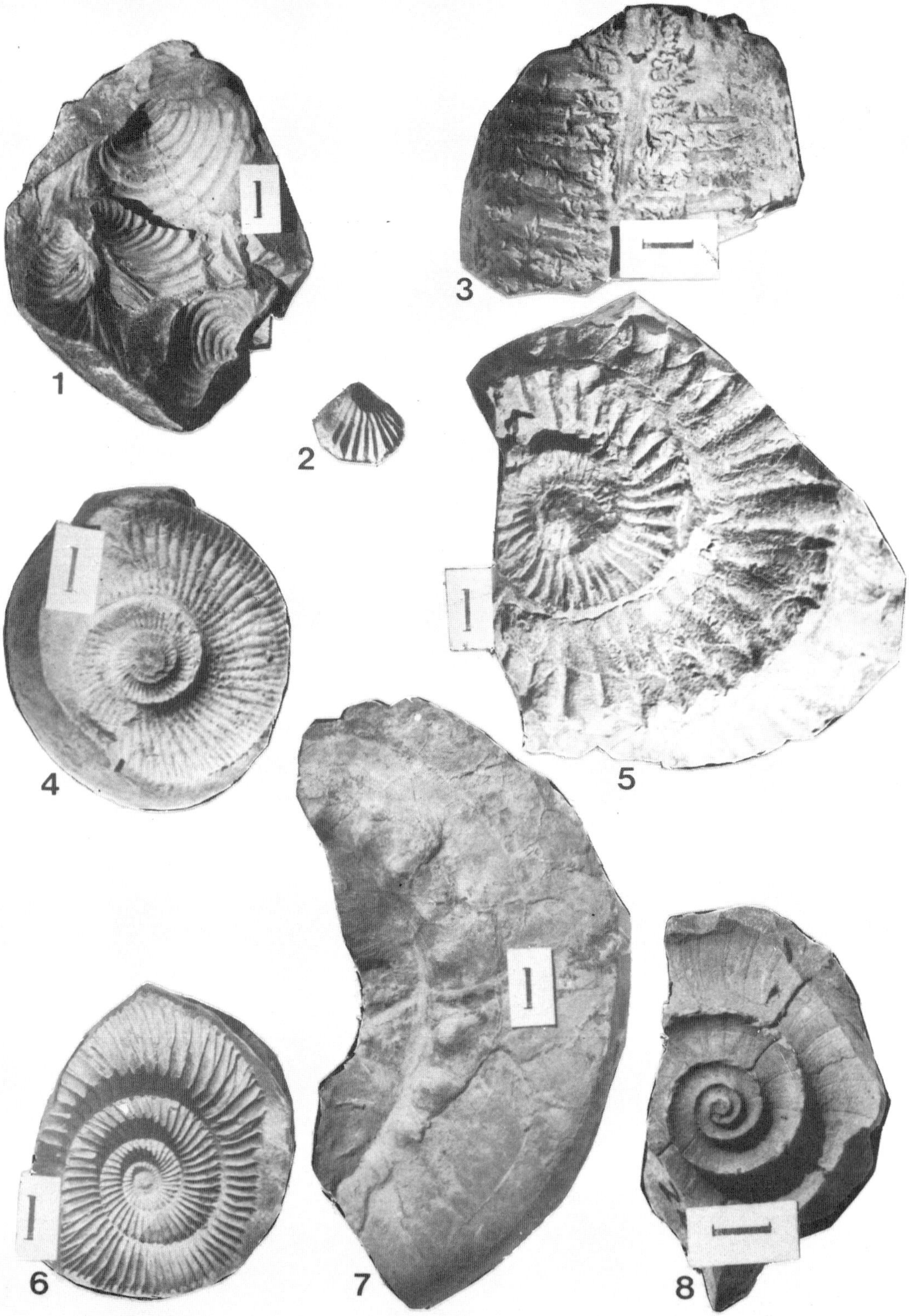

Figure 26 Fossils from the Spiti shale in Lapthal area of Malla Johar Tethyan zone. I. *Inoceramus* sp.; 2. *Rhynchonella* sp.; 3. *Perisphinctes* sp.; 4. *Macrocephalites* sp.; 5. *Hoplites* sp.; 6. *Perisphinctes* (*Aulacosphinctes*) sp.; 7. *Himalayites* sp.; 8. ? *Kossmatia* sp. (scale represents one centimetre)

Figure 27 1. Ventral view of *Cossmatia*; 2. Ventral view of *Aulacosphinctoides*; 3. Lateral view of *Kossmatia*; 4. Ventral view of *Boehmiceras*; 5. Lateral view of *Aulacosphinctoides*; 6. Lateral view of *Boehmiceras*; 7. Ventral view of *Aulacosphinctoides*/*Virgatosphinctes*; 8. Lateral view of *Aulacosphinctoides*/*Virgatosphinctes*; 9. Lateral view of *Aulacosphinctoides*

above-mentioned assemblage is a pointer to Upper Jurassic–Lower Cretaceous assemblage.

This series was also called Lower Flysch by von Kraft (1904) and it is according to the Spiti area dated to be Neocomian to Gault.

Flysch (Sangchamalla Formation)

Overlying the Giumal sandstone horizon without any unconformity, a radiolarian-bearing cherty horizon marks the base of flysch. This radiolarian-bearing local horizon is followed by shaly and feldspathic sandstone, chocolate reddish-shale with calcareous and cherty lenses, black silty shale, and greenish shale. The name 'Upper Flysch' was used by Heim and Gansser (1939). But it does not confirm the name given to the lower part of it which they called the Giumal sandstone. Thus a general flysch would be a more befitting name for this geosynclinal horizon of the Central Himalayan Zone. From the perusal of the literature on the geology of other areas of Kumaun, it is clear that this formation has its best development as Sangchamalla below Kiogad where complete sequence is exposed, and this is proposed as the type area of the Sangchamalla formation (*ca.* 1000 m or more). The details of the lithological sequence is given in Table 3.4 (Mehrotra and Sinha, 1981, modified after Shah and Sinha, 1974c). Purple marly shales have yielded the foraminiferal zone and determination has been done by Dr (Miss) P. Kalia of Delhi University: The assembly is indicative of Upper Cretaceous age.

The palaeontological investigation of the flysch samples has yielded a rich assemblage of dinoflagellate. The microplanktonic biota recorded has also been used to interpret the palaeoenvironmental conditions at the time of deposition of flysch sediments, besides the former and previously described Cretaceous flysch horizon. Studies of this aspect have been carried out in collaboration with Dr N. C. Mehrotra (Mehrotra and Sinha, 1981) (Figures 28 and 29).

Age and Depositional Environment

As far as the age of the Sangchamalla formation is concerned, it has been dated as Upper Cretaceous by Heim and Gansser (1939a). The above authors reported a few globular radiolarians (*Spumelliarians*) and imperforate siliceous foraminifera (Ophthalmididal). However, in view of the scanty palaeontological records this dating can not be considered as very precise. Dr (Miss) Kalia records some more foraminiferal

Table 3.4. Distribution of microplankton in the different levels of Sangchamalla formation (After Mehrotra and Sinha, 1981)

	Stratigraphic unit Exotic block	Characteristic fossils	Age
SANGCHAMALLA FORMATION	Greenish shales with bands of chocolate sandstones and dark grey radiolarian cherts (300 m)	Predominance of *Areosphaeridium diktyoplokus*, *A. arcuatum*, *Homotryblium tenuispinosum*; rare representation of *Oligosphaeridium* spp. and *Hystrichokolpoma unispinum*	Upper Eocene (?) to Middle Eocene
	Greenish grey greywacke and dark shales bearing fucoid markings, with bands of radiolarian cherts towards the top and a band of brown feldspathic sandstone near the base (650 m)	Predominance of *Oligosphaeridium complex*, *O. pulcherrium*, *Diphyes colligerium*, *Cordosphaeridum exilimurum*; first appearance of *Areosphaeridium diktyoplokus*, *A. arcuatum*, and *Homotryblium tenuispinosum*	Lower Eocene to Palaeocene
	Purple marly shale with brown bands of foraminiferal ooze (70 m)	Abundant and restricted occurrence of *Odontochitina cribropoda* and high representation of *Systematophora schindewolfi*	Upper Cretaceous
	Dark greenish shale with a few (palynological samples not collected) greywacke bands (50 m)		
	Guimal sandstone		

Thicknesses and lithological details after Shah and Sinha, 1974C.

genera—*Globotruncana*, *Heterohelix*, *Plummerita*, *Shackoina* and *Eouvigerina*—from the purple shale samples (Shah and Sinha, 1974c, p. 19) and supports the Upper Cretaceous dating of Heim and Gansser. It was only recently that Mehrotra and Sinha (1978, 1981) found that the upper age limit of the formation does not end with Upper Cretaceous but extends into Eocene. A large number of fossil dinoflagellate species were recorded from the different stratigraphic levels of the formation. The detailed taxonomy of them has been given in this paper (Mehrotra and Sinha, 1981). The lower part of the Sangchamalla formation consists of dark green and purple shales with a few bands of foraminiferal oozes and greywacke. The upper purple shale (70 m) level of the formation is marked by the abundant occurrence of *Odontochitina cribropoda* Deflandre and Cookson (1955). This species is well known for its restricted occurrence in Upper Cretaceous age. This is further strengthened by the representation of high frequencies of *Systematophora schindewolfi* (Alberti) Neale and Sarjeant (1966) in this level (Tables 3.4 and 3.5). This species has been reported from Germany and England and occurs in large numbers only in Upper Cretaceous sediments (Davey *et al.*, 1966); it is rare in Eocene sediments. The typical Eocene forms, *Homotryblium tenuispinosum* Davey and Williams (1966) *Aerosphaeridium diktyoplokus* (Klumpp) Eaton (1971) and *A. arcuatum* Eaton (1971) which occur abundantly in the upper levels of the formation, are totally absent here. All these fossil evidences are suggestive of an Upper Cretaceous lower limit of the Sangchamalla formation. However, it must be kept in mind that since no palynological sample has been collected from the lower limit then an Upper Cretaceous age for it cannot be completely ruled out. The 650 m thick sequence of dark green shales associated with radiolarian cherts, forming the middle part of the formation, also proved quite rich in microplanktons. Ten samples have been found productive from its different levels. The predominant dinoflagellate genera recorded are *Oligosphaeridium pulcherrium* (Deflandre

Table 3.5. Known stratigraphical range of some significant dinoflagellate species of Sangchamalla formation

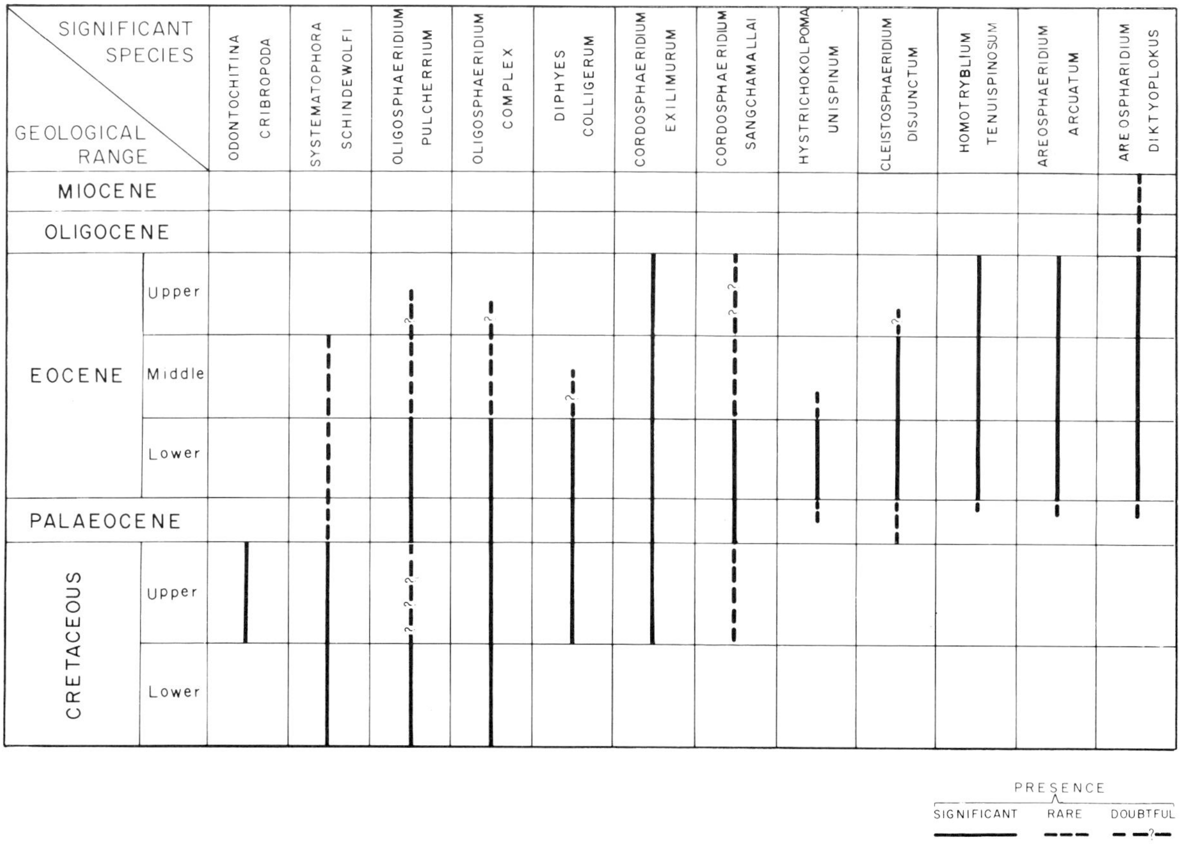

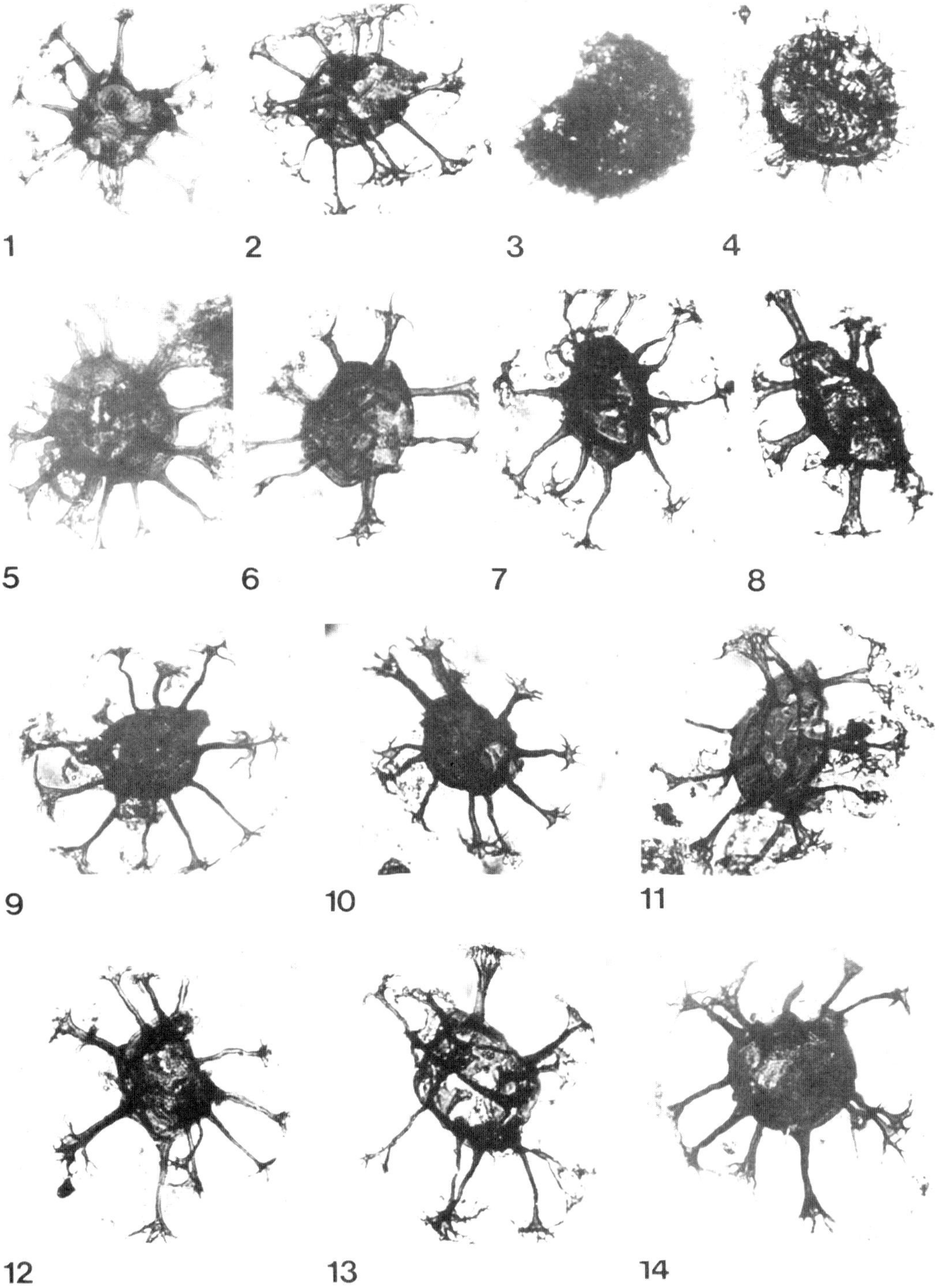

Figure 28 All magnifications × 490. 1,5,7. *Hystrichosphaeridium himalayensis* sp. nov.; 2,6,13. *Oligosphaeridium pulcherrium* (Deflandre & Cookson) Davey and Williams, 1966; 3. *Cleistosphaeridium aisjunctum* Davey, Downie, Sarjeant and Williams, 1966; 4. *Diphyes colligerum* (Deflandre & Cookson) Davey and Williams, 1966; 5,10. *Homotryblium tenuispinosum* Davey and Williams, 1966; 11. *Oligosphaeridium* sp.; 12,14. *Oligosphaeridium complex* (White) Davey and Williams, 1966

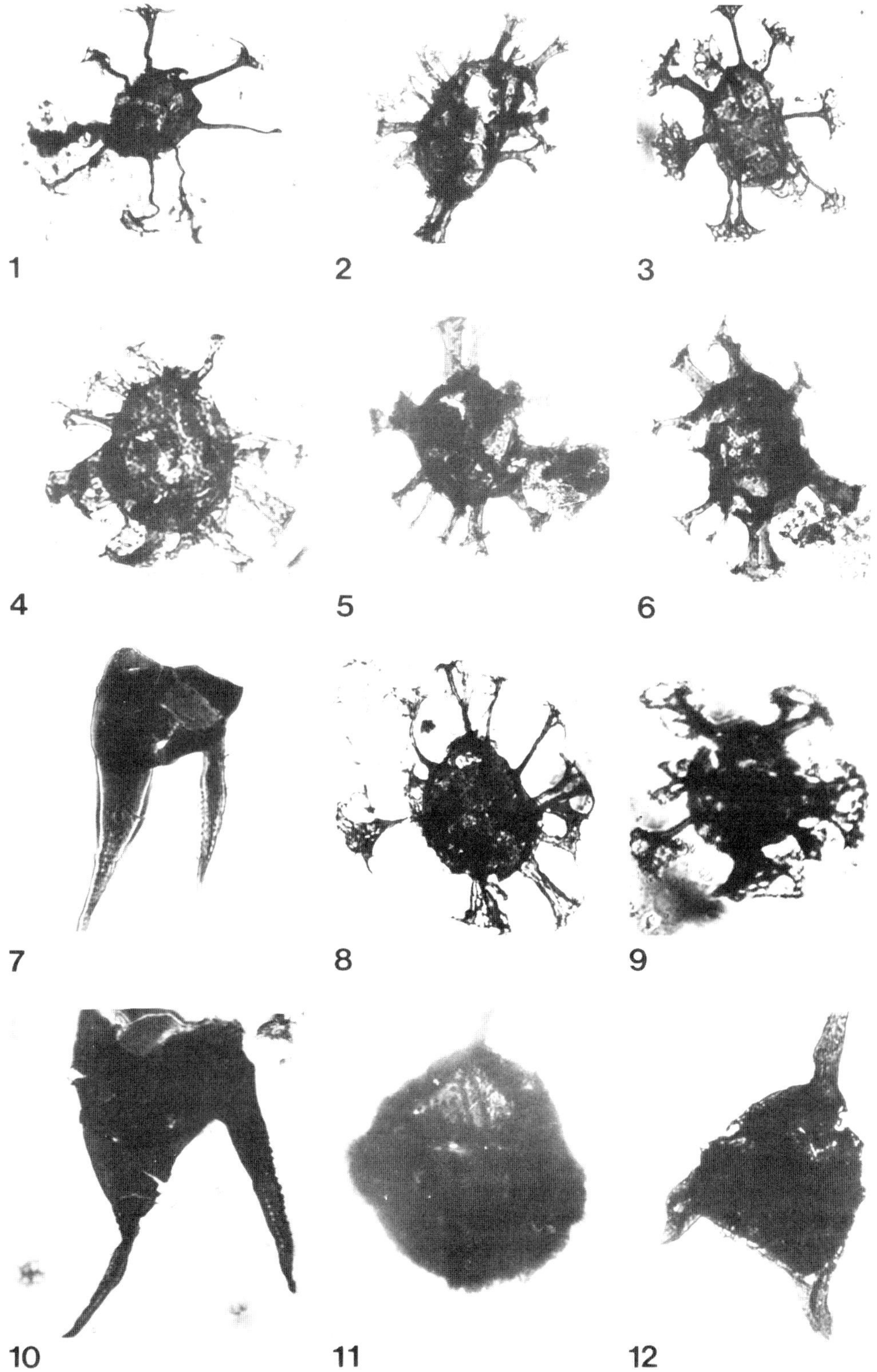

Figure 29 1. *Areosphaeridium arcuatum* Eaton (1971); 2. *Hystrichokolpoma unispinum* Williams and Downie (1966); 3. *Areosphaeridium diktyoplokus* (Klumpp) Eaton (1971a); 4. *Cordosphaeridium exilimurum* Davey and Williams (1966); 5,6 *Cordosphaeridium sanghamallai* sp.nov.; 7–10. *Odontochitina cribropoda* Deflandre and Cookson (1955); 8. *Adnatosphaeridium* sp.; 9. *Systamatophora schindewolfi* (Alberti) Neale and Sarjeant; 11. *Gonyaulacysta* sp.; 12. *Deflandrea speciosa* Alberti (1959)

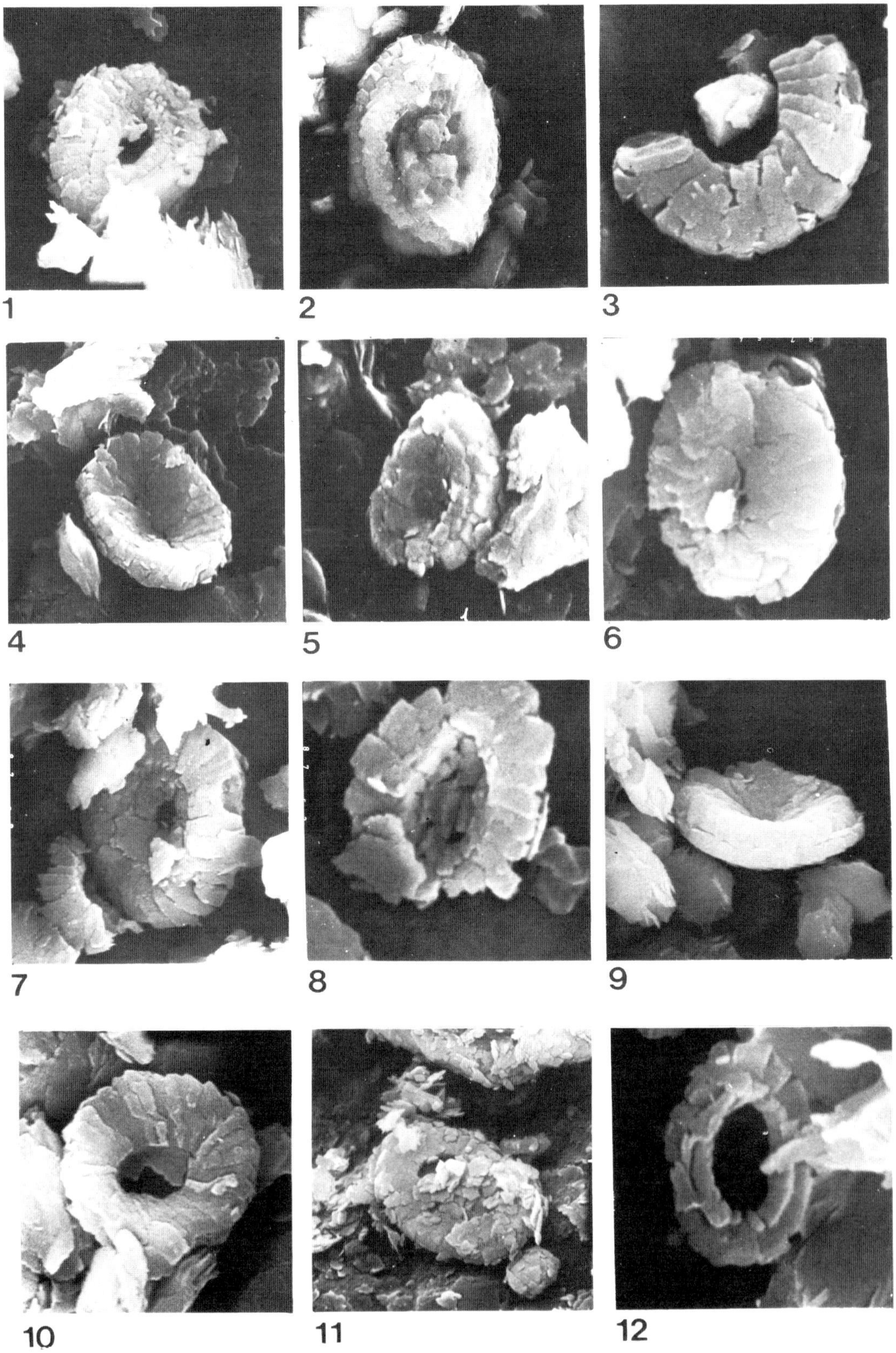

Figure 30 1. *Watznaueria barnesae* (Black) Perch-Nielsen, × 4750; 2. Proximal side of coccolith, × 2325; 3. Unidentified coccolith, peripheral disc fragment, × 6650; 4,5. *Watznaueria* sp., 4, × 4750, 5, × 6650; 6. *Watznaueria bioparta* Burky, × 8550; 7. *Watznaueria barnesae* (Black), approx. × 950; 8. *Biscutum constans* (Gòrka), × 10450; 9. *Biscutum* sp., × 4275; 10. Unidentified coccolith, peripheral disc, × 4750; 11. *Staurolithies* sp., × 4275

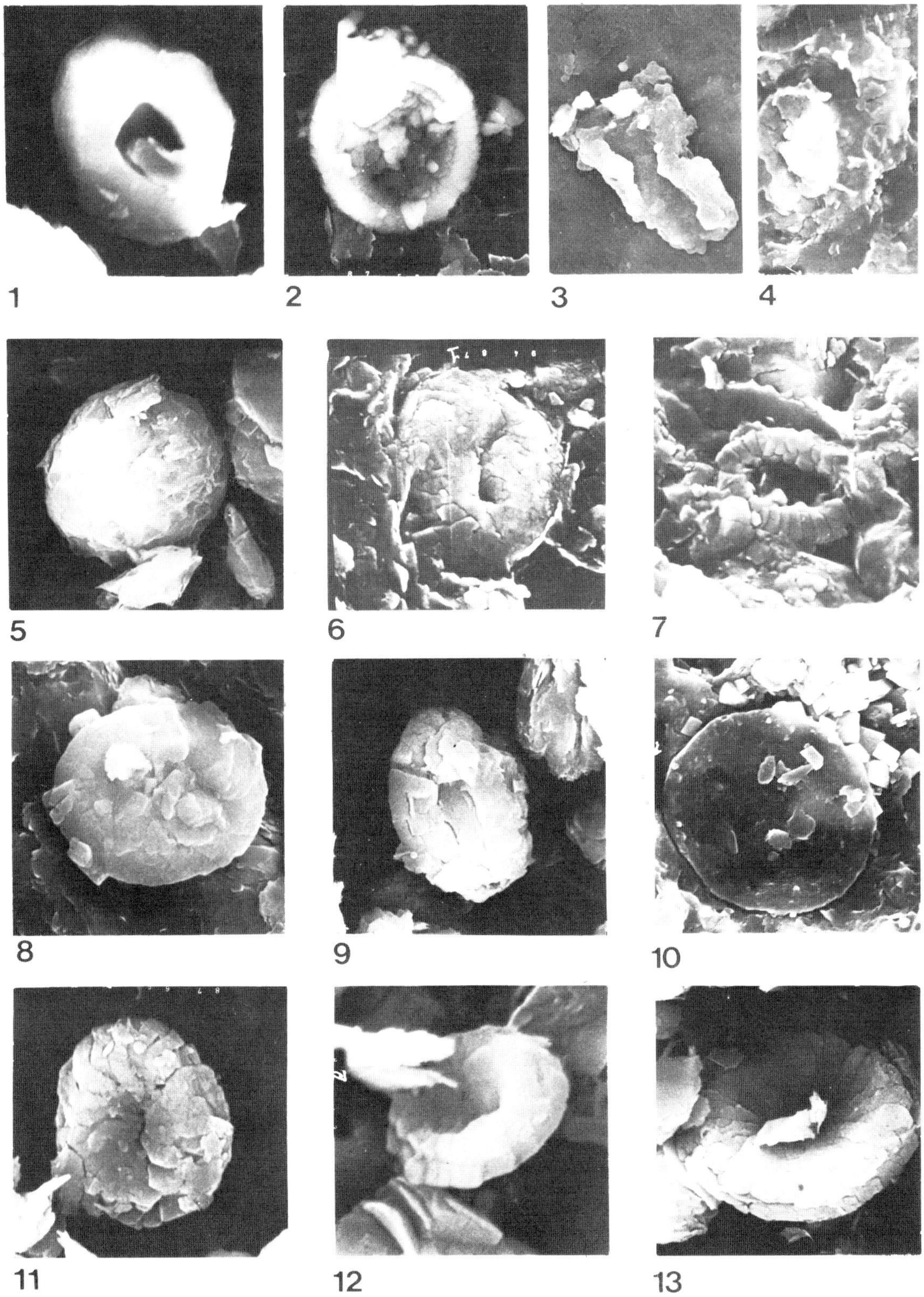

Figure 31 1. Obliterated unidentified coccolith, × 3255; 2. *Broinsoina* sp., × 4185; 3. *Lucianorhabdus cayeuxi* Deflandre, × 1395; 4. Print of outer disc of coccolith, × 3720; 5. *Thoracosphaera* sp., × 4185; 6,13. *Watznaueria* sp., × 4650; 7. Outer disc of coccolith, × 4185; 8. *Watznaueria* sp., × 4185; 9,10 Unidentified coccolith, 9, × 4185; 10, × 2790; 11. *Watznaueria bioporta* Burkey, × 5580; 12. *Biscutum* sp., × 9300

and Cookson) Davey and Williams (1966) *Cordosphaeridium exillimurum* (Klumpp) Davey and Williams (1966), *Hystrichokolpoma unispinum* Williams and Downie (1966), and *Diphyes colligerum* Davey and Williams (1966). All these forms are well known for their abundant occurrence in the Palaeocene–Lower Eocene sediments of England (Davey *et al*., 1966) and India. From Indian *Oligosphaeridium pulcherrium*, *O. complex*, and *Diphyes colligerum* are known from the Lower–Middle Eocene sediments of North Cacher Hills, Assam while *Oligosphaeridium complex* is quite common in the Lower Middle Eocene Subathu sequence of Lesser Himalaya (Khanna, 1978). *Cordosphaeridium exillimurum* and *Aerosphaeridium diktyoplokus* are the other stratigraphic species common in the Subathu and Sangchamalla formations. The appearance of typical Eocene forms like *Aerosphaeridium arcuatum*, *A. diktyoplokus*, and *Homotryblium tenuispinosum* in this level also supports its being Eocene in age. The above fossil evidence suggest definitely a Palaeocene–Lower Eocene age for this stratigraphic level of the Sangchamalla formation. Its upper limit may extend into middle Eocene also which is difficult to say with precision in view of absence of any dinoflagellate species which has been recorded from Middle Eocene sediments only (Table 3.5).

The topmost greenish shale–radiolarian cherts (300 m) sequence has also shown positive palynological results. The samples from green shales are found very rich in *Aerosphaeridium diktyoplopus*, *A. arcuatum*, and *Homotryblium tenuispinosum*. Representation of *Cordosphaeridium exilimurum* is also quite fair. The absence of typical Lower Eocene species *Hystenuispinosum*. Representation of *Cordosphaeridium exilimurum* is also quite fair. The further suggests that the level is definitely younger than the Lower Eocene, i.e. Middle Eocene, whether it touches the upper part of Eocene or not, is not possible to say as there has not been recorded any species of dinoflagellate which is known to occur only in the Upper Eocene. *Hence, Middle Eocene is the upper age limit of the Sangchamalla Formation.*

The predominance of dinoflagellates in the sequence is an indication of shallow marine depositional environment, within the flysch sedimentational cycle. It would not be preposterous to infer that flysch graben with loaded sediments has experienced pulsating vertical movements with shallowing and deepening of the basin. The study of glauconite appearing with radiolarian also affirms the shallowing of purple shaly horizon having marked oxidizing environment (Sinha and Srivastava, 1978, 1986). Thus the dinoflagellates and glauconite appearance are complementary to each other affirming a pulsationally sinking and rising flysch basin.

It is very interesting to note that a large number of species are common in the Sangchamalla and in the London clay microplankton assemblages. This is attributed to the existence of some sea link between the two regions during Eocene times. Reid and Chandler (1933) indicated the existence of this sea route which allowed the Indo-Malayan flora to migrate as far as Great Britain during the deposition of London clay sediments. Further, Singh (1952) on the basis of foraminiferal evidence it has been indicated that the Tethys sea might have extended from the high ranges of Sind–Baluchistan frontier to Iran, Asia Minor, North Africa, Turkey and Greece to western Europe as far as the Pyrenees. The close similarity between the microplankton assemblage of the Subathu (Khanna, 1978) and Sangchamalla Formations with that of London clay has further strengthened the above probability.

Rich assemblages of coccoliths have been isolated from the flysch sediments of Malla Johar in the Tethyan zone of Higher Kumaun Himalaya (Figure 33) with the help of Dr (Mrs) Olga Dmitrenko, Institute of Oceanography, Academy of Sciences of the USSR, Moscow (Sinha and Dmitrenko, 1983). The forms being reported: *Watznaueria barnesae* (Black) pereb-Nielson, *W. biporta* Bukry, *Biscutum constans* (Gorka) Black; *Staurolithites* sp; *Braarudosphaera bigelowi* (Gran, Braarud) Deflandre, *Thoracosphaera* sp., *Broinsonia* sp., *Lucianorthabdus cayeuxi* Deflandre, *Marthasterites fureatus* (Deflandre) Stradner, *Lithraphidites carniolensis* Deflandre etc. (Figures 30, 31 and 32). The coccoliths are described and illustrated using scanning electron microscope. They have been reported from the Upper Cretaceous horizons of different parts of the world. The upper limit is yet to be finalized in view of recent trend of world wide study.

Sedimentation study of the flysch and the occurrence of glauconite with radiolarites has given an important clue to its history of sedimentation and environment of tectonic phenomena in the flysch graben (Sinha and Srivastava, 1978, 1986). The occurrence of *Liosphaera* sp., *Staurocaryum* sp., and *Cyrtocapsa* sp. of radiolarites have been found in the glauconite bearing

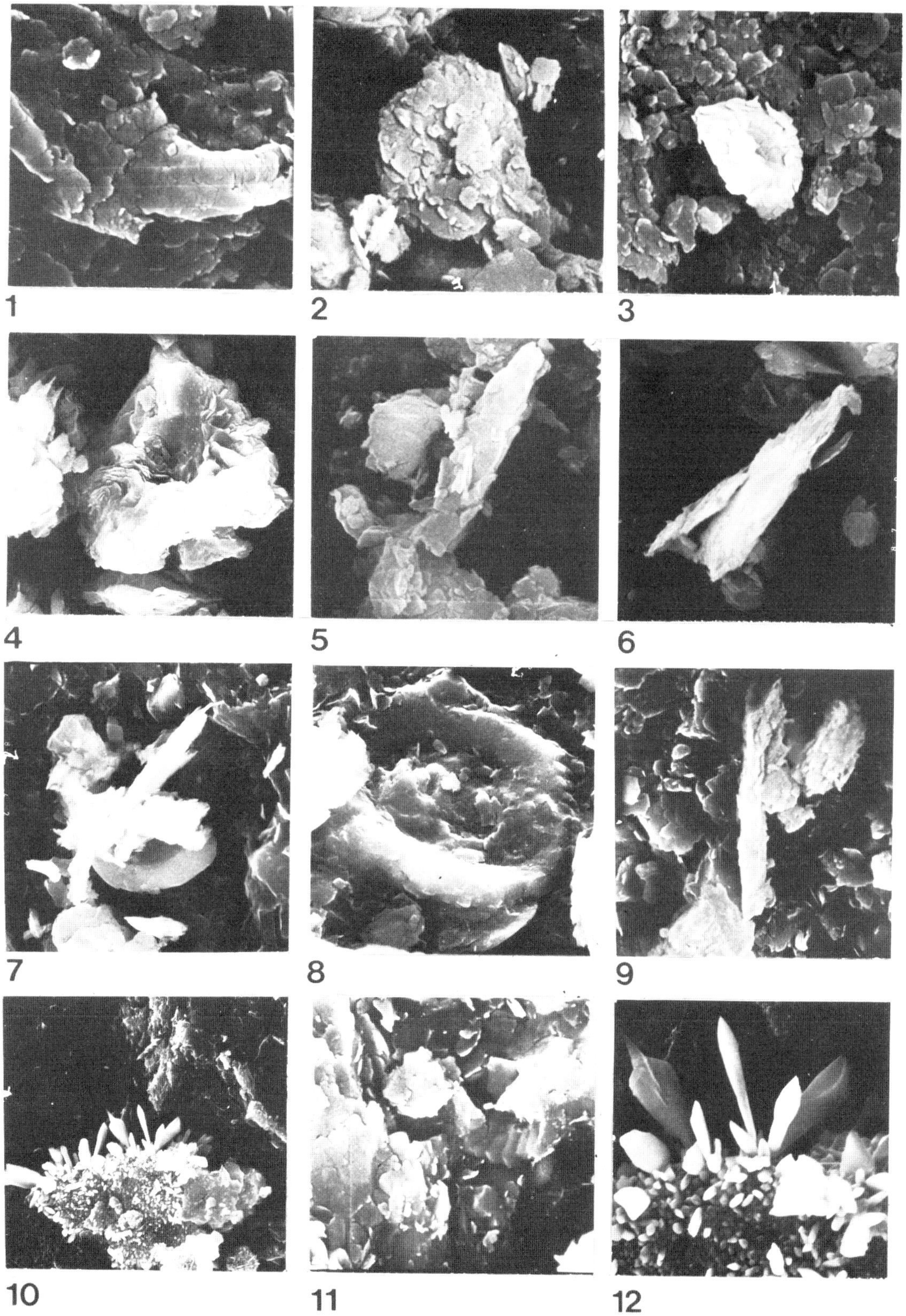

Figure 32 1. Altered fragment of outer disc of coccolith, × 4900; 2. Unidentified coccolith, × 4900; 3. *Biscutum constans* (Gòrka) Black, × 4410; 4. Deformed highly altered coccolith, × 2156; 5. *Lithraphidites* sp., × 4410; 6. Fragment of column of coccolith, × 4410; 7. *Lithraphidites* sp., and unidentified coccolith, × 4410; 8. Obliterated outer disc of coccolith, × 2450; 9. *Lithraphidites carniolensis* Deflandre, × 3920; 10,12. Crystal growth caught under SEM; 10, × 1470; 12, × 4410; 11. Fragments of outer disc of coccolith, × 4900

medium grained, moderately to poorly sorted greywacke units of Lower Flysch sediments in Malla Johar of Kumaun, exposed in the Sangchamalla of Kiogarh area. These radiolarites indicate a deep marine environment for the deposition of greywacke units while the associated glauconite indicates a shallow marine environment. This type of association is of marked significance and rare in the sedimentation history of the clastic horizons in Himalaya. Glauconite is found replacing the radiolarian bodies and also as grains indicating that it is a diagenetic product during the shallowing phase of the depocentre due to tectonic uplift of the deep marine sediments in this region.

Petrographic Description

The sandstone units consisting of radiolarians and glauconite are primarily composed of medium grained quartz, chert, feldspars with subordinate amount of clay pellets, biotite, limonite and haematite (Sinha and Srivastava, 1978) (Figure 34). The quartz grains constitute nearly 65 per cent of the total volume, glauconite 10 per cent, radiolarians 15 per cent, and the remaining 10 per cent is made up of the other accessory minerals. The rock as a whole represents a medium grained, moderately to poorly sorted greywacke in which quartz grains are dull, pitted, of undulating and non-undulating varieties. Polycrystalline quartz grains are also not uncommon. The quartz grains are angular to subrounded in shape and are mostly inclusion free. Secondary overgrowths of silica have also been noted on a few grains of quartz but it is not a common phenomenon. Sometimes the quartz grains are coated with limonitic material and have pointed, long, sutured and concavoconvex grain to grain contacts. The chert is found as fine mosaic in between the void spaces of the detrital grains. The grains of cherts are also commonly observed having the shape and size of the quartz grains. The occurrence of chert is significant denoting the derivation of some material from a sedimentary source. The feldspars are mainly microclines–albites and, except a few grains, are not fresh. They are mostly altered to sericite. The limonite is of secondary origin.

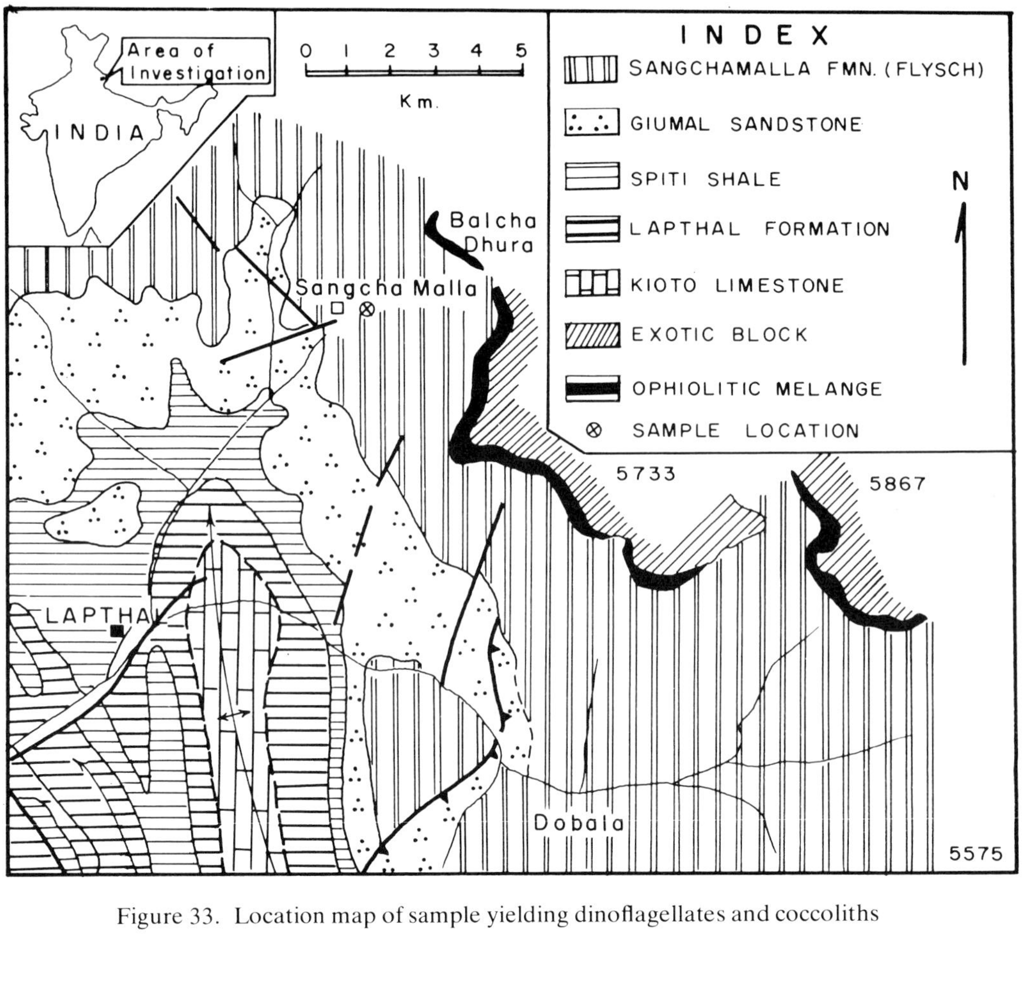

Figure 33. Location map of sample yielding dinoflagellates and coccoliths

Glauconite is present as granules and in the peripheral portions of the radiolarian bodies. The glauconite has replaced the radiolarians as is evident from the remains of radiolarians within the glauconite. This replacement must have happened during the diagenetic stage when favourable physicochemical conditions were available in the changing environment of the geosynclinal conditions. The glauconite granules are representative of current agitated environment for their deposition. The glauconite may have generated in the marine waters of normal salinity having reducing conditions from the available organic matter in the depositional basin. The presence of glauconite represents slow sedimentation in a depth of up to 10–400 fathoms (18 to 730 m). The origin of glauconite in such conditions may be attributed to a number of parent materials such as clayey substances filling cavities, foraminifera, radiolarians, and casts of other marine organisms. The replacement of radiolarians by the glauconitic material is believed to be early diagenetic before the overburden preserved them.

Radiolarians

The radiolarians are represented by circular or elliptical clear areas and sometimes they show spines or more rarely internal lattice characteristics of the organisms. But in most cases they are damaged by recrystallization of the matrix or

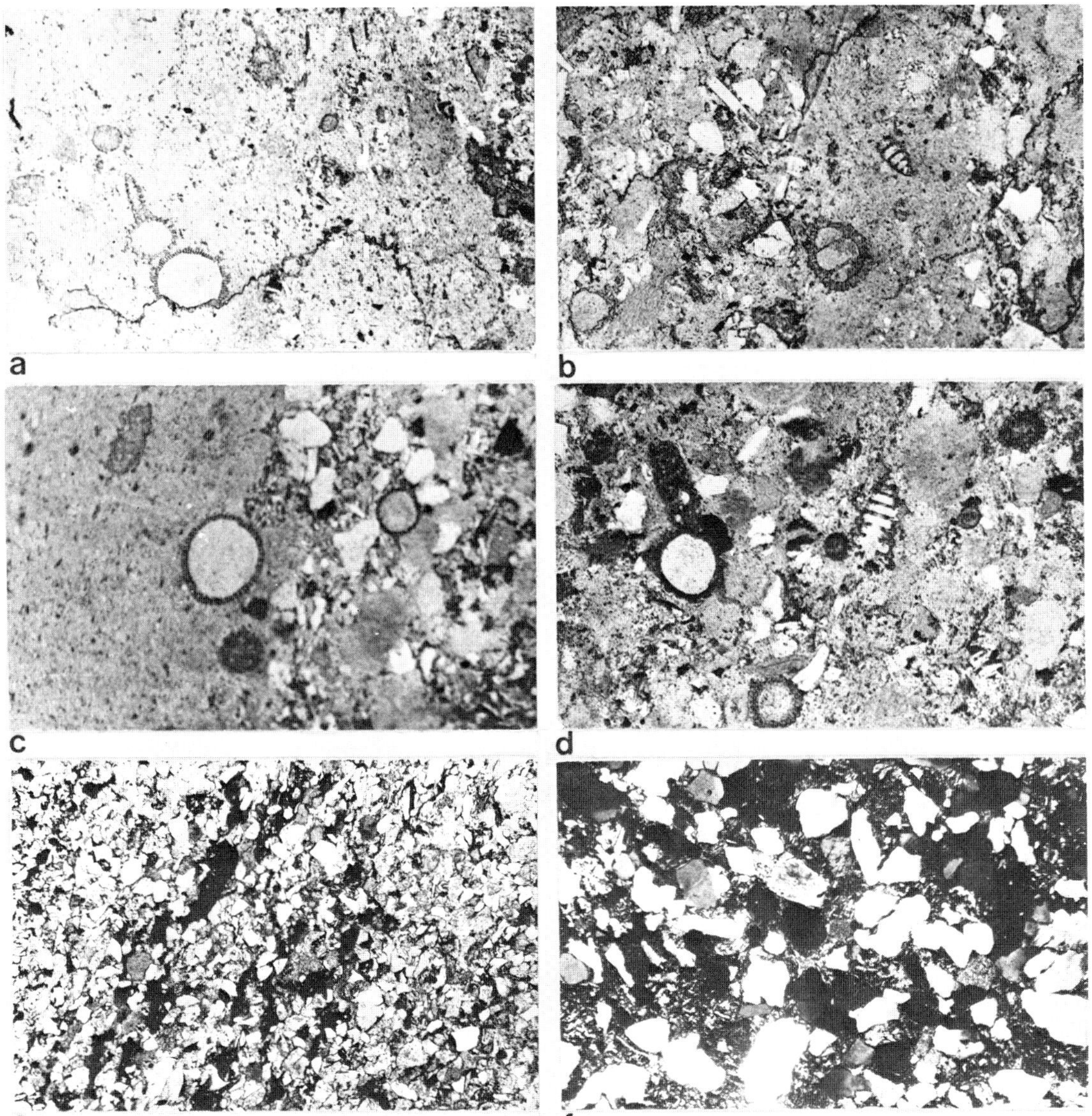
a b c d e f

Figure 34. Occurrence of glauconite with radiolarites in flysch sediments. (a) *Staurocaryum* sp.; (b),(d) *Cyrtocapsa* sp.; (c) *Liosphaera* sp.; (e),(f) glauconite grains in the medium grained, moderately to poorly sorted greywacke

replacement by glauconite. The occurrence of *Liosphaera* sp., *Staurocaryum* sp., and *Cyrtocapsa* sp. have been recognised in these rocks which are presumed to be a deep water assemblage. In the case of *Cyrtocapsa* sp., the apical horns have not been preserved (Figure 34, a, b, c, d; Sinha and Srivastava, 1978).

The *Liosphaera* sp. ranges in age from Ordovician to Recent, *Staurocaryum* sp. Cretaceous to Recent, and the *Cyrtocapsa* sp. from Jurassic to Recent. So these radiolarians are not indicative of precise age to these rocks because of their distribution in a long span of geological timescale.

Discussion

There are two significant observations derived out of this study: (1) Association of glauconite with radiolarian and (2) Occurrence of glauconite in greywackes. The glauconite mineral which is found in these rocks is certainly of secondary origin as evidenced by its petrographic characteristics. The organic matter, available at a suitable depth and suitable physicochemical conditions have been helpful in the genesis of this mineral which has later on replaced the radiolarian tests. There is another possibility that radiolarian tests also have played important role in the genesis of glauconite because such causes are found in the geological literature. This genesis must have taken place at a shallower depth which have been attained due to the uplift of the sediments in which radiolarians flourished in a greater depth of the marine water. This uplift facilitated the favourable depth, salinity, pH, E_h and other factors in the marine water to generate glauconite and to replace the radiolarian casts during the early diagenetic stage. Although radiolaria start appearing with the chert in the upper part of Giumal sandstone horizon (Figure 15) as a pointer to the deepening of the ocean floor, preparing the reception ground for flysch sediments, as a matter of fact the true geosynclinal sedimentary facies begin with the radiolarian-bearing cherty horizons followed by greywackes. Curiously enough the infilling of radiolarian chambers by glauconite, which is a marker of shallow basin, indicates that there was a marked uplift sometime during the Middle Cretaceous. This may be the hiatus of *Oregonian/Austrian* orogenic cycle to be recorded first time in the Tethyan zone of Himalaya (Sinha and Srivastava, 1986). It is to be noted that uplift and subsidence are typical features of a flysch bearing geosynclinal trough, and hence the appearance of glauconite may be attributed to the pulsational movements in the flysch graben-zone. However, the final picture to justify the Oregonian phase in Himalayan Tethys would emerge with more new data. In recent years the notable contributions towards the age of flysch have come from the finding of Lower Tertiary dinoflagellates (Mehrotra and Sinha, 1978) and coccolith under the scanning electron microscope (Sinha and Dmitrenko, 1983).

Kumar *et al.* (1978) reported the cone-in-cone structure in the Upper Flysch series (Upper Cretaceous) Malla Johar area. A comprehensive account of Upper Cretaceous microfauna of the Spiti–Niti Pass area of the Himalaya has been published by Mamgain and Sastry (1975).

3.2.4 Exotic blocks of Malla Johar

The ophiolitic melange with associated exotic blocks exposed in the Kiogad region of the Malla Johar region of the Tethyan zone of Higher Central Himalaya represents one of the best melange sequence in our globe (Sinha, in press).

In 1892, Griesbach, Diener, and Middlemiss discovered some isolated limestone blocks embodying Permo-Carboniferous and Triassic fossils near Chirchun in the border region of India and Tibet (earlier called Hunde's), in the upper Kiogarh valley in higher Kumaun region (earlier known as Malla Johar). Griesbach (1893) and Diener (1895c) recorded their observations in their geological maps. Griesbach believed them to be due to faulting while Diener left the 'Klippen' unexplained, while comparing with European ones. In 1899 Dr Walker was sent to examine the Chirchun area. Owing to ill health, he returned in a very short time, but collected 9000 Permo-Carboniferous fossils from the Peak Chirchun no. 1.

In 1900 A. von Kraft was ordered to Chirchun, but his plan to cross through Langpaia Lekh into Tibet was frustrated by orders from the then British Government, strictly forbidding him to enter Tibet. In those days particularly Europeans or any foreigner were strictly debarred to enter Tibet (then Hunde's). The orthodox belief of Tibetans was that their pious land would be made unholy by the entrance of non-Buddhist Europeans. A. von Kraft thus camped at Lapthal within the limit of British India.

In 1892, Griesbach near Sangcha Talla discovered loose blocks of a red and white limestone with *Jovites n.f.* ex *aff. J. boanensis*, E. v. Mojs. Mojsisovic concluded that the blocks belonged to the Carnic stage of the Upper Trias.

As mentioned earlier A. von Kraft in 1900 worked on this area for six weeks during the month of July and August. He systematically mapped the 'exotic block' area, and collected numerous fossils, chiefly ammonites. His conclusion that 'The exotic blocks were brought up to the surface by a violent volcanic outburst, and later disturbance having thrust the whole into confusion' could not hold good in the context of modern concepts. This brilliant geologist did not see the publication of his work in Geological Survey of India Memoir volume 32 in 1904—he died prematurely in Calcutta on 22 September 1901.

Later on Professor Carl Diener of Vienna University was entrusted to work over his collection. Diener (1898) commented 'In their occurrence amidst much younger sediments, and without apparent stratigraphical connection with the latter, which makes the structure of the Chitichun area one of the most intricate and most remarkable in the Central Himalayas'. It was the remarkable feat of young Swiss geologist Augusto Gansser in 1936 who came to work with his teacher, the late Professor Arnold Heim, and crossed into Tibet in the garb of Buddhist Lamas. It was established that the exotic blocks of Kiogar–Chitichun extended to more than 5000 km^2.

The researches of Arnold Heim and Augusto Gansser in 1936 in the Kiogar region after the master piece investigation by von Kraft in 1900 led them to infer that the 'exotic blocks' could be classified in to three categories in accordance to the physical association: (1) Exotic blocks associated with the Flysch; (2) Exotic blocks enclosed in the basic volcanites along the thrust plane, and (3) The exotic humps along the ridge in Kiogarh water divide (i.e. Kiogarh 1, 2, 3 etc.).

The systematic mapping and study of the exotic blocks could not be taken up by Indian geologists due to political and administrative reasons, since the border between India and Tibet (China) is not negotiable freely. Moreover, in the Kiogarh region the watershed is actually the line of the international border. However, the Kiogarh watershed was well studied from the southern slope and with determination a successful climb was made up to the base level of Kiogarh no. 1 where ophicalcite and basic rocks are exposed.

1. Exotic blocks in flysch have been elaborately described by von Kraft (1904) in the south of Kiogarh peaks, they all occur as exotics. They are almost all disturbed due to solifluction. A block of Liassic age was initially studied by von Kraft (EB4 pl. 3) and again in 1936 Heim (Figures 13, 35 and 36) made a detailed study with Gansser (Heim and Gansser, 1969, p. 125, Fig. 35). This large Lias block is about 300 m long and 60 m thick. It was von Kraft (1900) who first discovered *Arietites* and *Phylloceras* and he recalled the striking resemblance to the Adneth facies of the Salzburg area of Austria. The presence of *Sponga spicular* and radiolaria in the dense limestone prove the deep sea facies. Many other such Liassic exotic blocks (about 24 m × 10 m) in Upper Cretaceous Eocene flysch are known to be embedded associated with ophiolite as melange; these ophiolitic rocks are represented by serpentinites, spilites and amygdaloidal basic igneous complex. Sometimes exotic blocks of various compositions are clustered together with breccias and tuff of volcanic origin. In Fig. 130, p. 158 Heim sketched a composite exotic block of Permian age at 5407 m. Gansser (1964, p. 125) thought that it is not certain whether the blocks of volcanics are remnants of larger exotic block mass in now disintegrated and fallen due to gravity slide or represent single or clusters or blocks occurring with the flysch.

After close observation during our 1973, 1974 and 1976 expeditions, it is inferred that these blocks are the embedded remains of a broken mass from older rocks in younger Cretaceous–Eocene flysch along the deep-seated fault zone further north in the Tibetan region, along the Sutlej gorge.

2. A large number of exotic blocks have been noticed to be enclosed in ophiolite by workers in this region (von Kraft, 1904; Heim and Gansser, 1939a; Gansser, 1974a). The exotics in the basic and ultrabasic complex from a widespread sheet (Figure 35) showing enclosed limestone exotic blocks in the ophiolite complex.

There is a wide variation in the composition of basic and ultrabasic rocks called ophiolite. The igneous masses consist of large serpentine sheets with zones of ophiocalcite, some peridotites, diabases, basic, often amygdaloidal porphyrites and spilitic rock, and variations of basic tuffs intermixed with siliceous flysch sediments. Microscopically these spilitic rocks have a fluid texture indicating flowage, amygdales of calcite, and cirulitic zeolites. Gansser (1974a) perhaps mentioned them as amygdaloidal porphyrites. The

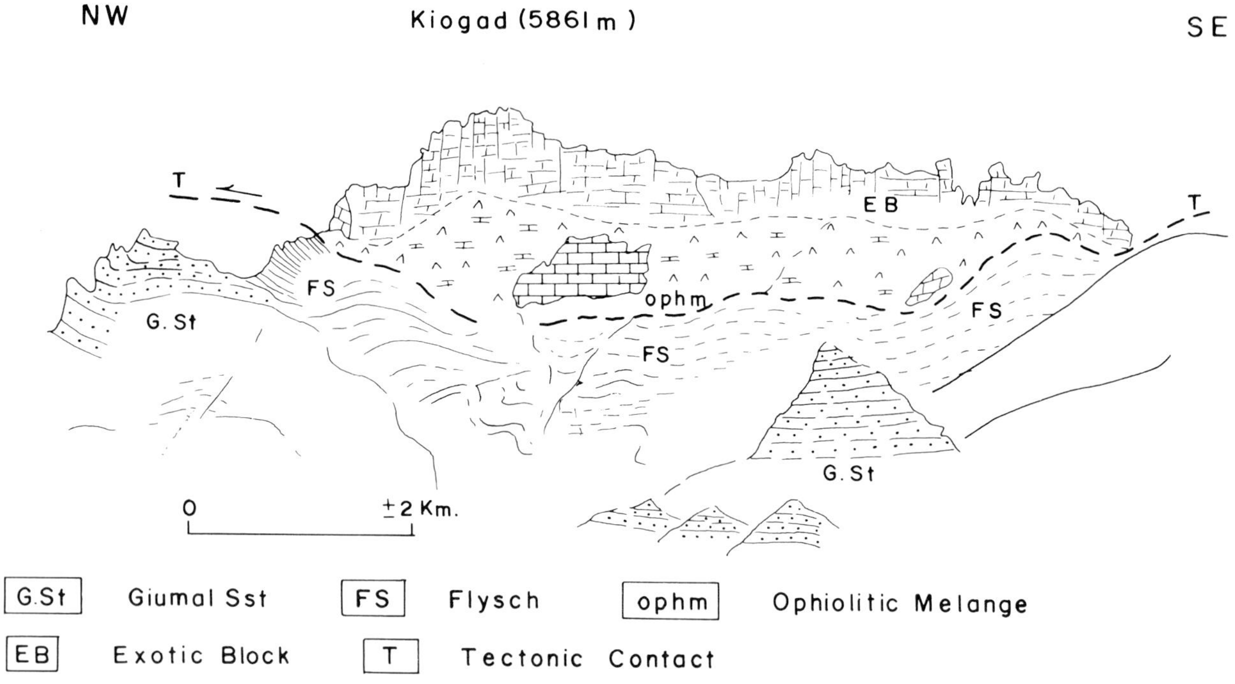

Figure 35. Exotic blocks of the Malla Johar Kiogad 1 peak

albite oligoclase microlites show alteration and rather small crystals of clinopyroxene (augite) are also present. The opaque minerals are scattered throughout the mass. Sometimes the brownish amygdales increase in volume up to 45–50 per cent, and are incompact whereas other specimens having a greenish-grey tint, are fine grained and compact. Out of the five specimens put under K/Ar dating process only three could furnish a valuable age data. The absolute ages recorded are: 107.5 Ma (boundary between Lower and Upper Cretaceous), 75 and 73 Ma (Companian–Maestrichtian) (Sinha and Bagdasarian, 1977). Thus the earlier 107.5 Ma phase of ophiolitic eruption seems to be followed by 75 Ma and 73 Ma eruptions. It has been observed that the most common exotic blocks are Triassic or younger and Permian are missing. The entire range of Kiogarh peaks (1 to 5) are within the ophiolitic complex. The other enclosed components are white Kiogarh limestones, red radiolarian cherts, dense brick-red limestones and marls (Gansser, 1964). A red limestone block furnished a rich Carnian cephalopod assemblage that is completely different from the Himalayan regime found in other similar blocks observed by von Kraft. A list of forty species of ammonites has been supplied by Diener (1912). Gansser (1964)

Figure 36. Kiogad 1 and 2 peaks photographed from the southern face

remarked that they are also some glauconitic sandstone blocks quite similar to Giumal sandstone.

3. In the third category comes the Kiogarh peaks (1 to 5) exotic mass. At a glance on the ridge with Kiogarh peaks, it looks like a continuous thrust-sheet, but with desiccated and eroded features disconnecting each broken sheet of dense and fine marblized limestone. The average thickness of the mass is 200 to 300 m and the total area covered is 20 km^2 (Gansser, 1964). In some places a vague plane of stratification could be observed which dips towards the north. The development of ophicalcite at the contact of limestone and basic igneous rocks is very conspicuous and its thickness varies at different places from a few metres to tens of metres. The summits formed by these limestone blocks vary between 5861 m to 5800 m.

Beyond the Balchadhura pass watershed the Tibetan region looks like an undulated rather flat country with reddish crumpled flysch rocks, and far beyond the serrated line of the Sutlej, flowing westwards from the Kailas region, is another white line of snow-capped peaks forming the Kailas range. A beautiful panoramic gleaming view of Tibet, the roof the earth, was observed at 15.45 on 23rd June, 1973. But a thunder shower and hailstorm forced a quick climb down from Kiogad peak 1. Shackleton (1981) provides a first hand account of the geology of the adjacent part of Tibet. It has observed that Zangbo (Tsangpo = Brahmputra) ophiolite belt is underlain by a sedimentary melange containing 'exotic blocks' ranging in age from Permian to Cretaceous. It has the appearance of an olistostrome but may have been scraped off the subducting crustal slab. To my mind these are the original facies remnant of the 'exotic blocks' of Malla Johar.

The lithostratigraphic framework of the Tethyan zone of Garhwal and Kumaun Higher Himalaya has been summarized in Table 3.6.

Table 3.6. Lithostratigraphic framework of the Tethyan Garhwal–Kumaun Higher Himalaya

Time unit	'Lithounit'	Lithology	Assemblage zones	Fossils
	Sangcha-malla Formation	Greenish shales with bands of radiolarian cherts		Radiolaria *Odontochitina*, *Oligospharium*, *Systematophora*, *Diphyes*, *Cordosphaeridium*, *Aerosphaeridium*, *Hystrichokolpoma* etc.
		Greenish greywackes and dark shales		
		Purple marly shale, with foraminifera ooze	*Globotruncana*-*Heterohelix* zone	Radiolaria Fucoid markings *Globotruncana*, *Heterohelix*, *Plummerita*, *Shackoina*, *Eouvigerina*
		Dark greenish shale with greywacke bands		
CRETACEOUS EOCENE	Giumal Sandstone	Greenish-grey sandstone and sandy shale with thick bands of massive radiolarian cherts		No determinable fossils except radiolaria
		Thick bedded glauconitic sandy shales and sandstones		
Portlandian	Saligram Member Spiti Shale	Black shales with phosphatic, ferruginous and calcareous concretions	*Perisphinctes*-*Hoplites* zone	*Macrocephalites*, *Inoceramus* *Belemnopsis gerardi*, *Perisphinctes*, (*Virgatosphinotes*) *frequens*, *Lytoceras* sp., *Hoplites*, etc
Callovian	Sulcacutus Member	Ferruginous oolite with coquina	Belemnopsis Sulcacutus zone	*Belemnopsis* sp., *Reineckites* *Macrocephalites*, etc.

Table 3.6. Continued

Time unit	'Lithounit'	Lithology	Assemblage zones	Fossils
JURASSIC	Lias Lapthal Formation	Dark blue to grey limestone with bands of coquina	*Pecten–Astarte* zone	*Pecten* sp., *Astarte*, *Cardium* sp., *Trigonia* sp., *Arca* sp., *Avicula* sp., *Belemnites*, *Rhynchonella*, *Lima*, *Ostrea*, *Pleurotomaria*, *Lima*
Rhaetic	Kioto Limestone	Grey limestone with numerous bands of coquina		
		Nodular and oolitic limestone		
TRIASSIC		Cross-bedded calcarenite and arenaceous limestone	*Megalodon* zone	*Megalodon*, *Spiriferina*, *Pecten*, etc.
		Grey and blue dolomitic limestone		
Noric	Kuti Shale	Alternating bands of black shale and limestone		Few pelecypod shells, *Juvavites*, *Lilangina*, *Pecten*, *Lima*
Carnic Ladinic	Kalapani Limestone	Nodular limestone	*Ptychites* zone	*Ptychites*, *Gymnites*, *Halobia*, *Arcestes*, etc.
PERMIAN		Grey massive limestone		
	Kuling Shale	Black crumply shale with thin bands of limestone with concretions towards the top	*Cyclolobus—Oldhami* zone	*Paramarginifera himalayansis*, *Spiriferella rajah*, *Linoproductus cancrini*, *Chonetes vishnu*, *Dielasma* sp., *Waagenoconcha purdoni*, *Geeitothyridium roysii*, etc.
DEVONIAN	Muth *Quartzite*	White sugary orthoquartzite *with bands of dirty white* quartzite	*Schelwienella williami* zone	*S. williami*, *Leptaena rhomboidalis*, *Strophomena*
		Chocolate brown quartzite and dolomitic limestone	*Pentamerus* zone	*Pentamerus* sp., *Camarophoria* sp., *Leptaena rhomboidalis*, etc.
SILURIAN	Variegated Formation	Purple limestone and shale with bands of quartzite	*Strophonella* zone	*Atrypa reticularis*, *Strophonella* sp. ?, *Orthis* (*Dalmanella*) *basalis*, *Favosites*, *Leptaena rhomboidalis*
	Yong Limestone	Green nodular biohermal and biostromal limestone	*Calostylis* zone	*Calostylis dravidina* ? *Streptelasma* sp. and massive stromatoporids

Table 3.6. Continued

Time unit	'Lithounit'	Lithology	Assemblage zones	Fossils
ORDOVICIAN	Shiala Formation	Grey to pinkish sandstone and quartzite with bands of limestone to the top	*Rafinesquina alternate* zone	*R. alternate*, *Leptaena halo*, *Favosites* sp., *Saffordia* sp., *Monotrypa* sp., *Strophonema*
		Alternating bands of sandstone and shale	*Monotrypa* zone	*Chaemaerops*, *Laptaena* ? *trachealis* ? *Rhinidictya* sp., *Asaphus*, *Lioclema* sp.
		Alternating bands of greenish shale and biostromal limestone. Green splintery shale with thin bands of arenite	*Rafinesquina aranea* zone, *Orthis testudinaria* zone	*Rafinesquina aranea*, *R. muthensis*, *Triplecia* sp., *Skendium* sp., etc.
	Garbyang Formation	Green needle shales with occasional bands of limestone		
		Alternating bands of sandstone and shale with graded bedding	*Eccliopteris* zone	*Eccliopteris kushanensis*
CAMBRIAN		Cross-bedded calcareous sandstone		
		Greyish green graded bedded sandy shales		
		Crinoidal and oolitic limestone	Horizon bearing indeterminate trilobite fragments and lingulids	
		Brown marl		
		Brown dolomitic limestone with alternating bands of shale		
PRECAMBRIAN	Ralam Formation	Arenaceous shale		
		Dark-coloured quartzite Conglomerate alternating with quartzite		
	Martoli Formation	Graded bedded blackshales; slates and phyllites		
	Vaikrita Group	Quartzite, quartz schist, calc-silicates, kyanite and sillimanite gneisses, migmatites, etc.		

CHAPTER 4

Tectonic Framework of Central Himalaya with Special Reference to the Higher Zone

The research area of the Indian Central Himalaya is divided from south to north along the width into subparallel structural facies zones. The Main Boundary Fault (MBF) plays the role of an important tectonic element dividing the Siwalik molasse along the foothills of Himalaya against parautochthonous and allochthonous tectonic units of Southern Himalaya (Sinha and Jhingran, 1977; Sinha, 1987).

In this research work the geologic division of Himalaya from south to north would be considered along the structural line known as the Main Axial Zone (MAZ), situated in the crystalline complex of Higher Himalaya. Since this very deep-seated structure in the Himalaya has played a crucial role in the history of its geologic development as well as incorporating the 'root-zone' of a thrust-sheet pushed southward, the territory of Himalaya at its southern end will be henceforth referred to as the 'Southern Himalaya' and that in the north extending into the Tibetan region as the 'Northern Himalaya'. This zone in the earlier conceptual work of the author was referred as the 'Main Central Axial Lineament of Himalaya' (Figure 2; Sinha and Jhingran, 1977; Ashgirei, 1977). Detail about the MAZ has been elaborated in Chapter 5 (Figure 37) and in a recent paper by the author (Sinha, 1987). The earlier proposed tectonogeomorphologic divisions (Gansser, 1964), *viz*. Sub-Himalaya, Lesser or Lower Himalaya, Higher Himalaya, and Tethys or Tibetan Himalaya, will be referred in their morphotectonic sense to perpetuate well-known terminology used in Himalayan geology for many years.

The following tectonic units have been identified in the territory of Southern Himalaya extending from southeast to northwest (Fuchs and Sinha, 1978):

1. Parautochthonous Unit comprising the Krol sequence and flyschoidal Simla formation. The existence of a 'Krol Nappe' derived from afar is denied.
2. The above-mentioned unit is succeeded by the Chail Nappe (thrust-sheet).
3. 'Jutogh Nappe (thrust-sheet)' together with higher 'Crystalline Nappe' represented by a lower unit of medium grade metamorphics and an upper unit composed of high grade crystalline, usualy disposed as allochthonous synforms.
4. Carbonate–terrigenous parautochthonous unit with the Riphean stromatolites, acting as gliding surface for higher thrust-sheets.

The Main Central Thrust (MCT) (Bist, 1986; Bist and Sinha, 1982, in press) tectonically separates the terrigenous–carbonate parautochthonous zone of Southern Himalaya from the metamorphosed crystalline complex of 'Vaikrita'. The Vaikrita Central Crystalline complex in its turn is divided from a huge pile of sedimentary Tethyan geosynclinal complex, deposited in a geanticlinal regime developing from Precambrian to Cainozoic, by the Tethyan Thrust (TT). In the central part the arcuate disposition of Tethyan basin helps to coincide the *en echelon* deposed Main Axial Zone (MAZ) with the Tethyan Thrust (TT) (Figures 13 and 37). In the higher

Himalayan terrain within the Vaikrita complex, there has been observed the zone of divergency with steeply dipping high grade metamorphic rock suites, called the Main Axial Zone (MAZ) (Figures 37 and 53) of Himalaya, incorporating the 'root-zone' of crystalline thrust-sheets (Sinha and Jhingran, 1977, Ashgirei, Sinha *et al.*, 1977; Ashgirei, 1977; Powell *et al.*, 1979; Sinha, 1987).

Over a decade of research in the higher terrain of Central Himalaya has led to the preparation of geologic-cum-tectonic map of more than 20,000 km^2 of inaccessible territory keeping the MCT as the southern boundary and the Tibet and Nepal borders as the boundaries in the north and east, respectively. The final map emerged in the scale of 1:125,000 (Figure 13). A portion of the area in the northwestern part was mapped in larger scale of 1:63,360. Another crucial tectonic window of Sirdang–Gammalla was mapped in 1:50,000 in the southeastern portion of area in the Darma and Kali valleys to explain its tectonic significance in relation with high grade flat thrust-sheet of crystallines. The detailed geotraverses described and illustrated in the next chapter on tectonofacies zonation will need to be consulted to help explain the structural and tectonic tangles.

4.1 Southern Territory of Central Lesser Himalaya: A Tectonic Synopsis

The outermost Miocene–Pliocene zone in the foothills of Himalaya is mainly composed of Siwalik molassic sediments which have been thrusted over by different tectonic units along the Main Boundary Fault (MBF). Adjacent southern terrain of Lesser Himalaya consists of parautochthons. Auden (1934) described the geology of the Krol Belt in his classical paper, and his zone

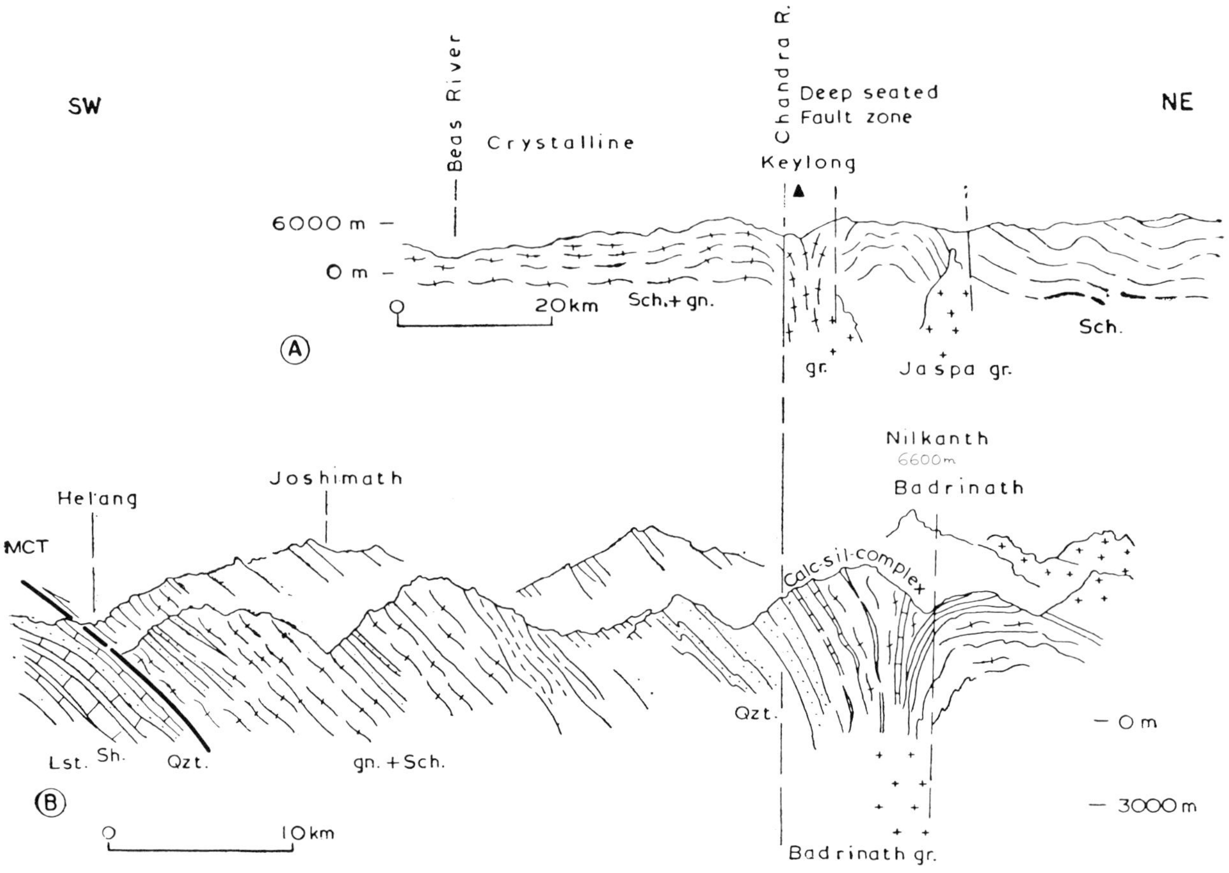

Figure 37. The Main Axial Zone (MAZ) at (A) Keylong and (B) Badrinath (modified after Powell and Conaghan, 1973a; Heim and Gansser, 1939a)

may be taken as the type area for the parautochthonous unit. This unit shows a series of anticlines and synclines, the first frequently sheared by reversed faults. Thus combined fold and wedge structures are characteristic features. There is a marked contrast between the often steeply dipping beds which build up that unit and the overlying nappes being horizontal or gently dipping. This contrast in tectonic style and the fact that only the southern frontal portions are proved by tectonic windows to be allochthonous shows that this belt is best termed parautochthonous. But this does not imply that the Krol thrust is translated as a thrust-sheet from afar north being deposited in a separate trough in the northern zone. In the Simla region the Krol Belt shows SW-directed folds (Auden, 1934; Bhargava, 1972; Sinha, 1972b, 1974a; Fuchs and Sinha, 1978). The Pachmunda and the Krol Synclines (Figures 8 and 9) are separated by an anticline at Solan; Simla slates and Subathu in the core of that anticline, according to Auden (1934), represent a tectonic window. The rocks exposed in that window seem to belong to the northern portions of the Tertiary Zone. The displacement of the southern parts of Krol thrust, proved by this window is, however, only 3 km. The structural plane separating the Tertiaries from Krol Belt rocks was called 'Krol Thrust' by Auden (1934). It is noteworthy that nowhere were the Krols observed to overlie the Riphean carbonate–terrigenous complex of the autochthonous Shali–Deoban Tejam formations. This Riphean carbonate complex was deposited in the adjacent northern structural facies trough. In the joint publication on the problem of tectonics of the Garhwal–Kumaun Lesser Himalaya with Dr G. Fuchs, author has maintained his view that the only uppermost C_3 unit (Figure 10) is the real 'Chail unit' (Fuchs and Sinha, 1978). The alternating sequence of carbonate and terrigenous quartzitic beds belonging to autochthonous Shali–Deoban Riphean sequence is to be attributed to tectonic consequence rather than stratigraphic normality. The Chail tectonic unit (C_3) should maintain its independent identity with quartzitic acid metavolcanic phyllitic composition representing early eugeosynclinal regime (Ashgirei *et al*., 1977a, 1981; Sinha *et al*., 1983). And this very Chail thrust-sheet physically cover the still lower units of parautochthonous series with conspicuous tectonic unconformable contact. Considering all the available data the depositional eugeosynclinal sediment catching basin of Chail with its root-zone now could be traced just south of the MCT zone. This is supported by the presence of acid metavolcanic porphyry rocks in some places as observed in the Mandakini valley between Agastmuni and Okhimath (Bist and Sinha, 1982; Bist, 1986).

The Chail thrust-sheet in its turn is covered by the Jutogh and crystalline thrust-sheets. The Jutogh thrust is well exposed and widely expansed in area around the Simla–Chor massif of Himachal Pradesh (Figures 7 and 8). These thrust-sheets of Chail and Jutogh are the responsible factors of 'reverse metamorphism' phenomenon besides those of Almora and Garhwal nappe (Gansser, 1964; Jain, 1971, 1972; G. Kumar, 1981). These have been horizontally translated for more than 30 to 100 km from higher Himalayan root-zones situated in the north. As a consequence of main phase of thrusting (post Dagshai), the whole pile of parautochthonous units and nappes was compressed, leading to folding and local wedge structures. In that phase perhaps the Main Boundary Fault (MBF) was reactivated, enabling the Lesser Himalayan units to move *en bloc* over the northeastern portion of the Tertiary belt. The folding led to the formation of windows in anticlines and of outliers of higher nappes in the synclines. These anti- and synforms are crossed by zones of axial depression or culmination roughly in north–south direction. The Chor area (Figure 10) is a typical zone of axial depression between Himachal and Garhwal–Kumaun Lesser Himalaya. A detailed work of the Simla area, adjacently northwest of Chor mountain, has been published by author in different papers (Sinha, 1972b, 1977a,b, 1978a, 1980a). To the east follows a culmination zone in the Tons region (Fuchs and Sinha, 1978). Further southeast there is a slight axial plunge and therefore more and larger Klippen outliers of higher nappes occur in that direction. At the same time the lowest unit of Lesser Himalaya—the parautochthonous unit—gets more reduced and finally disappears in the eastsoutheast direction near Naini Tal (Fuchs and Sinha, 1974, 1978). The last phase of vertical faults have influenced the structure of Garhwal and Kumaun, but to a much lesser degree. Faults probably traverse the fold structure in the Baijnath area. In the southwestern limb of the Lansdowne synform the transverse faults seem to be effective. The Aglar Fault (Fuchs and Sinha, 1978) in the northern limb of the Mussoorie syncline is approximately parallel to the strike trend of the rocks, and similarly the Naini Tal fault in the southern side of Naini Tal lake complicates the south limb of Naini Tal syncline (Fuchs and Sinha, 1974). Five generations of shortening structures B_1 to B_5 have caused corresponding deformations related to specific orogenic episodes in the southern Himalayan region (Schwan *et al.*, 1985).

4.2 Central Zone of Higher Himalaya: The Zone of Central Crystalline Complex (Figs. 13, 14, 37 and 53)

The higher Himalayan zone in question is situated between two great thrusts, *viz*. the Main Central Thrust (MCT) and the Tethyan Thrust (TT). This zone is constituted by the metamorphosed rock suites of different grades. They are intruded by granite, aplite and pegmatite.

4.2.1 The Main Central Thrust (MCT)

The term Main Central Thrust (MCT) was first proposed by Heim and Gansser (1939a) after their Central Himalayan Swiss Expedition in 1936, to describe the tectonic contact surface between terrigenous–carbonate autochthon and overlying metamorphic complex of mica schists and gneisses. Often along the contact there are thick layer (approximately 150–300 m) of schistose sericite–quartzite schists. This tectonic plane dips at an angle of 45° to 15° towards the northeast. Everywhere in the research area this thrust-surface is clearly marked, but in some places, *viz*. in the Mandakini valley, it is separated with Chail thrust which is mapped near Guptkashi, and later further downstream near Bhatwari. Mapping on either flank explained that again in the east and west the MCT is duplicated by the Chail thrust surface (Valdiya, 1980a,b). Pati and Rao (1981) expressed their view that two thrust planes occur in the major area of Uttar Pradesh Himalayan region (Garhwal and Kumaun). The lower thrust plane is correlatable to the Chail Thrust and the upper to the Jutogh Thrust of Himachal Pradesh, i.e. the MCT. At places where both the thrusts are present the lower or the Chail Thrust has been called the MCT, whereas in the area, where the Chail unit is tectonically overlapped and the Jutogh comes in direct contact with the sedimentaries, it is the Jutogh Thrust which has been called the MCT. Valdiya (1980a) while redefining the MCT proposed another thrust further north and called it the 'Vaikrita Thrust'. The MCT is clearly expressed over the surface in the mountain slopes and deep cut river gorges of Alaknanda, Gori Ganga, Kali and Sutlej, etc. The contact zone is characterized by a wide zone of crushing (Bouchez and Pecher, 1981), striations and mylonitization, accompanied by retrograde metamorphism. This thrust surface has been responsible for the translatory movement of crystalline sheets causing the Klippen of Almora, Askot, Baijnath, Dudatoli and Chor, etc.

The MCT is characterized by another interesting feature of schuppening or overlapping system of parallel thrust planes and reverse faults (Bist and Sinha, 1981). This feature is vividly observed near the MCT contact zone in the gorges of Alaknanda, Mandakini, Pindar, Gori Ganga, Darma and Kali, etc. On the exposed wall surface it could be seen how the sedimentary terrigenous–carbonate is thrust over by crystalline thrust and forming by a schuppen structure.

The final Middle Miocene major movement led to colossal crystalline thrusting. Bist (1986) and Bist and Sinha (in press) explain the system of thrust planes in the Mandakini valley through a duplex system of imbricate pattern.

Shackleton (1981) after traversing from Tibet to Nepal observes that the MCT, taken as the limit between the High and Lesser Himalayas, is mapped about 2 km south of Nepal/China border, within garnet mica schists some 200 m below the lowest horizon of augen gneiss. There is no metamorphic break, but a steep metamorphic gradient. The MCT must be at least largely premetamorphic and a shear zone rather than a single thrust, as recognized elsewhere.

4.2.2 'Divergence structure': 'Root or Würzel zone'

The late Professor Arnold Heim (Heim and Gansser, 1939a, p. 72) in his review of tectonics in the Upper Alaknanda gorge observed: 'seen from the distance, on the high walls of the western side, the injected series with lime–silicate layers continues and gradually steepens to a vertical and even over-vertical position, in the shape of a synclinal fan. Its vertical axis, striking W20°S, passes the river below the bridge of Bamani 1.5 to 2 km south of the temple of Badrinath'. He plotted the structure (Heim and Gansser, 1939a, Plate IV, section 7b) but the detailed significance of the structure was not discussed even when Gansser reproduced this section in 1964 (Plate III, Section 4), in the classic monographs 'Geology of Himalaya'. The author while synthesizing the work on the deep seated structures in Himalaya and the Caucasus reproduced this structures in his lecture at Tbilisi (USSR) in 1975 (Figure 37), and compared the same produced by Powell and Conaghan (1973a) from Rohtang Pass area. A similar divergency in the crystalline complex with lime–silicate, gneiss quartzite intruded by younger granites was observed in the upper reaches of Pindar river before reaching the snout of Pindari glacier. All these divergence (or bivergence) structures could be aligned in a linear NW–SE Axial Zone of Himalaya, hitherto referred in this work as Main Axial Zone (MAZ) (Sinha, 1987) of Himalaya (Figures 37, 51, 52 and Photo Panorama No. 1). Further east in the gorges of Giri, Darma and Kali this divergency could not be plotted by author in his sections, perhaps because the last phase of thrust to push Himalayan arc into an arcuate bend obliterated it. Moreover, in this very zone the crystallines are squeezed as compact as 10 km or less in width with tight isoclinal folds (Figure 38). Perhaps, the almost vertical fault zone further east near Lilam in Gori valley and Urthing in Darma valley might be the remnant of this zone. Another possibility also could not be ruled out that this zone coincides with the Tethyan Thrust (TT) further north. There are other sublongitudinal or strike-faults with steep angle inclinations of 70° to 80° NE with the hot water springs dotted along this zone, like *Lata–Tapovan fault* (Figure 13) following the course of Dhauli Ganga upstream of Vishnuprayag. Similarly the fault planes at Baling in the Darma valley and Lilam in the Gori valley are difficult to trace on the surface, as it is a snow covered area, but by their morphogenic nature it seems that they join with each other. At Dar in the Darma valley we also come across a vertical fault zone with younger granitic intrusions.

Amongst the quartzitic zone from west to east there are vividly seen tight isoclinal folds with axial plane parallel to the general trend of schistocity dipping northeastwards.

4.2.3 Gam Malla–Sirdang—tectonic window

In the gorge of Kali and Darma rivers we come across within the high grade crystallines a zone of sedimentary complex with inverted anticlinal with axis dipping north (see Figures 102, 103 and 104) and on the southern side a zone of crushing along a fault of reverse nature is prominent. After detailed mapping it became clear that this tectonic window is bound by the MCT in the north and a reverse fault in the south. It gets closed by overlapping crystalline thrust westwards and further southeast in Nepal (Fuchs, 1981). The *Jhamarigaon–Sosa* reverse fault (see Figures 103 and 104) at 50–60°NE inclination is responsible for the formation of the entire overthrown southern crystalline chunk up to Dharchula. Most striking feature is the emplacement of amphibolitic bodies along the tectonic zone of their region and their diversified age relation.

4.2.4 The Tethyan Thrust (TT)

In the higher zone of the research area the northern contact line between central crystalline complex of the Vaikrita group and the sedimentary pile (13,000 m) of the Tethyan zone extends far more than 200 km (Figures 12 and 13). In all the sections traversed along the gorge of rivers Dhauli, Pindar, Gori, Darma, and Kali, the tectonic contact between crystalline high grade metamorphic and sedimentary or very low grade sericite schists was observed. In the northwestern region this tectonic contact was first mapped (Shah and Sinha, 1974c; Figures 12 and 54), and in successive work (Sinha, 1980b; Fig. 14) in the

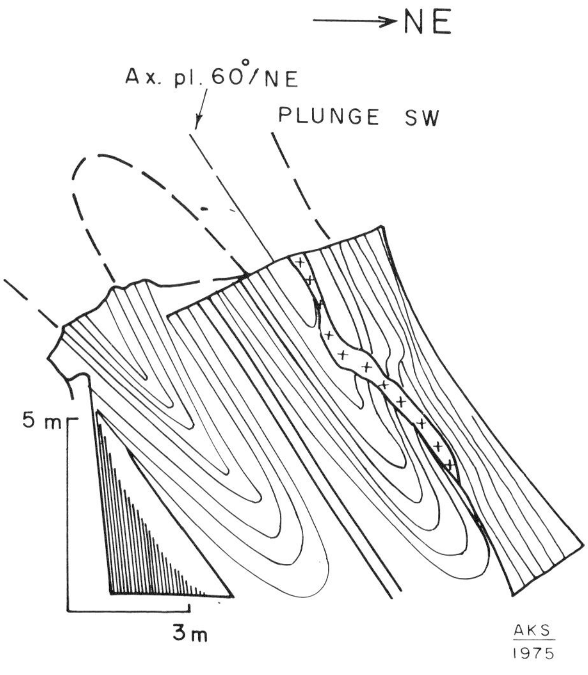

Figure 38. First generation isoclinal fold involving calc-granulite and mica-schists, on the right bank of Kali river

northwestern area, it was called as *Malari–Niti deep seated fault*, behaving as high angle thrust (45–50°) with upthrown northern wing. Owing to its regional character this thrust in the present research work would be named as the *Tethyan Thrust (TT)*.

In the southeastern part of central area, where the two peaks of Nanda Devi (7820 m) and Nanda Devi East (7434 m are situated, the TT assumes the character of a transcurrent fault and echelonment occurs, causing the star pattern glacier valleys. With a shift of 10–20 km the TT continues more as a moderate angle thrust 35°–45° separating thick Martoli phyllites and crystallines. In the Gori Ganga valley at Laspa (see Figures 71 and 72) it forms rather loop type geomorphology and shows profuse intrusion of younger granitic body. Along the tectonized zone develops the biotite porphyroblasts schists of Rilkot (300 m) at the contact between kyanite–sillimanite schist and sericite–phyllite.

Further southeast of the mapped area near Daling in the Darma valley we come across a tectonic break with low inclination of 30–35° due northeast, forming rather a system of schuppen zone. Here along the TT and Baling Fault also there is profuse intrusion of post-tectonic granites, and pegmatic–aplitic suites.

In the next profile along the Kali river gorge near Budhi there is a clear cut exposure of a steeply-dipping fault, accompanied by a system of parallel faults. It is noteworthy that here thick Martoli series is completely cut out and in contact with the crystalline are resting argillaceous carbonate suite of Garbyang formation. This zone is also intruded by two phases of Neogene granites.

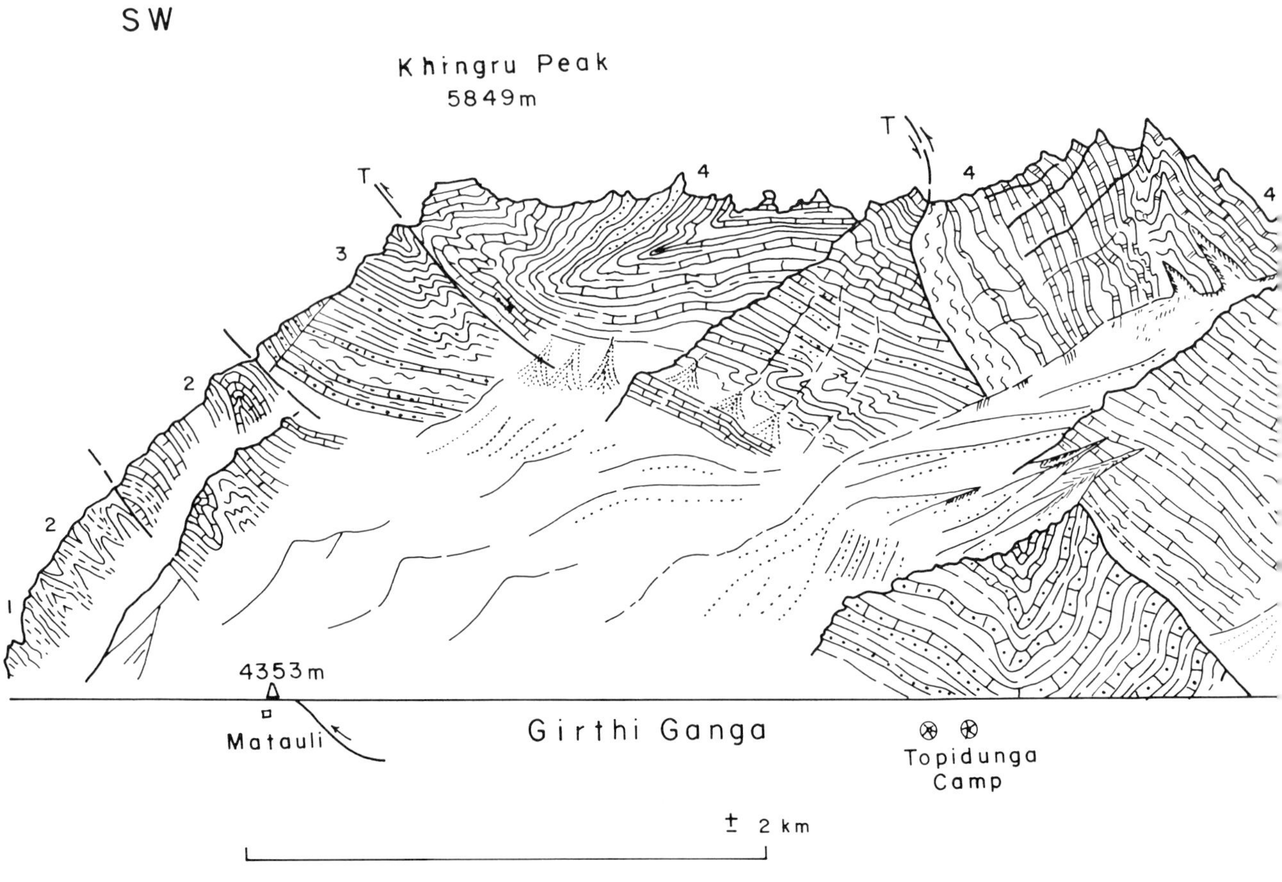

1. Kalapani limestone,
2. Kuti shale,
3. Kioto limestone (Lower Unit),
4. Kioto limestone (Upper Unit),
5. Lapthal Formation,
6. Spiti shale.

In the far eastern area in the territory of Tibet and Nepal, the work of Chinese geologists Jixiang *et al*. (1980) was cited by Shackleton (1981) calling the TT the ***Chiatsun Thrust***. This supports the author's conclusion of Tethyan Thrust (TT) of Tethys.

4.3 Northern Zone of Higher Tethys Himalaya

The tectonic framework of the Tethyan zone maintains its separate identity and independent history of geologic development in the broader realm of Himalaya. The regional direction of fold-axis in the Palaeozoic sediments shows NW–SE prolongation. Obviously the Precambrian (?) sediments are deformed in its own style, but with common NW–SE fold direction of Palaeozoic rocks. In the Gonkha Gad, a tributary to Gori Ganga the Martoli Group phyllite show steep inclination and sharp angular unconformity with overlying the Ralam conglomerate forming the base of Palaeozoic sequence. Thus an early Precambrian phase of orogeny is definitely present in the Tethyan zone which may correspond to the *Baikalian phase*. In the sediments of Mesozoic age the general deformation axis direction turns to assume a submeridianal trend. The Precambrian Martoli group has undergone an early phase with a NW–SE general axis besides other unidentifiable trends, but the cleavage partings assumes a NW–SE trend and dipping steeply across the bedding plane towards the northeast as it is observed along the right bank section between Martoli and the Milam glacier (see geo-

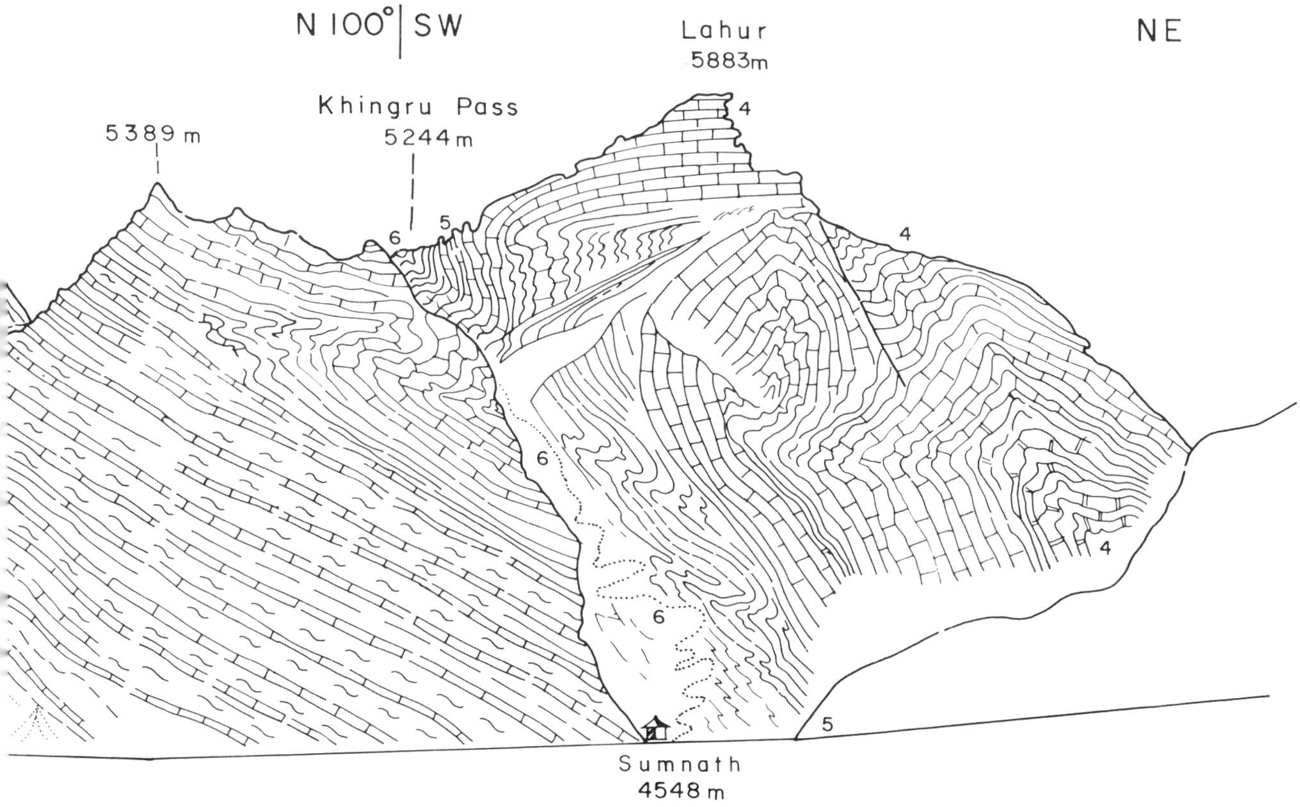

Figure 39. Panoramic geologic sketch view of the escarpment between the Khingru (5449 m) and Lahur (5883 m) peaks on the right bank of Girthi gorge. Sketched from the southern flank of the Topidunga composite. For detail see Figures 40 and 41 and Photo panorama no. 4.

traverse Figure 72). The above-mentioned deformation style is present in the younger beds.

Main anticlinal structures accompanied by synclines are:

1. Anticlines of Lahur and Bampanag with axial trend NNW–SSE and plunging NNW with 10–15° below the flysch horizon of Kiogarh water divide (Figures 39, 40, 41, and 42). Figure 41 sketched from Chidamu camp depicts the complication and inversion the conjugate anticline and syncline structures.
2. Anticline of Unta Dhura Pass (NW–SE trend) (Figure 13).
3. Anticline of Bedang (NW–SE trend) Figure 87).
4. Inverted anticline with synclinal core along the Jolingkong and Dawe pass ridge (see Figure 50, the section across Bedang camp along Dhauli Ganga traverse).
5. Inclination of axial plane towards the southwest at 45°.

Besides the above-mentioned regional fold structure in the Tethyan zone, in different formations and in rock types with different lithological compositions there appears the later and higher degree of deformation, depending upon their competency.

The lower part of dolomitic limestone and phyllite of Garbyang suffered gravitational toothpaste type folding with structural N and NW trend in the left bank of Girthiganga opposite the Barmatiya campsite and barite mineralization district (Figures 43 and 47; Sinha, 1980a). Such gravitational slumping structures have been found to be accompanied by flat dipping or almost horizontal displacement of not larger amplitude. A typical synclinal fold pattern in the right bank of barite mineralized district of Barmatiya (Fig. 43) shows the axial plane direction towards 280°N with 40° inclination having a plunge of 25°/200°N. The lithology of the rocks complicated by above-mentioned fold is siltstone and arenaceous argillite.

Higher up in the Mesozoic sequence of Kuti the

Figure 40. Lahur antiformal structure plunging NNE. For details pertaining to the structurally complicated limbs, see Figures 39 and 41

style of folding deformation completely changes to a disharmonic type (Fig. 45), due to incompetency of alternating thin layers of limestone and predominating argillaceous component. The Kuti horizon is caught between massive quartzite of Muth and the limestone of Kioto. Gradually upwards the folding style changes in the carbonate and argillo-carbonate formations of Kioto and Lapthal, and the fold axis direction changes to be submeridian. It has been suggested that the fold pattern in the Lapthal formation (Figure 46; Photo Panorama No. 3) is reminiscent of the *Jura type* open folds of the Alps (Gansser, 1964; Shah and Sinha, 1974; Sinha, 1980a). At Lapthal we also come across sublatitudinal *microfold axes*.

The major type of faults in the Tethyan zone should be divided into two categories:

1. Strike-faults which sometimes assume the nature of thrust or transcurrent fault. Their orientation changes from NW–SE to NNW–SSE.
2. Faults across the dip with orientation of NE–SW to sublatitudinal.

4.3.1 Strike-faults

The following thrust faults belong to the first category (Figures 12 and 13; Geological maps).

1. Shilong–Siruarch Thrust

Along this thrust, the older Ralam conglomerate is thrusted over the younger Garbyang as observed in Gonkha Gad, and Girthi valley section (Figure 12). It is a low angle (20–35°) thrust plane clearly seen in exposure due to the characteristic red colour of the Ralam conglomerate horizon over dirty grey dolomitic phyllite of Garbyang. As marked on the map this is the very thrust which is parallel with the Tethyan Thrust (TT) right from Niti in the Darma valley situated northwestern end of the area to middle zone in Gonkha Gad, a tributary of Gori Ganga. Tracing further southeast in the Darma valley it is perhaps the same thrust which forms a single plane due to mergence with each other, showing the barite–copper mineralization near Dantu at the

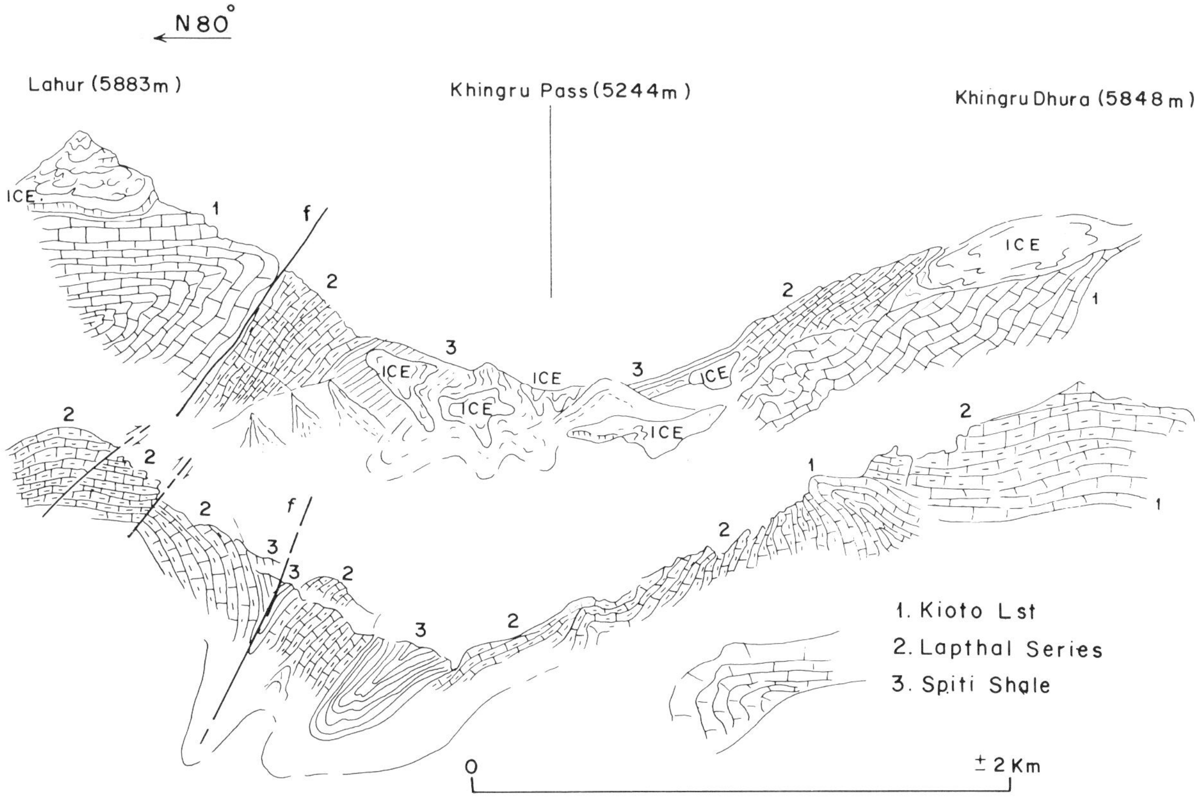

Figure 41. Inverted fold structures in the complicated limbs of Chidamu–Lapthal syncline

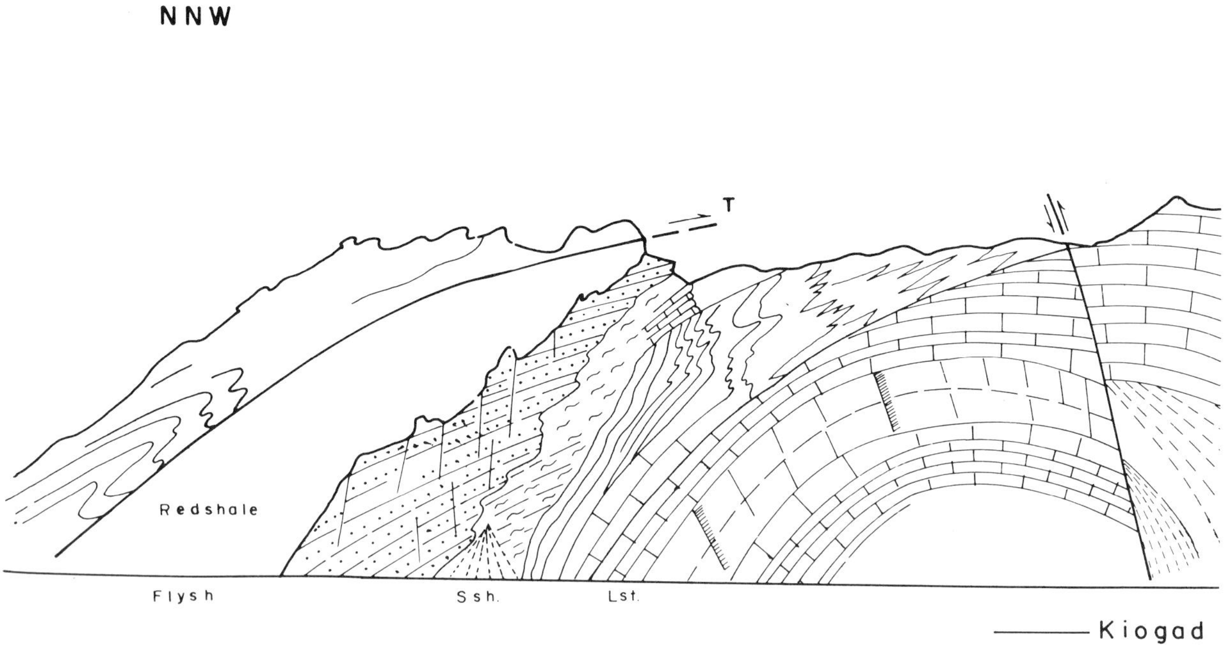

Darma–Lasser river confluence. Under the thick snow cover it was not possible to trace it further southeast and, moreover, in the Kali gorge near Budhi, the Ralam is completely absent due to abutting against the fault and the TT.

2. *Nihal–Bedang–Dung Thrust* (Figures 47, 48 and 49; for others figures refer to traverse chapter 5; Fig. 89)

As seen in the figure, in the right bank of Kali river the Silurian is thrust over Kuling shale–*Fenestella* shale (?) sequence. Other most remarkable feature is the reverse fold pattern with axial plane dipping towards south with thick Muth quartzite core (Figure 47). This type of 'back-folding' is a rather uncommon phenomenon in the Himalaya. In Alps this has been called as 'Ruckfaltung' in Heim's Fig. 67, p. 100 (Heim and Gansser, 1939a). Axial pitch is towards the northwest.

In the geological map (Figure 13) and Traverse map of Darma and Kali (Figure 88) this thrust is clearly traced. Especially near Bedang its sole gets complicated and further northwest it becomes obscure due to survey difficulty with Tibetan boundary as well as extremely difficult access. The help of aerial photographs is the last resort. However, it has been elaborately depicted in sketch figure across Kuti valley near Nihal (Figure 48). Near Dung in Gonkha Gad it has also mapped by author in 1976 expedition (also refer to Figures 71 and 72).

3. *Lipu Lekh–Wilsha: (Muth): Thrust* (Figure 13; see also Figure 88)

This thrust-zone could be seen and mapped standing in the south of steep vertical mountain wall near Lipu Lekh on the way to Kailash–Manasarowar. Thrust was marked by Heim and Gansser (1939a) in their sketch geologic map along with the other four parallel thrust planes. However, the author could not physically map the other two beyond the last one since they fall within Tibetan territory. Since the hanging wall is entirely composed of a massive escarpment of Muth Quartzite resting over Triassic-Jurassic with low angle northerly dipping plane, hence the word 'Muth' is bracketed with its name. The extreme northeast border of India with Tibet is the location of this thrust extending NE–SW almost parallel with the border line (Figure 49).

4. *Lamphia–Dhura Thrust*

It forms the part of four parallel thrust planes situated north of earlier described Lipu Lekh

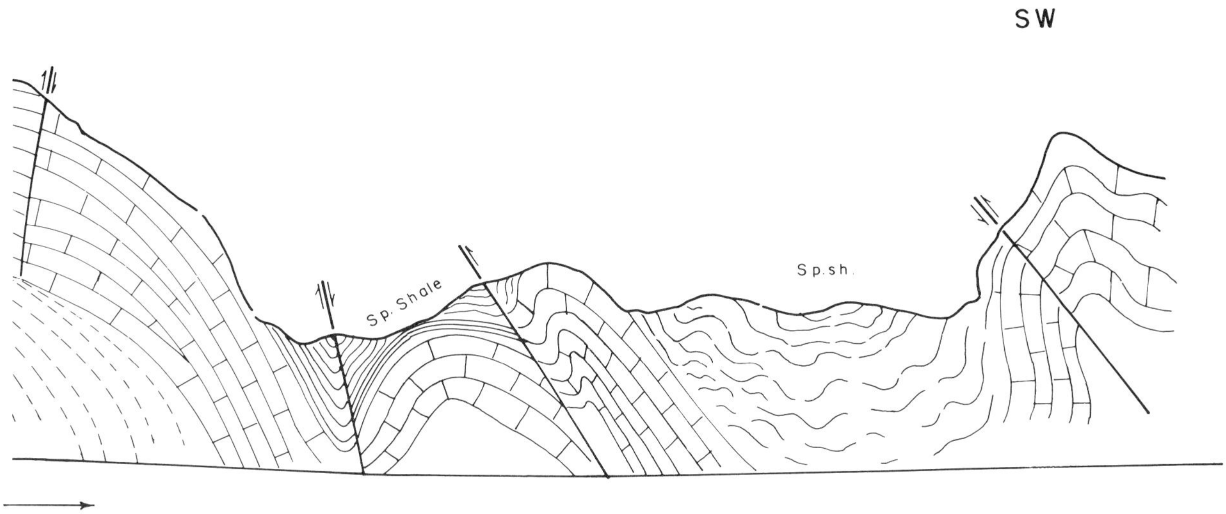

Figure 42. Plunging Lahur anticline dissected along Kiogad gorge, Note the thrust T in flysch and younger faulting in the limbs

Thrust, bordering the Tibetan border. Silurian along this plane is thrusted on the Devonian.

5. *Schuppen Thrust Structure of Bedang–Dawe* (Refer to Figures 88 and 89; Photo Panorama No. 2)

This is conspicuously exposed in the north of Bedang camp site (3850 m (the extreme end of traverse section Figure 89) along the northeast slope of escarpment. It was sketched by the author in 1975 from the trail Joling Kong pass (6000 m). Along this schuppen thrust system the whole sequence of Silurian variegated with the Yong limestone and the Devonian Muth quartzites are first folded in a recumbent fold and subsequently sliced and thrust parallel to the major axial plane. Finally the whole structure complex is also faulted transcurrently. The complicated tectonic sketch of this zone is shown in the traverse map (Figure 88) and geological map of the author (Figure 13).

6. *Bampang Thrust* (see Figures 38 and 71)

A sketch of the southern face of the escarpment in the right bank of Girthi Ganga forge over the Topidunga campsite. This thrust is intraformational in character and complicates the quartzitic horizons of the Kioto formation.

All the above-mentioned thrusts are mappable in the field by knowing the lithological features and the colour of the rock type. It is not possible to map the terrain by climbing the heights of 5000–7000 m. They are evidently not a large

Figure 43. Folding pattern in Garbyang dolomitic arenite and intercalated phyllite in Barmatiya in mineralized zone. Axial plane 40/280N, Plunge 25°/200°N

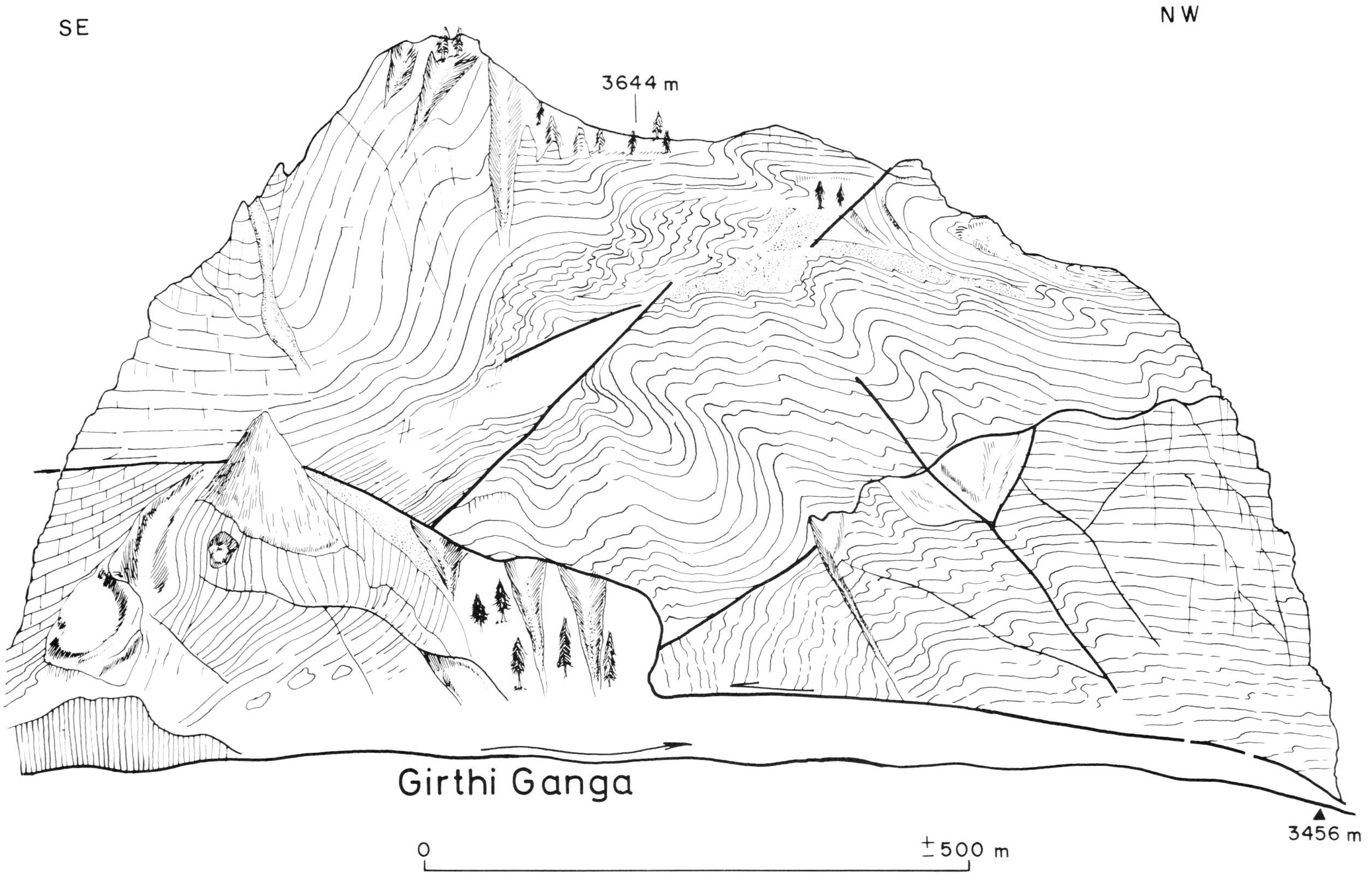

Figure 44. Gravitational back and toothpaste folding pattern in the arenaceous dolomitic Garbyang formation in the right bank of Girthi Ganga gorge near Barmatiya

Figure 45. Disharmonic fold pattern in the Kuti shale on the left bank of Kiogad near Rewlibagar. Photograph facing southwest (width of photographed structure approx. 500 m)

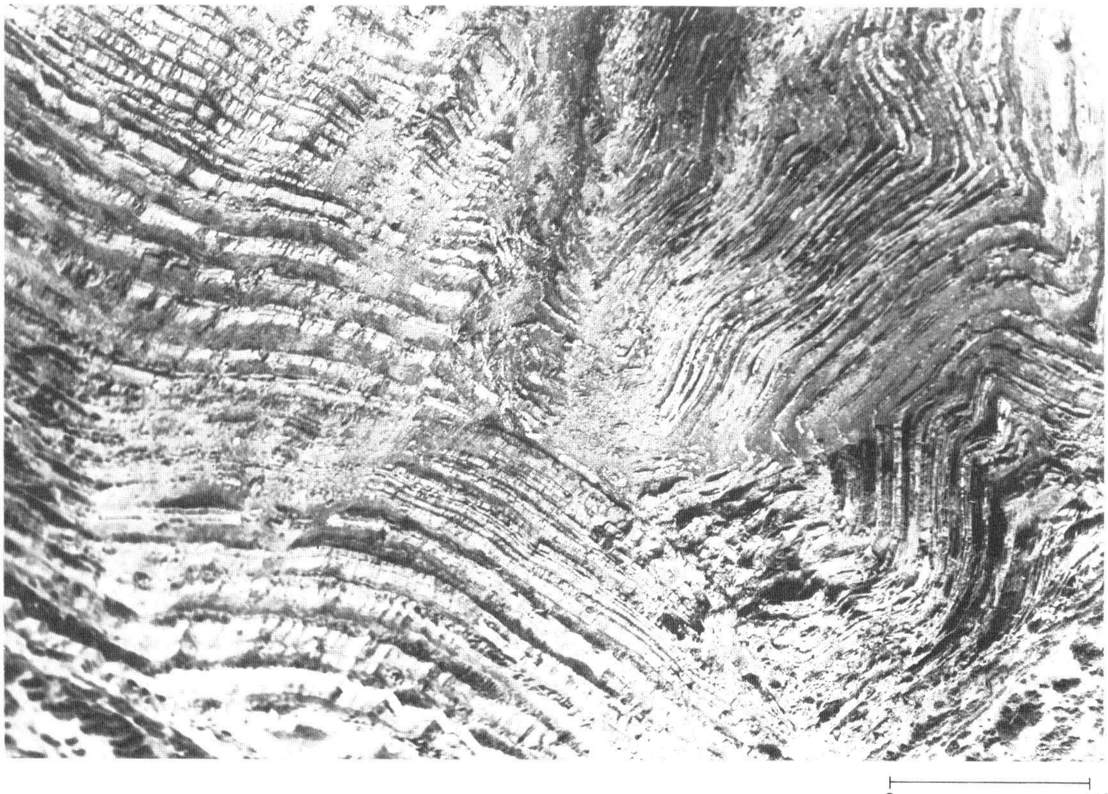

Figure 46. Box type folding, complicating the Lapthal formation and the Spiti shale. The broad style of tectonics is similar to Jura type

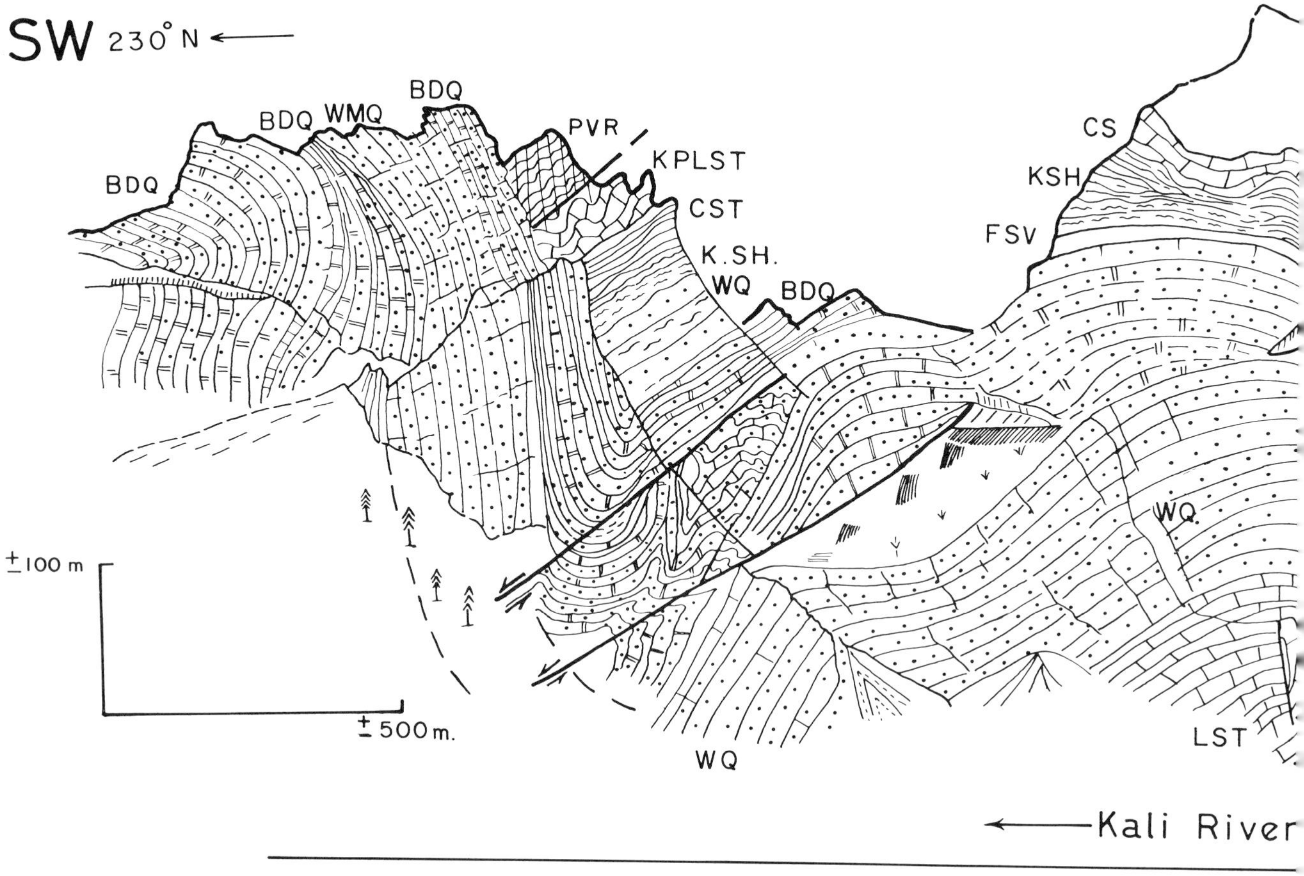

Figure 47. A detail geologic profile of back folding and overthrusting (T—Nihal Thrust) on the right bank of the Kali Zone. Explanation: Shiala Fmn, YLST—Yong limestone; PVR—purple variegated Silurian; BDQ

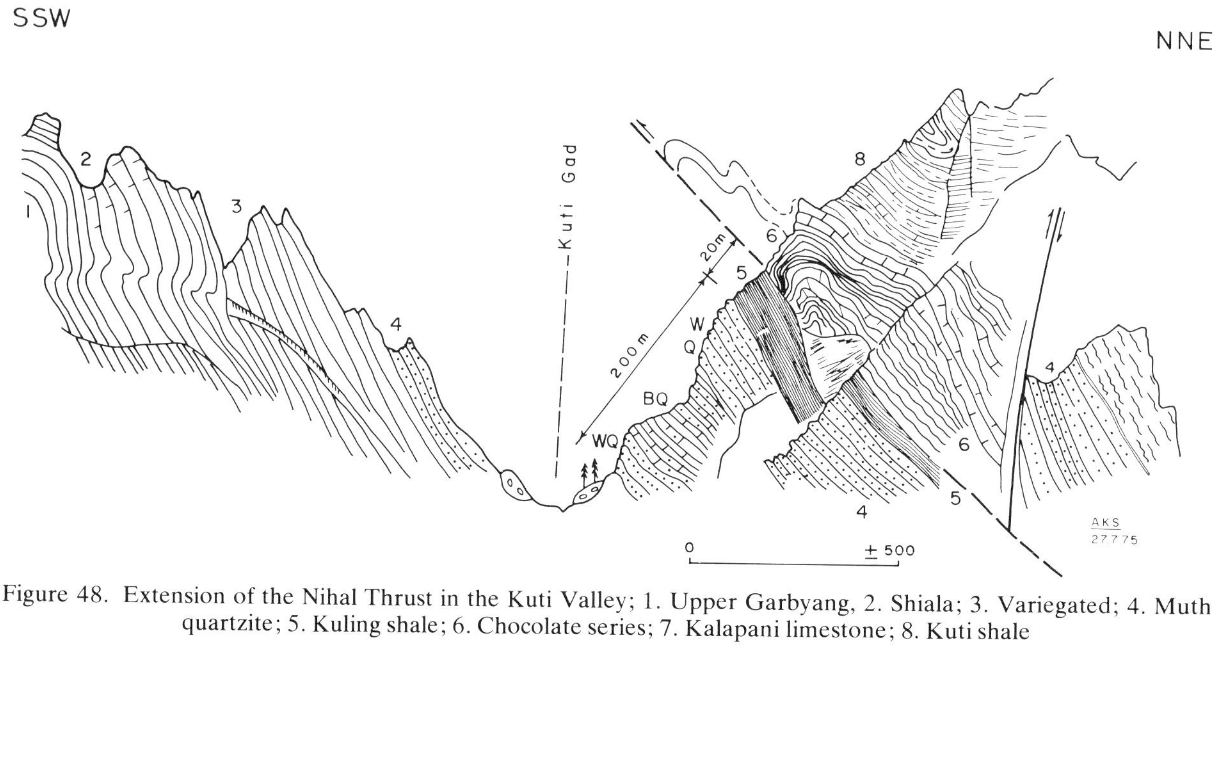

Figure 48. Extension of the Nihal Thrust in the Kuti Valley; 1. Upper Garbyang, 2. Shiala; 3. Variegated; 4. Muth quartzite; 5. Kuling shale; 6. Chocolate series; 7. Kalapani limestone; 8. Kuti shale

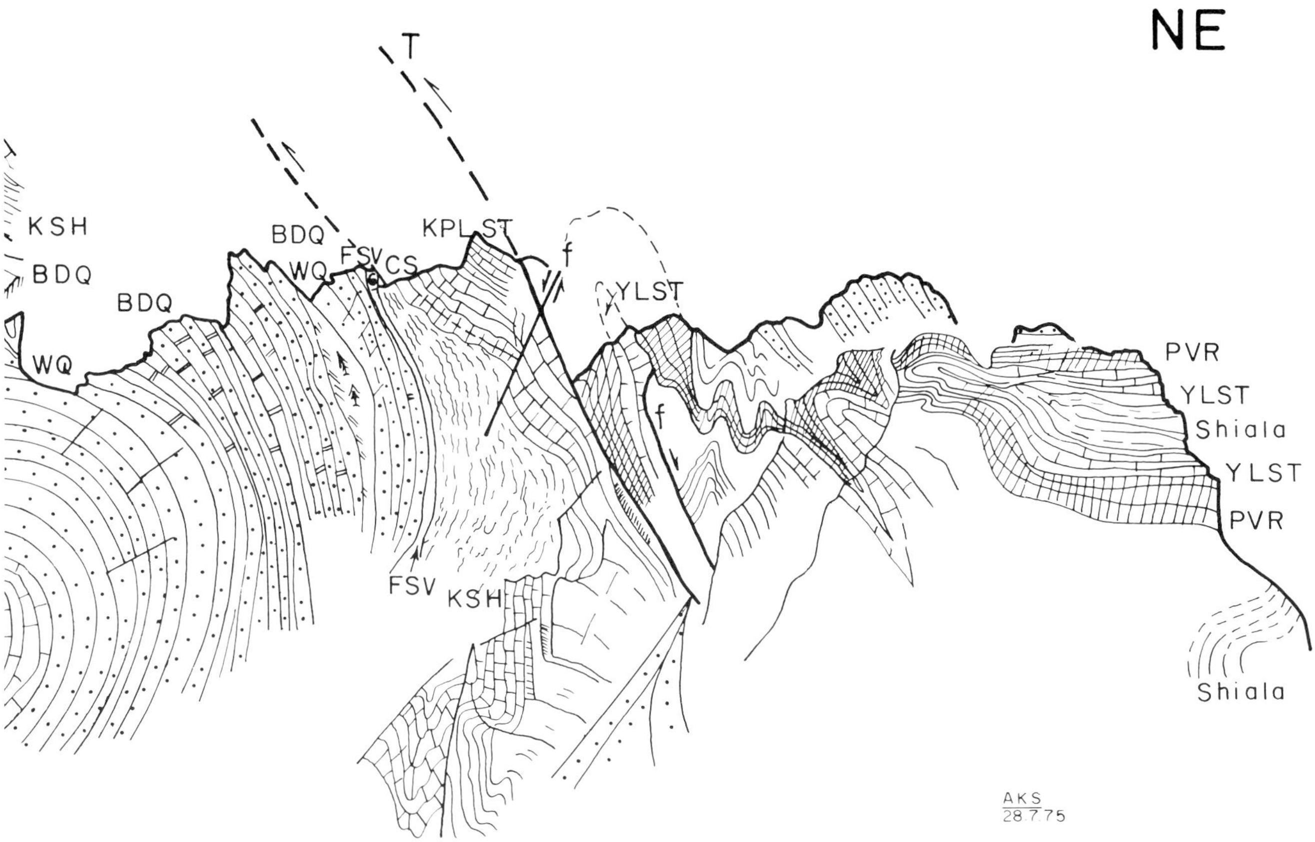

gorge between Gunji and Kalapani. This phenomenon of back folding is rather uncommon in the Himalayan Tethyan —brownish dolomite; WQ—*Fenestella* shale; CS—Chocolate series; KPLST—Kalapani limestone

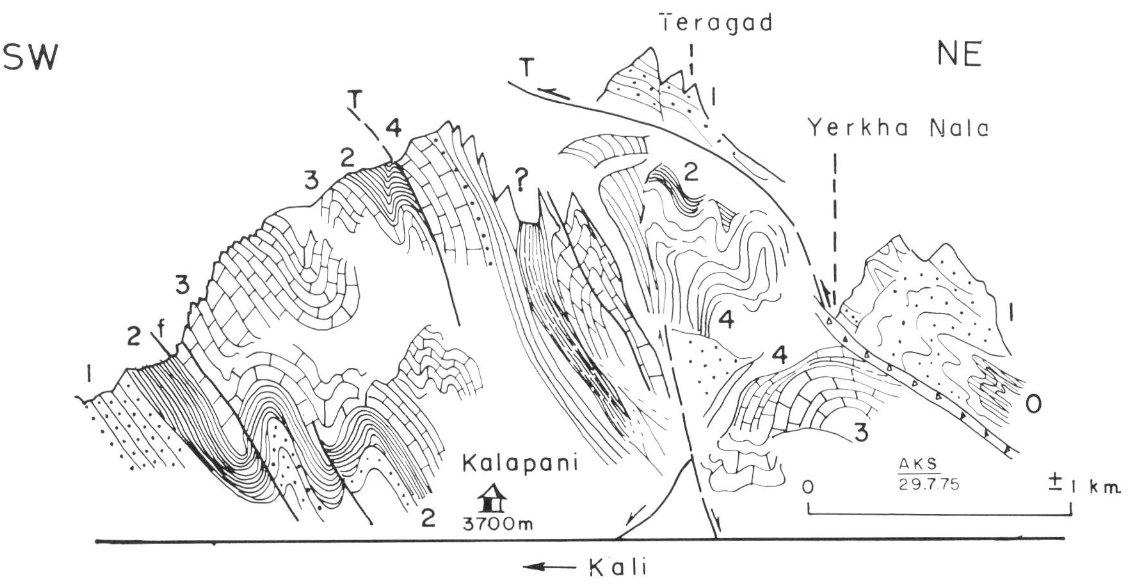

Figure 49. Repeated thrusting on the way to Lipu Lekh and fan fold in the background of Teragad: 0—Variegated Silurian, 1. Muth quartzite; 2. Kuling and Chocolate; 3. Kalapani limestone

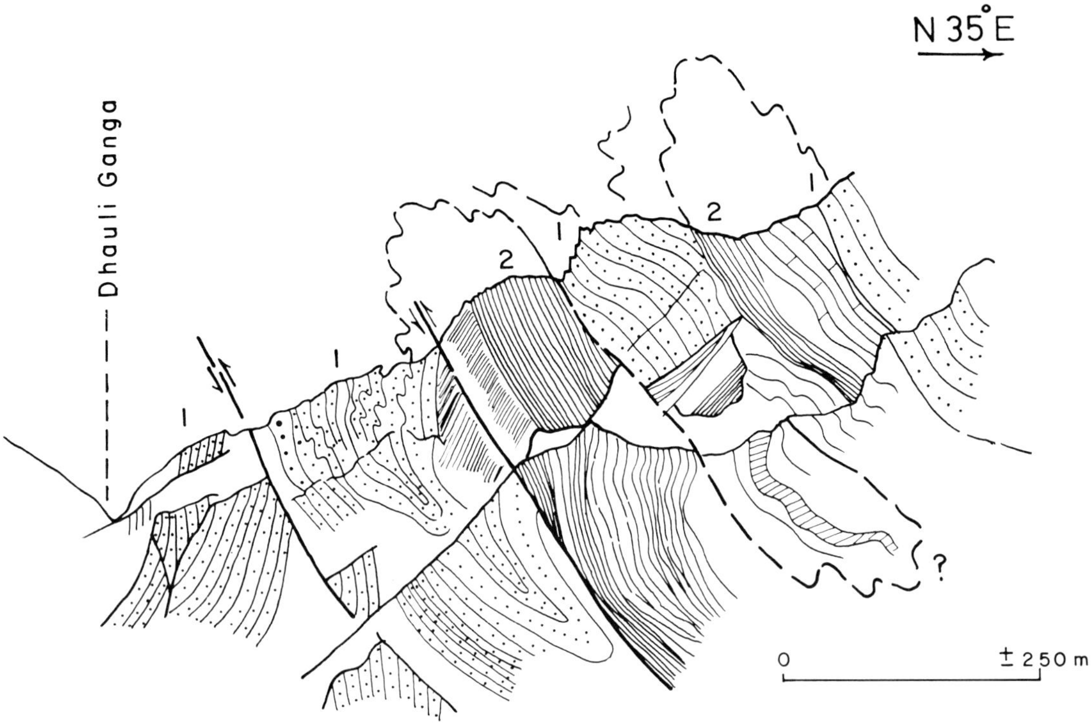

Figure 50. Repeated thrusting and isoclinal folding in the Upper Palaeozoic sequence of Dhauli valley near Bedang, 5 km upstream to Puingrung; 1. Muth quartzite; 2. Variegated Silurian

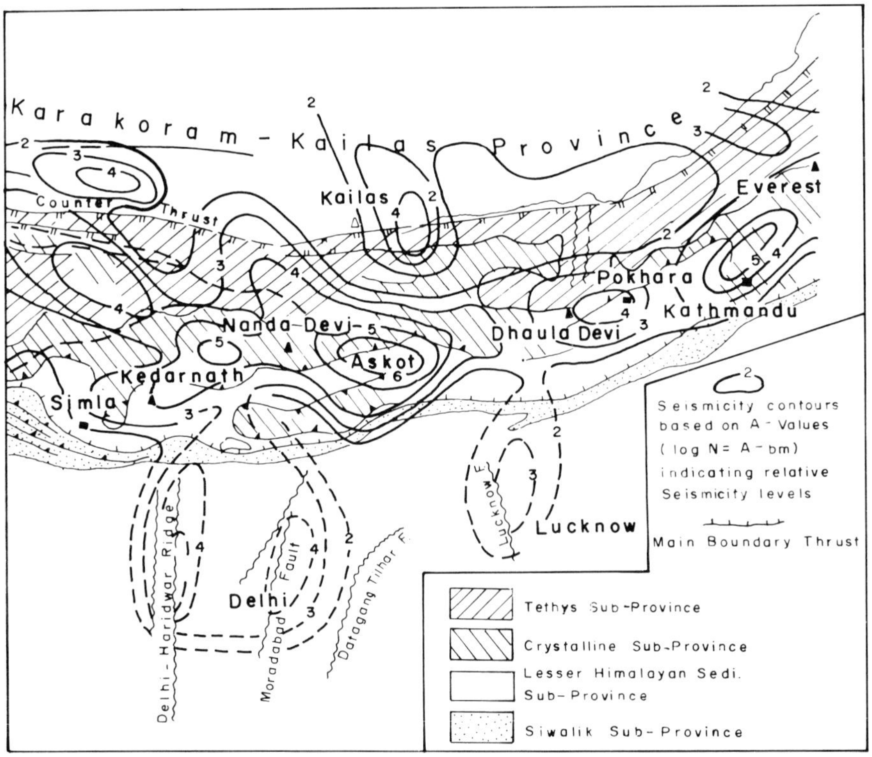

Figure 51. Seismotectonic map of Central Himalayan region (adopted after Kaila and Hari Narain, 1976)

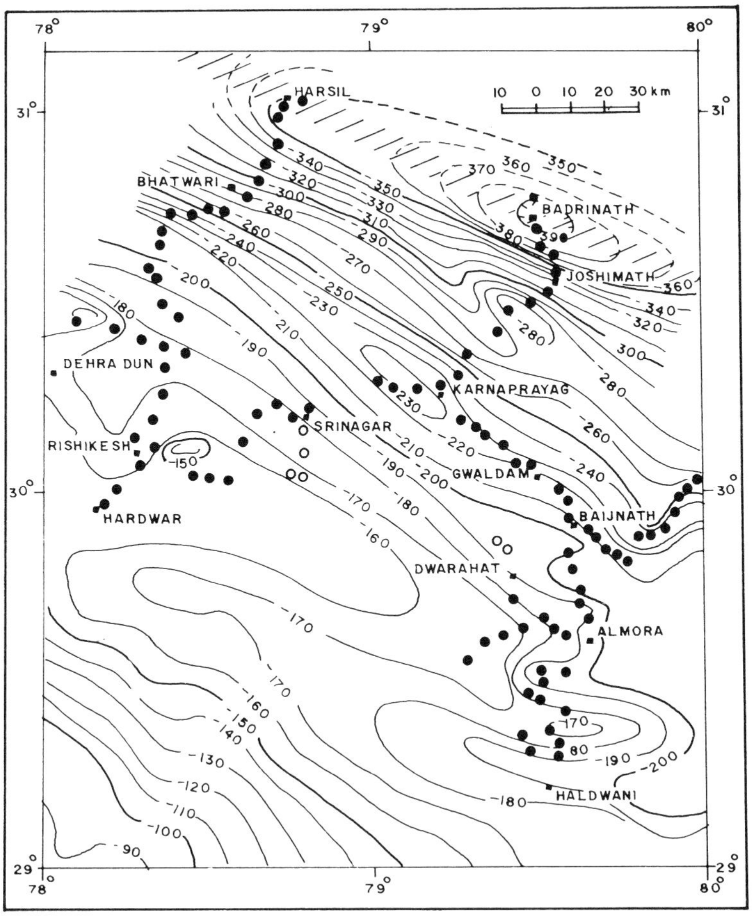

Figure 52. Bouguer anomaly map of the Kumaun–Garhwal Himalaya. Note the anomaly at Badrinath where the Main Axial Zone (MAZ) is situated (adopted after Qureshy *et al*., 1974)

scale thrust compared to southern Himalaya, but average amplitude displacement should vary between hundreds of metres to 5 or 10 km.

7. *Sumna–Marchok Fault*

This is mapped in the northwestern part of area in Dhauli and Girthi gorges having NW–SE trend could be traced more than 40 km (Sinha, 1980a) (Figure 12).

4.3.2 Dip faults

Faults to be grouped in second category are those having trend in NE–SW direction (Figure 13).

1. *Pachu–Lowe Dhura Fault*

This is mapped mainly on the basis of aerial photographs after noticing the conspicuous dislocation of the eastern limb to the northeast about 5 km or more and respective southwest direction in a transcurrent fashion. Except in the Gori valley, the upper reaches of this fault is purely interpretative as stated earlier on the basis of aerial photomapping. Such movements are in conformity with the tremendous thrust movement which this part of higher Himalaya has experienced from Tibetan highland area.

2. *Bedang Fault* mapped in the Darma valley (Figure 50).

3. *Haldawa Fault*

Observed in Gonkha Gad in the amphitheatre structure of Haldawa–Pinglawa Khola, are in unison with the nature of the last one and parallel to each other in their disposition (Figure 13).

4.3.3 Kiogarh Nappe (Figures 12, 13, 53 and 54)

This colossal nappe is responsible for bringing the 'exotic blocks' of Malla Johar in association with ophiolite suites of rocks over Cretaceous–Palaeogene flysch disposed along the Kiogarh heights (from 1 to 5) situated in the northeastern extreme Central part of author's map area (Figures 12 and 13). The ophiolites take part in the thrusting movement as decollement facilitating the gliding movement. Its root-zone is obviously along the great Himalayan suture in the Tibetan territory which has been profusely squeezed and completely obliterated the *in situ* position of 'exotic block' facies with a displacement of approximately 80 km (Gansser, 1974a, 1976, 1979), Gansser (1964) has drawn that the 'Jungbwa: Amlang-La: Kiogarh' nappe system with the flysch (Sangchamalla formation) having

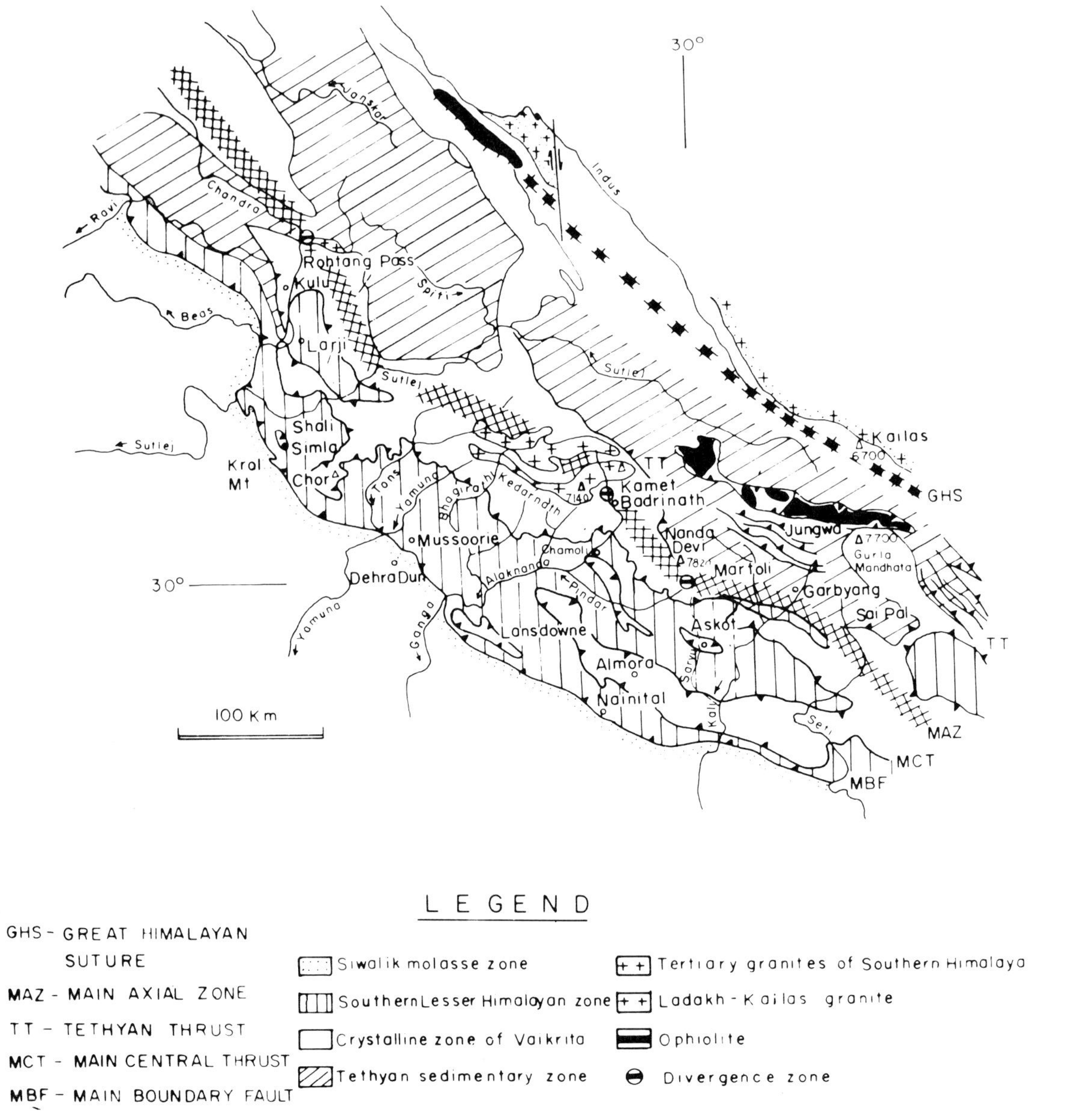

Figure 53. Major tectonic features of Central Himalaya

with the flysch (Sangcha Malla formation) having olistromal 'exotic blocks'. These 'exotics' were first discovered by Griesbach in 1891, 1893 and Karl Diener in 1898, and studied in detail by von Kraft in 1904. The author during the 1973 expedition studied these blocks by climbing the Kiogarh no. 1 peak (5861 m) with Dr Shah (Shah and Sinha, 1974c; Shah, 1977; Sinha, 1980a) and established the detailed geologic–tectonic aspects including radiometric dates and geochemistry of the ophiolitic rocks (Sinha and Bagdasarian, 1977). K/Ar age was found to be Companion–Maestrichtian: 75, 73 Ma and 107.5 Ma.

The expedition in 1976 helped the author to map the thrust-structures over Giumal but within the flysch formation (Figure 42) extreme NNW wing showing thrusting (T) over faulted and plunging Lahur anticline.

CHAPTER 5

Geotraverses through the Central Himalayan Segment

5.1 Geotraverses through the Higher Zones of Central Himalaya

Geotraversing in the Central Himalayan region was started by the author as early as 1969. In continuation of the trend of the work, the author took up the difficult job of preparing the geological traverse profile of each possible section from south to north. Simultaneous investigations of lithology, stratigraphy, structure, petrography, petrochemistry and geochemistry, geochronology, and mineral deposits were taken up. The river valleys and gorges are the only possible tracks across the Higher Himalaya and all possible traverses along them with their branching sub-traverses were covered in course of eight long years involving many expeditions.

5.2 Chamoli–Helang–Joshimath–Badrinath (Alaknanda Valley)

A distance of 55 km from Chamoli to Joshimath forms a segment of the pilgrim route to Badrinath. Badrinath is a further 40 km from Joshimath by road along the deep gorge of the Alaknanda. Chamoli itself is situated on a thick body of quartzites. These quartzites with all their inherited characteristics are sedimentary in origin. There occur two prominent bands (*ca.* 1–5 m) of augen gneiss concordant with the quartzites. One is about 4 km upstream from Chamoli towards Birehi stream in the left bank of the Alaknanda, while the other is situated beyond Birehi. A few bands of such gneiss could also be noticed downstream along the road towards Nandaprayag. Petrologically these gneisses are medium to fine grained banded rocks with quartz, plagioclases, micas, iron ore, and ilmenite. The rubidium–strontium isotope study of these gneissic bands provide an age of ±1800 Ma and increasing rubidium isotope content as we come downstream towards Nandaprayag (personal communication Professor V. P. Bhanot). A similar age picture has also been obtained from the Kulu valley (Frank, 1976).

It would be clearly understood with the sketch figure (Figure 54) along the Alaknanda gorge that the quartzite horizon monotonously dipping southwest and embodying concordant bands of gneissic rocks are also dotted with the emplacement of basic rocks.

At the confluence of Birehi Ganga, a vertical fault could be traced with marked crushed zone. The tributary Birehi follows the depression caused by this fault. Further upstream along this tributary there is a landslide-dammed lake—'Goṇa'. There has occurred two historical devastations with the outburst of this lake. The earlier event dates back to 26 August, 1894, and the recent one occurred on 20 July, 1970. In both these events water broke through, damaging the valley far down the stream. The later event of flash flood of Gona lake was accompanied by the landslide causing the damming of the Patalganga river further upstream. Breaching of the dams caused a flash flood in entire Alaknanda valley, which completely washed away Belakuchi village, with the loss of 35 lives, 15 vehicles, and a 15 km section of the Badrinath–Rishikesh Highway. Besides the loss of life there was heavy silting in all the *hydel* schemes connected with this river system.

Geologically Pipalkoti hamlet is situated on the apex of anticlinal structure (Figure 54a) composed of dolomitic limestone and slaty rock complex below quartzites. These dolomitic limestone are magnesite bearing due to hydrothermal alternation. Geologically in comparison to adjacent region these are the Riphean complex of Calczone described in the entire zone of Lesser Himalaya.

Observation of the road section from Pakhi village after Pipalkoti to Joshimath along the left bank of Alaknanda gorge provides an opportunity to cross through the sedimentary anticlinal domal structure to high grade metamorphic thrusted sequence (Figure 54a). Near Pakhi we first encounter slaty rocks of greyish dark colour which are in places severely deformed and showing microfolds and cleavages. These slaty rocks are calcareous and the limestone bands can be often encountered as we go ahead upstream. The limestones show massive nature and in turn they are overlain by thick white and greyish quartzites (*ca.* 500 m). Basic sills are emplaced in the quartzites. There has developed schistose structure in them. The quartzites grade to talc schists. The thickness of these basic sills is approximately 50 m but the similar type sills in quartzites of 50 cm to 1 m are more than 5 m in number in the neighbourhood. These quartzites show sericitic nature and also the schistosity (flaggy nature).

At Helang a pronounced tectonic break is met with marked by layers of basic rocks and overthrusting of catazone metamorphic gneiss and schists. This tectonic lineament is known in the entire Himalaya as Main Central Thrust (MCT) (Figs. 13 and 54b).

Close to the thrust plane which is represented by a system of parallel thrust surfaces dipping moderately at an angle of 30°–45° towards the northeast. Over the thrust plane immediately the crystallines are represented by chlorite schist interbanded with quartzites. There are two prominent bands of quartzites (Heim and Gansser, 1939a, pl. IV, Sect. 7b). Further upstream the gneisses alternating with schists bands become more prominent. Near Joshimath the gneisses acquire augen texture, and they are dotted with garnet. The schists become more biotite rich.

As we descend in the gorge of Vishnuprayag (Fig. 13) there is the exposure of biotite gneiss with minor flow folding structure. The general trend of dip is 45° due N 10°E. Heim and Gansser (1939a) estimated the following thickness to approximation of the rock types exposed in the narrow gorge of Alaknanda: (a) 1200 m of mainly mica schists with garnet; (b) 400 m of mainly gneiss; (c) 700 m of mainly mica schists with garnet; (d) 9400 m of quartzites with subordinate mica schists.

Auden (1935) observed that northward from Vishnuprayag (30°N 34′: 79°E 37′) is found a great thickness of bedded paragneisses consisting of recrystallized quartzites, garnet–mica schists and massive granulites containing biotite, muscovite, microcline, plagioclase, quartz, and garnet. At Pandukeshwar it could be observed that the quartzite series is characteristically arenaceous and massively bedded. Within the quartzites we come across pockets of gneisses (Fig. 54b) which is produced perhaps due to selective metamorphism of different source rock while deep burial metamorphism. The original current bedding structures are still preserved in spite of the complete reformation of the rocks. It is observed that with the slight change in strike the dip of quartzite series varies from 40°–70°. There could be found at places the tight isoclinal folding but this is more pronounced in the Dhauli Ganga valley near Jumagwar.

Along the waterfall of Kaliankoti near Hanumanchatti the rock changes into high grade metamorphite of sedimentary origin with augen gneiss and lime silicate complex. The dip becomes 60°–70° due north 25W (Figure 54a). In the photo panoramic view of Nilkanth peak (Figs. 55 and 56) can be seen the general geological set up and landscape view (see Photo Panorama No. 1).

Further upstreams the Badrinath 30°45′: 79°29′) region (the extreme right hand of Figure 54A, profile after Heim and Gansser, 1939) is a complex of quartzites, lime-silicates rocks, pelitic schists, migmatites and intrusive granites, and pegmatites (Figures 57 and 58). Heim and Gansser (1939a) by trekking along the old pilgrim route in 1936, as follows: (a) Biotite gneiss and mica schists, about 80 m, dip 80° to 53°E; (b) Banded lime-silicates, about 100 m; (c) Gneiss and gneiss-quartzite, dip 60° to S25°E; (d) highly metamorphic sediments with injected gneiss and calc-silicates forming a flat anticline of 10–20° pitch to S15°W, thickness about 1 km. The Alaknanda section, in general, from the zone of the MCT up to Badrinath was visualized as: (1) 7.8 km of mica schists and gneiss; (2) 9.4 km

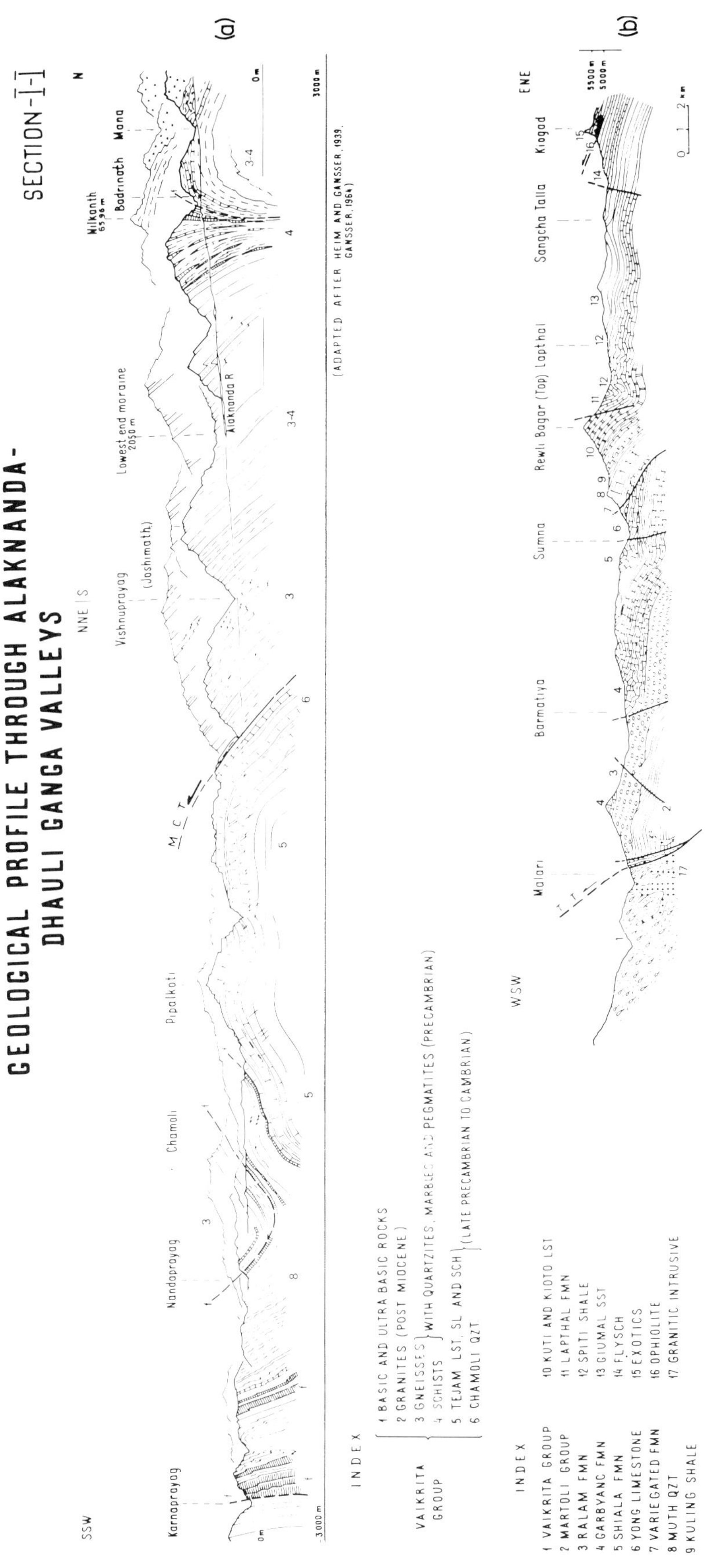

Figure 54. Geological profiles through Alaknanda–Dhauli Ganga valley (a) SSW–N; (b) WSW–ENE

Figure 55. A view of the holy shrine of Badrinath. According to Hindu mythology this is the seat of *Lord Vishnu*. This pilgrim hamlet is situated on the right bank of Alaknanda gorge close to the Main Axial Zone (MAZ), dotted with hot-water sulphurous stream

quartzite ('granulite'); (3) Approximately 4 km of injected paraschists characterized by lime-silicate layers and granite; aplitic dykes and sills.

Most striking structural feature of this zone is the presence of well exposed divergence of metamorphic complex. The section onwards to Badrinath shows a regular dip to the north, gradually becoming completely vertical (see Figures 37 and 54a), and finally a flat-dipping schistosity due south. Such tectonic style is called '*divergentio*' (Latin) corresponding to '*Root-zone*' or '*Würzel zone*' known from literature. Misra and Sharma (1973) attempted a division and suggested a two-fold division of this region: (1) the Helong–Pandukeshwar zone, and (2) the Badrinath zone. According to them the latter has been shown to consist of the following rocks types: (1) Quartz–muscovite–biotite–staurolite; (2) Quartz–oligoclase–orthoclase–actinolite–calcite; (3) Calcite–tremolite–quartz–tourmaline; (4) Hornblende–plagioclase–almandine; and (5) Calcite–scapolite–diopside. Thus 2, 3, and 5 would belong to the calc-silicate suite. L. N. Gupta (1978a) describes the occurrence of reaction skarns from Badrinath area resulting due to interaction between compatible and incompatible rocks during regional metamorphism. There has been mentioned about the mineral assemblages like forsterite, brucite, chondiodite, diopside, tremolite, green spinel, serpentine, etc. Down the temple of Holy Badrinath shrine of 'Lord Vishnu' (Figure 55) there flows a rich sulphurous hot water spring. The flowing water is collected in the concrete water pool for the purification baths of pilgrims before they enter the Holy Hindu shrine. The water temperature is approximately 60°C.

Figure 56. Nilkanth peak (6600 m), composed ot Tertiary granite, is situated behind the Badrinath Temple. The metamorphic rocks are in bivergence toward the northeast

This rich hot water stream is about 15 m above the river water level of Alaknanda, and the flow rate would be about 4 l s^{-1}. Although there is not much H_2S content in the water, the lime content results in the deposition of sinter downstream causing the stone to grow. This phenomenon makes the prejudiced pilgrims think this is a miracle of the Hindu God *Vishnu* who is enshrined in the temple. The chemical analysis of water is as follows: CaO—2.5 mg/100 ml; MgO—2.0 mg/100 ml; Na_2O—4 ppm; K_2O—33 ppm (analysis by Dr P. P. Khanna, WIHG, Geochemical Laboratory). Some smaller springs are scattered around the area along the tectonic line, and one at 50 m southward has the flow rate of 5 l min^{-1} at temperature of 30°C.

5.3 Joshimath–Malari–Niti Pass (Dhauli Ganga Section) (Figures 12, 13 and 54b)

In the preceding pages, a section up to Joshimath and its continuation upstream Vishnuprayag through Badrinath up to Mana has been described. Now we take up the section following the bank of Dhauli Ganga upstream to Niti Pass (Figure 13) via Malari, the confluence of Dhauli and Girthi Ganga.

Joshimath township is located mainly on the outcrop of biotite–muscovite mica schists and subordinate gneisses. Rb/Sr geologic dating and isotopic studies have revealed a Precambrian age with varying ratio of radioactive isotopes of strontium in the sequence. Upstream from Lata, the Dhauli Ganga flows in an almost E–W direction controlled by a regional fault of NW–SE direction (Figure 13). This reverse fault runs down from the Nanda Devi massif towards Vishnuprayag. Approximately 15 km upstream from Lata (Figure 13) there occurs a massive zone of the same type of quartzite that runs near Pandukeshwar described in the last section. The intense metamorphism and occurrence of garnet and kyanite in the upper zone near Jumagwar is very conspicuous. Along the deep gorge and steep cut wall the tight isoclinal folding in quartzites with their axis dipping northeast and a general NW–SE trend is quite conspicuous. Right up the Helang as one enters the zone of Central Crystalline the development of overturned and recumbent folds could be observed. The strike of the foliation surface varies considerably from NW–SE to NE–SW and some times E–W (Gairola, 1975). The direction being apparently northward, the dip of foliation plane varies from 20° to 60°. In folded quartzites the axial surface is inclined to 10–15°. Folds are having acute hinge and are of recrystallized material, suggesting a

Figure 57. The holy shrine of Kedarnath, seat of *Lord Shiva*, is situated in the valley of the zone of Vaikrita metamorphic complex and younger granites. In the left-side author is standing with his son Priyanshu (3 years)

synchronous folding and recrystallization. The schists and gneisses exhibit complex folds with the development of S_3 foliation at the fold hinges of second generation (F_2) is well marked. There are flow type folds and the thickness of layers varies from limb to limb. The typical S and Z-shaped folds develop on the limbs of microscopic folds. The general orientation of the fold axes is NE to E and a general plunge towards east. The axial surface is mostly inclined to the north and folded and in such cases S_3 foliation cut across S_2 foliation cross folds are often seen in the vicinity of South Joshimath township. The S_1 foliation is parallel to the S_2 foliation developed during the F_1 in the limbs. However, in the hinge zone S_2 foliation is found to cut across S_1 foliation, and S_3 cuts across S_1 and S_2 foliations. Sometimes refolding has resulted in the foliation of mushroom structure having (Gairola, 1975) sickle-shaped outcrop. In such cases fold axes of F_1 generation mostly plunge to the southeast, while the F_2 fold axes plunge in an easterly direction. The lineations are also found well developed in schists and quartzites. Interference of lineations is observed, in such cases the crenulations have caused the deformation of earlier generation mineral lineations. Lineations appear to have developed during the successive phases of F_2, when the crenulation lineations and lineations due to intersection of S_2 and S_3 (L_2) were superimposed on L_1 shows orientation variation whereas L_2 has a general NE to E trend. The boudinage is quite often, either in the limbs of folds constituted by quartz layers. Sometimes their folded layers are also boudinaged suggesting the continuation of deformation even after fold formation. The small scale shearing effect is also observed in Joshimath and subsequent zone near Malari.

As mentioned earlier the complex of schists, gneisses and migmatites occur after ending of quartzite series near Jumagwar and continues till village Malari where it gets truncated by a major tectonic lineament called the Malari Deep Seated Lineament or the Tethyan Thrust (TT) in general abutting against sedimentary pile of Tethyan zone (Figures 12, 13 and 53). This marks the beginning

Figure 58. The background view of Kedarnath shrine. The protruding massif of Tertiary granite could be seen in the high grade crystallines

of another structural facial zone. This continues up to the Niti Pass along the Dhauli gorge. The continued traverse enables a geologist to see the Martoli black pyritiferous schists unconformably overlain by the Ralam Conglomerate about 1.5 km upstream of Malari in the left bank.

The Ralam is conformably overlain by Garbyang, Shiala, Yong limestone with variegated Silurian, and a long fault (Figure 12) brings the repeating strata of Garbyang over it again near Geldhung (Figures 12 and 13) before Niti Pass.

5.4 Malari–Barmatiya–Sumna–Rimkhim Pass (Girthi Ganga, Yong Gad and Shal Shal Gad Section)

In the traverse description of previous subsection an attempt has been made to give a generalized geologic picture from Malari to the Niti pass. But starting ahead of Malari along the left bank of Dhauli Ganga about 2 km upstream, one comes across the unconformable contact between greenish grey argillite with silty component belonging to the Martoli Group and purple–red conglomerate horizon of the Ralam Formation. The village Malari is situated on the fluvioglacial morainic material brought by glacier from eastern high peaks against the Dhauli Ganga river. This mushroom-like perched block embodies unsorted material of 3 m × 1 m to rather smaller cm size. The unsorted Martoli sequence rocks material, as well as the young igneous intrusive, is scattered all along the contact zone of the fault. The Martoli sequence consists mainly of dark grey slate and quartzite. The slates are pyritiferous. All the sedimentary features are typical in the Martoli sequence to be compared with the black shale facies. If this sequence is to be compared with the euxinic environment condition in Caucasus, it reminds me of the so-called aspidic black shale sequence with pyrite cubes in the southern slope of Caucasius synclinorium. However, it would be difficult to establish homotaxiability between these two facies in geochronological sequence, since the Martoli sequence in the Tethyan part of Himalaya goes down to Precambrian, whereas the Caucasian sequence is of Jurassic age. The black shale is intercalated with the silty material in which the normal graded bedding sequence could be established. The shales show perfect development of cleavage inclined to bedding planes. After a 40 m shale sequence, a white and grey quartzite sequence follows. Pyrite is invariably present in quartzite sequence. The general dip is 40° due north 38°. The Martoli sequence near the village Malari is intruded by tourmaline aplite–pegmatite of 57 Ma (Sinha and Bagdasarian, 1977) (Figure 12).

At the confluence of Girthi and Dhauli Ganga, a greenish white quartzite of the Martoli sequence appears and further upstream along the left bank the conglomerate of the Ralam reappears. The conglomerate itself consists of unsorted quartzite pebbles and they are cemented by siliceous cement. Conglomerate which may have an approximate thickness of 30 m is followed by monotonous quartzite sequence of dirty colour showing sedimentary features, like tool marks, graded bedding, cross bedding etc. At the 8 km point, the traverse route is taken along the bank of Girthi Ganga over a temporary bridge. And a transverse fault with a steep angle could be seen in the right bank escarpment of the gorge. Further along the track up to 16 km to Barmatiya Camp, a permanent hut has been erected.

The rock sequence exposed at the contact zone of Ralam and Garbyang is marked by limestone with intense quartzification. The quartz veins are dotted with pyrite and copper minerals. Climbing along the track approximately 2 km upward along the hairpin bend, a geologist cannot miss the ancient copper mining adits on the left bank steep wall of Girthi Ganga gorge. At Barmatiya it becomes highly mineralized with the mobilized veins of hydrothermal barites and polymetallic minerals. Right on the track near the temple built by the road builders (Director General of Border Road Building Organisation of the Government of India) one may come across white barite vein (1 m thick) cross-cutting the country rock, and which may be mistaken for the quartz vein seen very commonly in the area. The whole limestone country rock is mineralized with veins and veinlets of copper and barite and calcite with copper iron sulphide ore also found disseminated in the country rock. Further up, the limestone becomes more and more arenaceous.

It is to be recorded that the cross and deformation exposure near Martoli village representing the tectonics shown of 'Malari–Bhamari' deep-seated fault, a northwest sector of the Tethyan Thrust (TT) is intruded by the syntectonic tourmaline pegmatite. The caught up block of crystalline intruded by pegmatite also appear on the eastern bank of Malari village (Figures 12 and 13). The intrusion has taken place in black shale as well. This sequence of black shale, quartzite and silty argillite is unconformably overlain by the conglomeratic horizon.

Between Kailashpur to 8 km DGBR camp a quartzite band appears with a shale having graded bedding and deformed with open type of folding. There the Ralam conglomerate reappears against the fault contact with the Martoli quartzite. A distinct fault of NE–SW direction having downthrow to the southeast separates the conglomerate with quartzite. Beyond this point the conglomerate appears and grades into quartzites of a greyish colour. Quartzites upwards becomes calcareous with the graded bedding, cross bedding ripple marks, and other sedimentary feature. At the 8 km point there is a fault with 40° hade and north–south extension. The dark, bedded dolomitic limestone with quartz veins and a patch of conglomerate with quartzite pebbles and siliceous cement of *ca*. 5 m thickness is encountered climbing up along the nala in northern slope in the contact. This contact is faulted with pyrite and copper mineralization. The oxidized gossan is also observed to be confined in the tectonized zone. Perhaps this mark the end of the Ralam sequence being followed by the Garbyang sequence of the lower Cambrian.

Let us come back to the mineralized zone of Barmatiya as described in the preceding paragraphs. This mineralized zone was subsequently investigated in detail in the 1974 expedition when emphasis on large scale geological mapping was followed by other activities of prospecting and exploration of the area for the evaluation of deposits of barite and polymetallic mineralization for commercial purposes. A detailed description of these mineral areas are given in Chapter 9.

Thus we proceed further seeing the geological sequence. Opposite the Barmatiya campsite, the dolomitic limestone sequence of Garbyang have been found to display back gravity folding of toothpaste type (Figure 44). It is to be noted that such back folding is typical and seen everywhere in the Tethyan sequence. The general trend of the rock is monoclinal dip towards the N-NE. Perhaps the movement along the tectonic dividing line between the crystalline and the Tethyan zone accompanied by the phenomena of gravity sliding in reverse direction of thrusting movement caused the back folding. Such phenomena has been reported by Bordet in Nepal Himalaya (1955).

The upper part of the Garbyang sequence shows more arenacous nature with typical sedimentary features and a shallower water depositional environment (Figure 15). Repeated horizons of marl occur as persistent yellowish bands which could be taken as marking the horizon of the upper part of the Garbyang sequence. In the upper horizon, fragments of trilobite have been observed 2 km upstream of Kiogad and the confluence of the Girthi Ganga river along the track to Sumna. Following the sequence within the quartzite with lingoid ripple marked surfaces (Figure 59) well preserved *Eccliopteris kushanensis* Garabu has been observed. This first occurrence of dateable fossil marks the Middle Ordovician age. This flat gastropod is the first fossil find in this normal sequence. The Garbyang sequence transitionally passes into greenish shale intercalated with a biostromal algal horizon intraformational breccia (Figure 60) in the upper part. This formation in previous literature (Heim and Gansser, 1939a) has been recognized as the Shiala formation and here I would also like to continue the same nomenclature. The sequence before we reach Sumna campsite, situated at the confluence of Kiogad and Yong gad, is observed to be that of

Figure 59. Ripples in Upper Garbyang enclosing flat gastropod *Eccliopteris* exposed in the Girthi gorge near Dobala

a biostromal environment accompanied by intraformational breccia and conglomerate, indicating unstable depositional environment (Figure 55b). With the turning along the Yong Gad and Shiala Formation is followed by a prominent reefal horizon which we name the Yong Algal limestone (Figures 61, 62, 63a,b and 64). On the basis of fossils, the Shiala has been divided into four zones in the Girthi valley, *viz*. *Orthis* (*Dalmanella*) *testidunaria* zone; *Rafinesquina areanea* zone; *Monotrypa* zone and *Rafinesquina alternate* zone (Figures 16 and 66).

The discovery of reefal algal limestone (Figure 63a,b) marks the first time identification of the biostromal zone in the Tethyan Himalaya. In the Yong Gad ravine, it represents a rather domal shape character and mixed-up feature. The horizon is characteristic of biostromal debris with the biota consisting of calcareous algal stromatoporids, corals and bryozoans. Further up the Yong Gad to Rimkhim campsite, the reefal domal structure with diverging dip on either position of exposure is overlain by the Variegated formation with purple and green calcareous shales and arenite. Previously it has been dated as the Silurian Shiala horizon. There is a marked unconformity with limonitic patches and it is followed by brownish white massive quartzite horizon of the Muth Formation. The lower portion of Muth horizon (Figures 15 and 64) of quartzite are doloarenite and are dated by fossils. But these remains are identifiable. Following upstream just before the confluence of Yong Gad and Shal Shal Gad (about 400 m from Sumna to Rimkhim) there is a fault of NNE–SSW extension causing the Muth quartzite to abut against the series of buff and greenish colour calcareous shale. This fault can be traced down towards east (see Figure 63) and it is observed that over the Muth quartzite the dark black shale of arenaceous micaceous nature appears with characteristic *Productus* and *Spirifer* fossils. This calls for Permian affinity without any trace of *Fenestella* of Carboniferous age. This, a prominent Hercynian hiatus is observed here (Figure 15). Looking at the Rimkhim ridge panoramic sketch (Figure 65) we could find that the shale of 20 m is followed by Triassic horizon. It is, in higher horizons, overlain by Lapthal and Kioto formation of Jurassic age. In the Barahoti plain a typical fossil *Cyclolobus Oldhami* marking the presence of Permian age was mapped (Figures 12 and 13). The Rimkhim is situated at a height of 5200 m. The Barahoti Plain is a beautiful pasture with green patches and water fountains. This flat ground has been used as grazing from ancient times along the Indo-Tibetan border. The campsite at Rimkhim faces the peculiar weather of high ultraviolet radiation with strong wind blowing the whole day from 10 o'clock from the northern side of Barahoti. The wind velocity is so strong that it is not be possible to stand erect. It makes it very difficult to work in this area and to map it properly geologically. Through the Chorhoti pass in the west of Rimkhim (see Figures 12, 13, 65 and 66) being bounded for the exploration to go to Niti pass side due to inclemency of weather and snow bound heights.

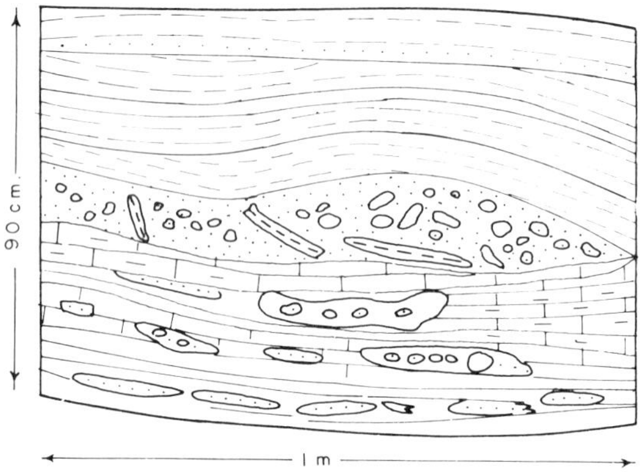

Figure 60. Intraformational breccia in the Shiala calc-argillitic sequence, in the Girthi Ganga gorge

Figure 61. Exposed sequence of Muth quartzite–Kuling shale–Kalapani limestone and Kuli shale in the Kiogad gorge. Photograph facing southwest

5.5 Sumna–Rewlibagar–Lapthal Sangchamalla Balcha–Dhura Pass–Kiogad Valley (Figs 45, 46, 54A, and Photo Panorama No. 3)

Following the right bank of Kio Gad upstream from Sumna (Figure 12), the section exposed up to Rewlibagar campsite before the ascent starts, a continuous fossiliferous succession of quartzitic, slaty, calc-argillitic, black shale, white calcareous and quartzite limestone with algal growth, and intercalated limestone and shaly horizons are met with. This section exposes rocks of the Shiala Series (Ordovician), algal greenish-nodular limestone of Yong formation, Variegated Silurian series, the Muth formation, black carbonaceous Kuling shale horizon, and Triassic of the Kuti formation (Figure 45).

The Shiala rocks are constituted by quartzitic shale, limestone bands, shaly and arenaceous shaly rocks. The collected fossils in the *Rafinesquina* zone have been described. Over the Shiala horizon a characteristic greenish biostromal horizon of algal limestone (30–40 m) is exposed on both the sides of the valley.

These algal limestones present a horizon which in places exhibits a domal appearance and occupies normal stratigraphic unit between the Shiala and the Variegated series. The angular unconformity between the Muth quartzite and the Variegated series is pronounced in this zone. About 4 km from Sumna below massive white, sugary quartzite of the Muth formation appears a horizon of approximately 20 m thick dolomite quartzite of rusty brownish colour with fossil remains. Obviously this represent an infra or transitional horizon between the Devonian Muth and the Silurian Variegated. In this section (Figure 54) there is a marked absence of typical Chocolate series which has been observed by earlier workers in the Kalapani area of the Kuti river. Triassic limestone bands with shale partings of the Kuti Formation have undergone disharmonic phase of folding during the last orogenic cycles, since it has occupied a sandwiched position between two competent horizons of the Kalapani limestone and Muth quartzite below and the Kioto limestone massive sequence on the top. The thickness of limestones bands in the Kuti shale varies 20 cm–2 m, whereas the calcareous shale bands are varying in thickness from 3 to 4 m and in some places giving up to 50 m (Figure 45).

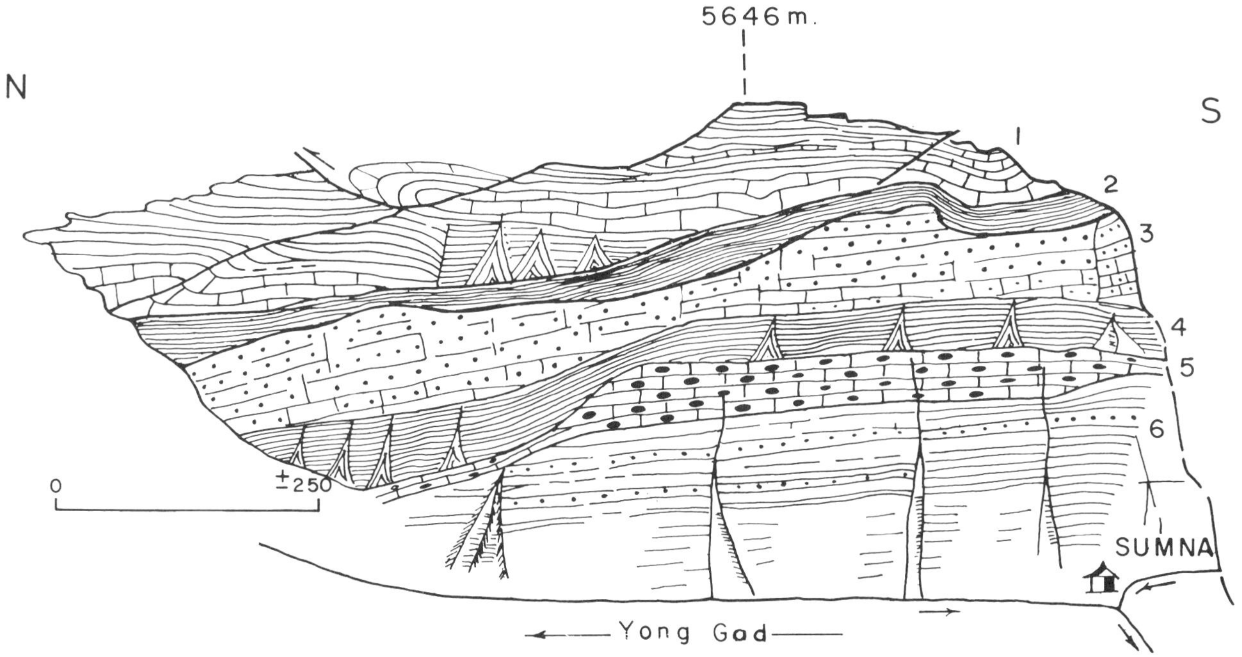

Figure 62. The right bank of Yong ravine, Exposures are: 1. Triassic Kut shale; 2. Kuling shale; 3. Muth quartzite; 4. Variegated Silurian; 5. Yong limestone; 6. Shiala formation

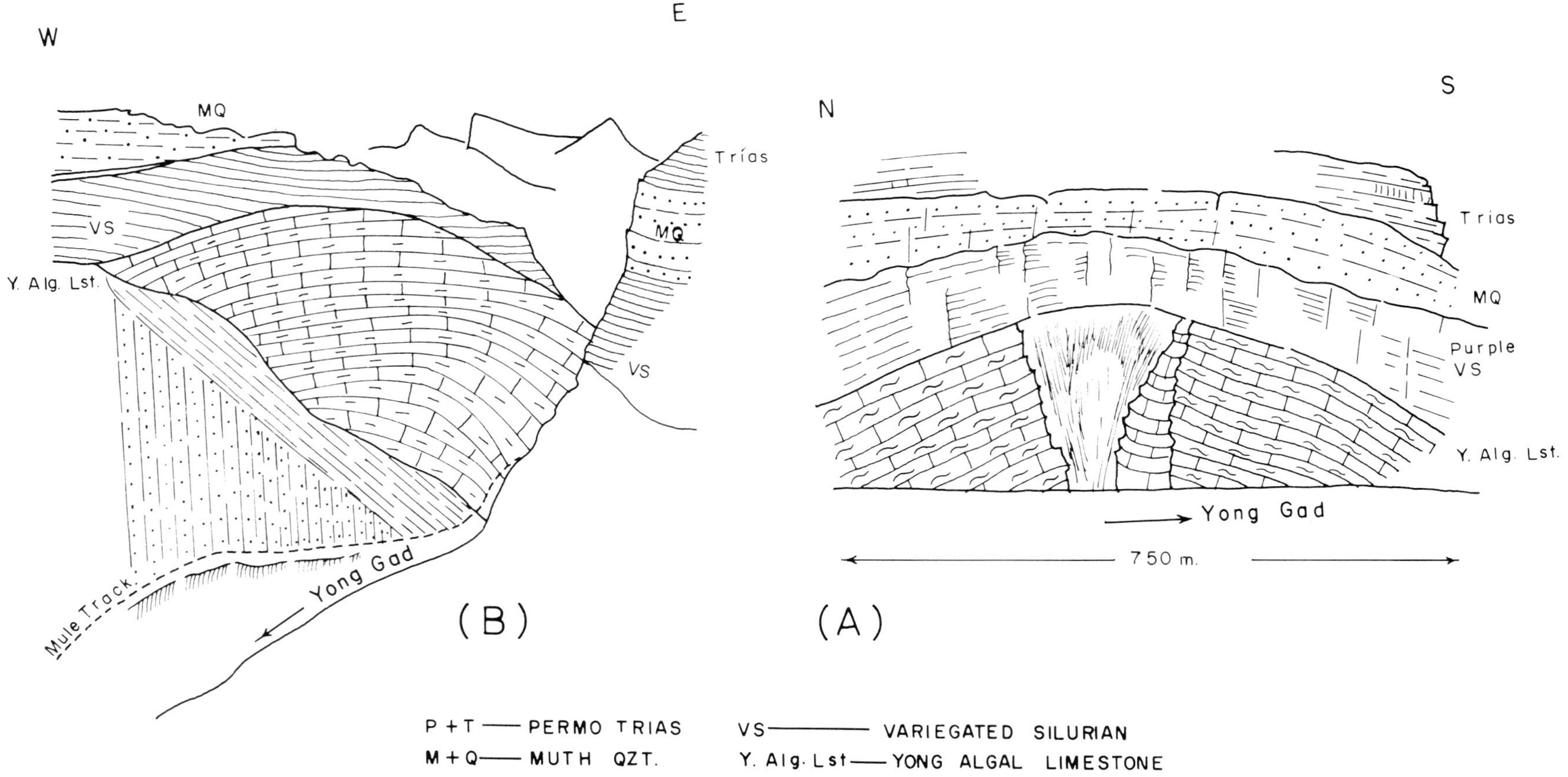

Figure 63. (a) Geological sketch view along left bank of Yong Gad about 3 km upstream of Sumna. (b) East–west cross-section of 'a' showing the extension of algal reef

Climbing up all along the hairpin bends, an arduous climb along the slope of Rewlibagar ascent in the right bank of Kiogad, one has to cross an approximate thickness of 300 m to 400 m of Kuti shale before come across the basal quartzitic oolite, and arenaceous grey–blue limestone with *Megalodon–Pecten* horizon situated about 300 m from the current of the Kio Gad stream. After climbing a height of 5645 m (Photo Panorama No. 3), we go down to the broad and wide Lapthal valley. The Lapthal series of rocks conformably lie over the Kioto formation. They consist of limestones and subordinate shale with rich fossils assemblage of Liassic stage. The rocks are folded and faulted in typical Jura type of folding. In the northern flank (see Figure 46) a back type folding with an NNW–SSE axial plane and plunging NNW could be seen. The southern limb is being truncated by a fault line with almost vertical displacement. The actual cross-section of the fold could be seen in the photographic panorama. The northern limb of box type folding due to severe deformation is cleaved to the bedding plane.

Traversing further northward an undulating topographic feature is produced due to weathering of the shale horizon of the Spiti shale (Figure 15). These horizons of the crumpled Spiti formation overlie the Lapthal formation with a transition bed of ferruginous oolite with Sulcactutus fossils. The base camp was pitched on the flat ground of this Spiti horizon, which is rich in ammonite fossils. The ammonites are enclosed in the concretions which could hardly be seen in any places *in situ* but usually they are rolled down in the gullies and Nala. In the basal part of this section we come across a rich *belemnite–hoplite* assemblage of fossils.

The following traverses from the base camp towards Balchidhura pass were taken in the next few days. The hills of sandstones with cherty components have been called Guimal sandstone. This horizon does not show any mark unconformity with upper part of the Spiti shale. After reaching the top of the ridge of Rewlibagar and looking north, a magnificent rampart of the Kiogad water-divide dotted with humps of exotic

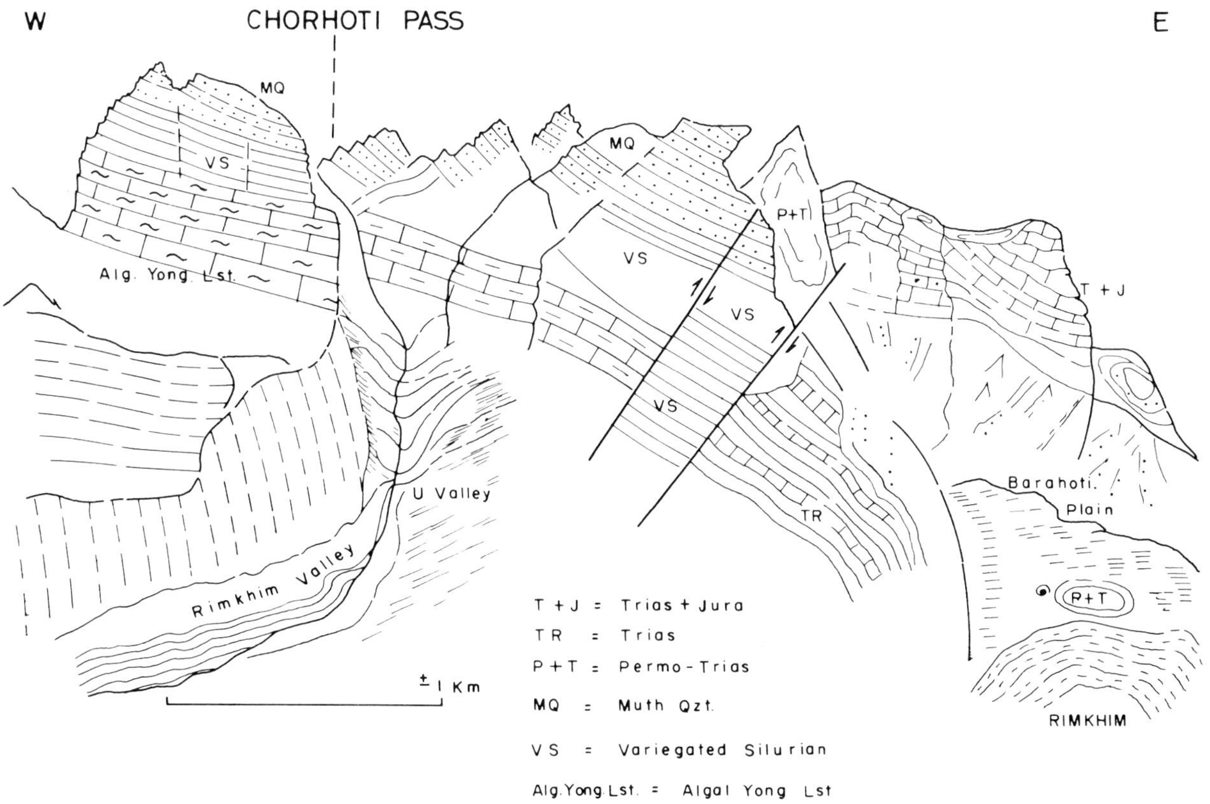

Figure 64. Disposition of Mesozoic strata in the Rim Khim water-divide with *hoplite* fossil location (P + T) in the Barahoti plain

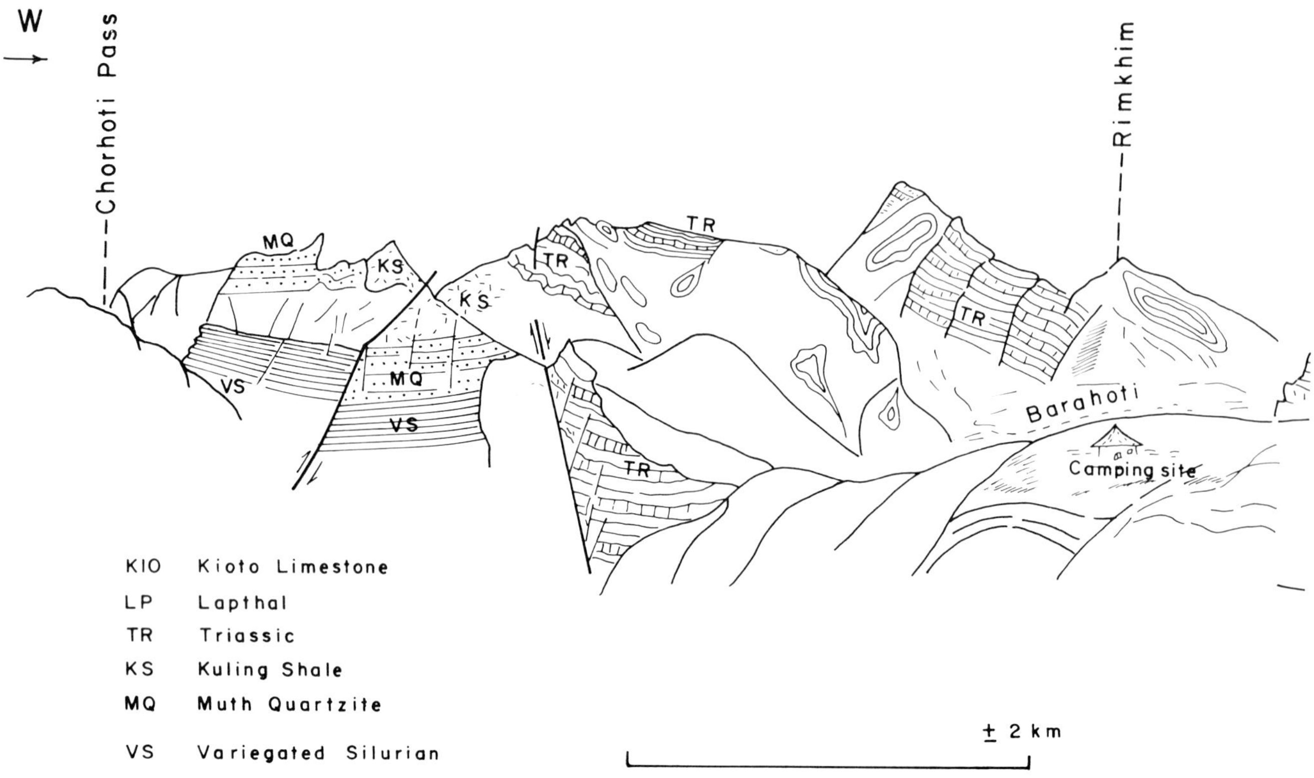

blocks of Malla Johar, known to the world through monumental work of von Kraft (1904), could be observed. Coming down in the valley of dark crumped shale, a flat pasture with springs and green patches could be seen. The northern border of the valley is constituted by a sandstone horizon with chert. The apparent thickness of Giumal sandstone is *ca*. 400 m and over this cherty sandstone horizon a typical flysch sediment appears. Along the track to Balchadhura pass an attempt was made to climb the Kio Gad No. 1 exotic block peak in the 23 July 1973 (Figure 36). With this aim we four geologist, Drs S. K. Shah, S. Kumar, V. K. Gairola and myself, started earlier in the morning hours from base at Lapthal and reached the Sanchamalla camp by mid-day, crossing about 5 km, but, finding it difficult to work due to the rarified air, Drs Kumar and Gairola dropped out on the way due to physical exhaustion. Further climbing was taken up only by Dr Shah and myself with local porters along the scree slope to reach to Kio Gad. It was too difficult for us to go up over 5000 m and also feet could not be supported on the loose scree materials. At last our efforts to touch the historic 'exotic block' of Himalaya was accomplished. We reached the top of it by 3.35 PM (IST) (Figures 12, 13 and 36). Between the flysch sediments below and overlying pell-mell block of limestone the ophiolitic melange mixed up with the calcareous rocks and basic to ultrabasic lava, known as ophicalcite (1–15 m) is formed. Besides the serpentinite all rock types could be seen as rolled blocks along the valley slope. Notable amongst them are spilite with amygdales of two different types: one having massive with less amygdales of light greenish colour and other with non amygdales of grey and brownish colour. These spilitic rocks were dated and their age varies from 107.5 to 73 Ma (Sinha and Bagdasarian, 1977). The panoramic view of the disposition of the 'exotic blocks' of Kio Gad could be seen, Kio Gad No. 1 (5861 m) (Figures 35 and 36). The flysch below the 'exotic blocks' could be divided in distinct zones with total thickness of *ca*. 700 m (excluding 'exotic blocks') having black shale with interbedded chert (120 m), red purple calcareous shaly (250 m) with a few to 10 bands of shaly limestone black shale sandstone and total of 500 m with shale and grey black colour. Since after the flysch the greywackish

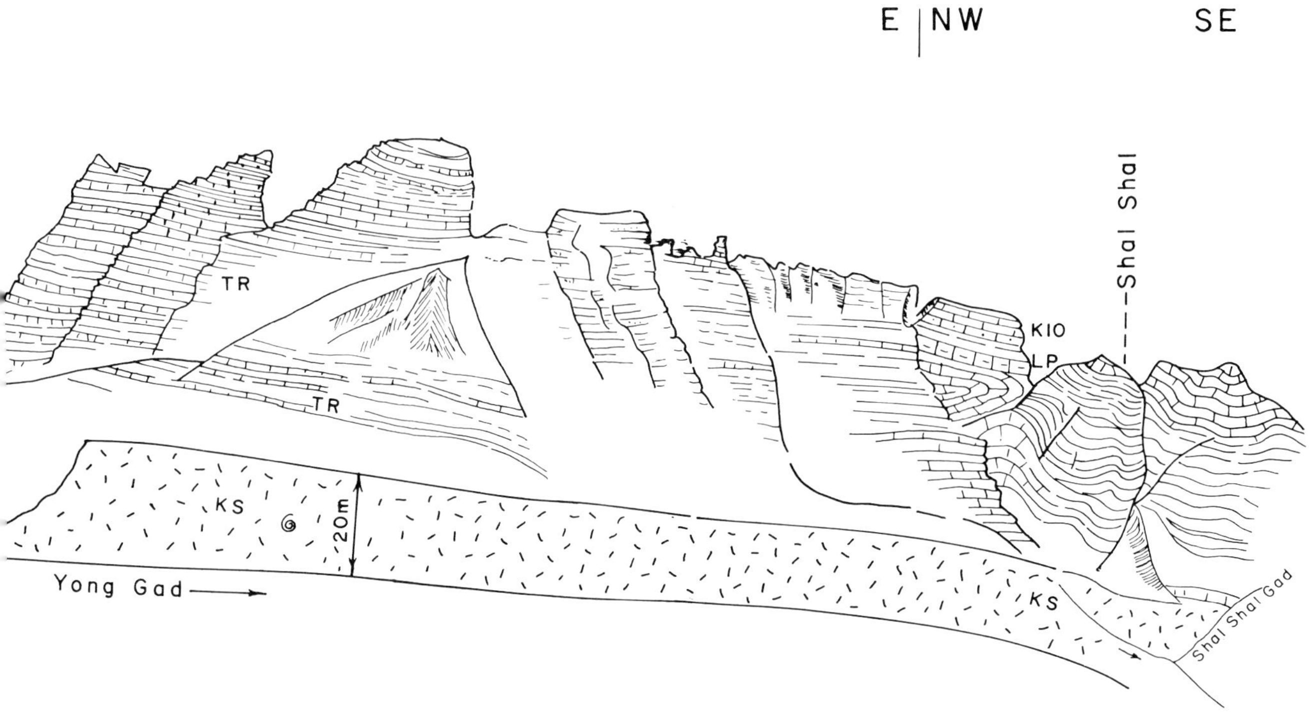

Figure 65. Sketch of panoramic view of Chorhoti Pass–Rimkhim watershed

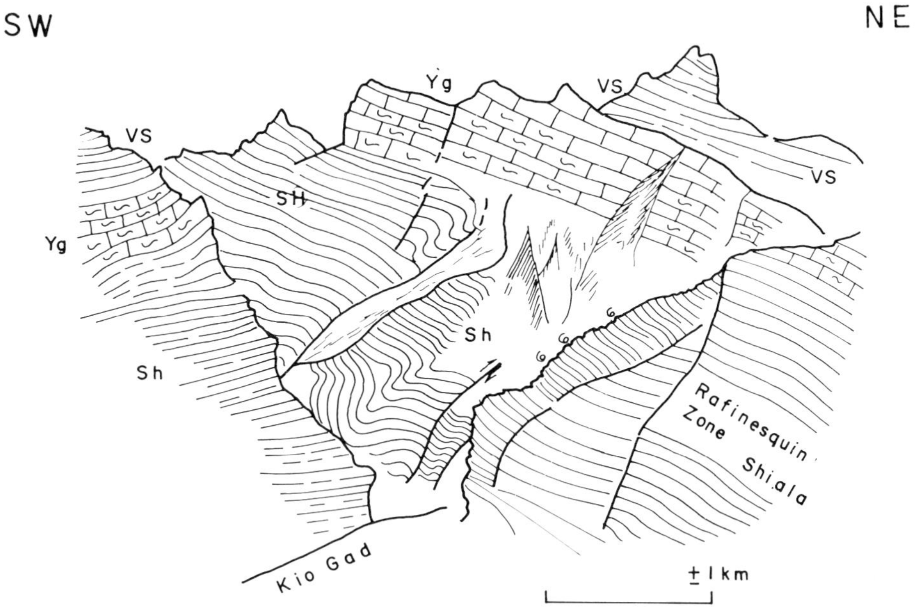

Figure 66. Structural view of rocks 500 m east of Sumna in Kiogad gorge. Sh—Shiala, Yg—Yong limestone; Vs—Variegated Silurian

sandstone followed. Our climb was made difficult by the sudden outbreak of hailstorm. Hence it is difficult to give further details. However, this problem could be taken up in the following chapter of stratigraphy. The limestone lenses in the red and purple horizon have yielded a rich assemblage of coccoliths and dinoflagellates. While climbing the upper part of flysch it was observed that the rolled 'exotic blocks' of marblized and cherty limestone with ophicalcite are embedded in flysch. It seems that these blocks were brought in the flysch sediments while the deposition was in progress in the flysch. But the upper part of the flysch were definite Triassic deposition with marked ophiolitic melange and ophicalcite complex at the base of thrust contact which facilitated their movement along the decollement plane from far northern zone somewhere in Tibet. Since this watershed represents the boundary of India and Tibet (China), no geological observation could be taken further north of it.

5.6 Bageshwar–Kapkot–Loharkhet section (Saryu Valley) to Dhakuri in Pindar River Valley (Figure 70a)

The traverse along the right bank of Saryu river were taken and the geologic features were documented in the way to cross over to the Pindar river valley section through Loharkhet–Dhakuri ridge. The township of Bageshwar is an old religious place with temples. It was made a base camp for traversing in adjacent area of Kumaun Himalaya. The traverse distance between Bageshwar and Kapkot is 24 km. The geology exposed on the way is mainly composed of calcargillite, calc-sericite schists; and chlorite-phyllite. Kapkot camp was made in the premises of Government Block development office. A jeepable road goes up to Bharari, and Bharari onwards one has to take a traverse on foot. Traverse sketch (Figure 70a) has been drawn from Bharari onwards along the left bank of Saryu river. At Bharari the dolomitic limestone is highly deformed and a local inverted fold could be observed in the left bank wall of the hillock over which minor irrigation canal has been dug. Further onward the strata dip 40° along a direction of 2–20° north. Dark grey limestone forms an anticline, as one reached Kharbagar, where the rocks are faulted and thrusted along northward-dipping thrust plane with marked crushing in the limestone folded along the zone dotted with sulphur hot-water spring. The analysis of the water is as follows: CaO–9.1 mg/100 ml; MgO—22mg/100 ml; Na_2O—54 ppm; K_2O—26 ppm. The water sample was taken from the spring no. 1 (Figure 67) situated 50 m upwards on the northern slope of Kharbagar village. The flow-rate is very low (*ca*. 5 cc per minute). Gas seeps out at 10 visible edifices along the E–W fault line (see Figure 67) in an area of 3 m × 2 m. The small spring flow water is collected in a northern pool 2 m × 2 m inside. The spring water temperature is normal *ca*. 20°C. pH paper indicate brick red colour between 7.6 to 7.9.

Water sample no. 2 was collected from spring no. 2 situated about 200 m towards west on the same level. The flow rate is much higher than previous one: *ca*. 2 l per minute. This is only zone source. The water with a pungent smell of H_2S is collected in a cemented water pool (5 m × 4 m). Another concrete cement pool (5 m × 4 m) has been built but it remains dry due to the flow from the new sources coming out through a low level seepage position. There is sufficient white crustation under the springs and profuse smell of H_2S around. The villagers informed that the earlier flow rate of water was quite high and this concrete water pool was constructed for the irrigation purposes. But after a few years time, the springs were getting dry. Local people were also informed that there are five such H_2S water discharge sources along the slope and one in the southern slope with red colour discharge, perhaps due to rich iron contents. Analyses of water sample no. 2 from spring no. 2 is as follows: CaO—9.0 mg/ml; MgO—18.1 mg/100 ml; Na_2O—52 ppm; K_2O—26.1 ppm (analyst Dr P. P. Khanna, Geochemical Lab., WIHG). Passing over the Kharbagar, village climbing further through Naseti village, we come across the zone of massive grey dolomitic limestone and up to the village Tamta where thinly laminated limestone and slate are exposed. The approximate thickness of the exposed limestone is about 400 m which is again followed by thick grey massive limestone down through Basi village and until we cross the Saryu river and come to the right bank of it near Kharbagar village where the thin limestone argillite is conformably overlain by a marked horizon of sericite schist. It is bright in colour and shining in character. This zone is similar to the zone below the Main Central Thrust horizon along the Girgaon Munsyari thrust in the Gori Ganga valley. The set-up is different here as there is no crystal-

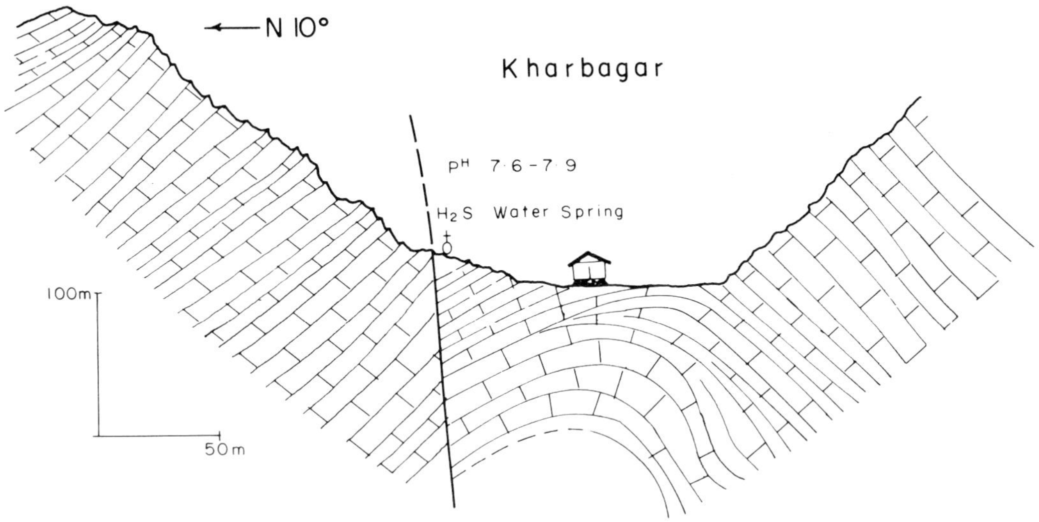

Figure 67. Structural set-up of the Kharbagar fault showing the sources of sulphurous hot-water springs

line overlapping, immediately, but rather again the laminated grey limestone follows with the normal stratigraphic column along the right bank. This sericite schist contain pyrite and quartz. Passing along the massive limestone (*ca.* 700 m) which is having a growth of algal stromatolites we come over to a phyllitic and calc-slate zone. Over village Shaling the estimated thickness of this slaty zone would be *ca.* 800 to 1000 m. Northwest Loharkhet Dak Bungalow it becomes more carbonaceous and assumes more dark slaty character. A block of limestone with beautiful algal growth of *Colonella* assemblage of stromatolite zone was found transported near Loharkhet Dak Bungalow (Figure 68). The tracking path to Pindari glacier over the Loharkhet passes through Parthali Dhar (2166 m, 2896 m, 3212 m) through the Chaur Gaon village climbing up we touch the zone of quartzite of white sugary variety including quartz sericite schist, and amphibolite schists which assume the character of chlorite schists due to metamorphism. Since the track almost goes along the strike of the exposed rocks, there is not much change in the geology and one has to walk along quartzite and quartzite sericite schists. Below the point 2896 m, a zone of chlorite–mica–quartzite schist appear marking the appearance of crystalline schists brought by Main Central Thrust (Figure 70, 2896 m Dhakuri Inspection Bungalow). This is followed by quartzose gneiss zone down up to Dhakuri PWD Dak Bungalow, through Dhakuri khal (3213 m) in the valley of Pindari river. Near the village Dhakuri there occur the crystalline schists over the sericite schist zone and metamorphosed basic amphibolite sills. The rocks are more granito-gneissic type. But in general these thick rock series *ca.* 500 m could be named as porphyritic. The porphyritic gneiss type are not noted in the Kali and Dharma valley sequence characteristic of Pangu–Khet type. Near the Dhakuri pass overlooking the Dak Bungalow, the rocks belonging to crystalline schist become rather coarse representing quartz–biotite gneiss.

The rain in this part of Himalaya is unpredictable. It was drizzling during the day time on 5 June and later on raining heavily before we could reach our camp at the Dhakuri Dak Bungalow.

The distance between Dhakuri and Loharkhet is 11 km, and makes one day walk to camp for the normal trekker.

5.7 Dhakuri–Pindari Glacier–Nanda Kot Section (Pindar Valley) (Figure 70b)

Trekking to the Pindari Glacier continues from Dhakuri Dak Bungalow onwards down the stream along the left bank of Pindari river. The next tourist camping halt is at Kati Dak Bungalow of the Public Works Department of Uttar Pradesh Government. The distance between Dhakuri and Khati is 8 km.

Coming down about 1 km along the spur from Dhakuri Dak Bungalow on a descending path, we cross through porphyritic gneiss followed by a mica schist zone (see the detailed section Figure 70b). The descent is beautiful, dotted with colourful flowers such as yellow wild roses and the others as described in the introductory chapter on flora. About 500 m before the Umula village along the track Khati on the left bank of the Toroni nala, porphyritic biotite gneiss is exposed with a rich development of micas and showing some schistosity. Above the mica schist there are amphibolitic sills. The thick rock sequence following the sills is represented by quartz–biotite–sericite schists. In turn the quartz–biotite–sericite schist series is succeeded by a quartzite sequence. The average thickness over the Main Central Thrust is as follows:

Quartzite (approx. thickness 2500 m)
Quartz–biotite schist (*ca.* 2000 m)
———— Tectonic Break ————
Amphibolitic sills (*ca.* 50 m; the upper part of mica schist is quite different)
Mica schist horizon
Porphyritic gneiss (more than 500 m)
——— Main Central Thrust ———

There is all probability to trace a wide and quite extensive tectonic break inclined at an angle of 40° NW between Khati and Umula village which in right bank of Pindari river and could be traced across the Sunderdhenga Gad (see right extremity of Fig. 70b). This lineament structure is also clearly visible in the aerial photographs of this area. This thrust plane seems to be parallel with the Main Central Thrust passing near Dhakuri. Over the Jignigaon (Fig. 70b) this tectonic line up to Jignigaon village sequence repeats with quartzitic gneiss and quartzitic biotitic schists and mica schists, in general. Crossing over the Khati, the crystalline rocks are represented by the dark porphyritic biotite gneiss which near 2 km milestone from Khati is succeeded by quartzite sequence with mica schist association. The quartzites are of a typical sedimentary nature, showing degree of metamorphism and inversion of strata. A beautiful folding structure (Fig. 70b, see inset) within the quartzite could be observed while crossing Pindari river over a wooden bridge at Maliadhaur, 6 km from Khati. In the sequence after Maliadhaur up to the 11 km rest camp of Dwali Dak Bungalow (2734 m) the quartzite rocks are rather different with wide isoclinal fold showing repeated inversion. It could be observed that a

Figure 68. Columnar stromatolites from calc-zone limestone near Loharkhet in the left bank of Saryu river

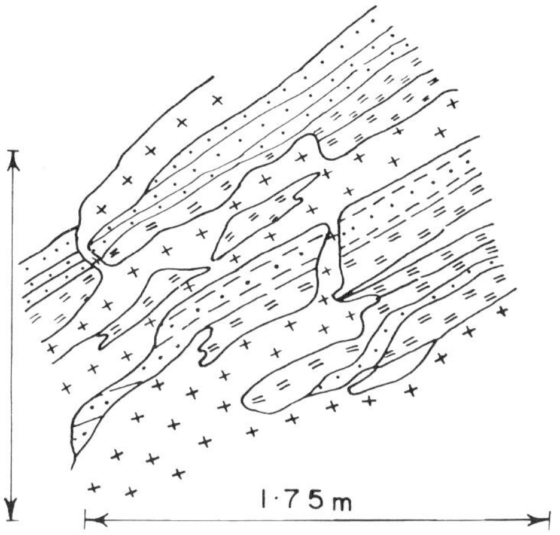

Figure 69. Cross-cut interrelationship between crystalline county rock and younger intrusives near Phurkia in Pindar valley: 1. Quartzite, 2. Quartz–biotite schist, 3. Garnetiferous aplitic granite

subordinate amount of calc-silicate rocks are associated with this quartzite schists sequence between Maliadhaur and Dwali. Upwards sequence is dominated by calc-silicates with quartzite but degree of metamorphism increases successively. The 5 km distance from Dwali and Phurkia Dak Bungalows is represented by a typical assemblage of kyanite-bearing schists, quartzite biotitic schists, and calc-silicate assemblages. Deformations in this high degree of metamorphism is intense with the increase in grade of metamorphism. The whole deformation further up from Phurkia up to the ruins of the old rest house about 2 km upstream is dominantly tight isoclinal type (see detail sketch in Figure 70b), the axial planes of which become a steep angle, and a zone of extreme disturbance and irregular deformation of the crystalline sequence follows. Reaching 3 km point along the track the massive pile of quartzite biotitic schist and calc-silicate rocks again assumes a dip from vertical to flat. This zone between the 2 km to 4 km from Phurkia Dak Bungalow represents a zone of divergence as shown in Figure 70b. This reminds us of the most significant divergent structures described in the Badrinath section (Figures 37 and 54) and the section of root-zone shown by Powell and Conaghan (Figure 37) and Sinha and Jhingran (1977). Although there have been sketches of this traverse described earlier, this significant divergent structure has remained unnoticed. This zone shows another important geological phenomenon which is characterized by the tourmaline granitic intrusion (Figure 69) of different phases of synkinematic origin (extreme left of Figure 70b). The first appearance of such granite is noticed near the 9 km point from Khati to Dwali rest camp. This granite is profuse in the zone of divergent structure and continues up to the right wall of Pindari glacier further up to extreme end of traverse. At the 7 km point to Pindari glacier from the calc-silicate quartzite mica schist horizon with a low angle between 10 to 20° dip due northeast appear the transition zone porphyritic biotitic schist, which we call here the Pindari schist equivalent to the Budhi and Rilkot schists observed in the Kali and Dharma valleys respectively. The description of these horizons will be taken up in the next sections. The cross-country picture of the intruding granites in quartzite schist and quartzite biotitic schist rocks 2 km before Phurkia rest house has been sketched and photographed (Figure 69). This sequence of Pindari quartzite biotitic schist visibly passes below the tongue of Pindari Glacier flowing down from the Traill's pass or Pindar Khala. It remained an ambitious task for us to climb up to the exact contact point between the crystalline complex and the sedimentary of Tethyan zone. At the exposure of sericite schist and phyllite rocks, a very short and quick glance of observation before the cloud covered the exposure made it clear that the sedimentary sequence is faulted against highly weathered crystalline rocks below the Chhangu peak (6322 m) is exposed below Traill's pass (see Figure 70 the upper part view of Pindar Glacier), on the extreme southwestern flank. The gate and tongue of glacier were visible (Figure 70b) for a momentous few seconds during the early hours of morning at 7 o'clock, and simultaneously I could snap the visible peak of Nandakot (6861 m), Nanda Bhanar, Nanda Khani (6029 m) in the eastern side of Nanda Ghunt, and the group of peaks in the western side.

Fig. 70a
Fig. 70b

Since the days of the British Raj, it has been popular with tourists to visit this part of Himalaya to see the glacier. This is the only glacier in the Central Himalaya that is approachable without many difficulties or a long-distance trek. Nowadays, the tourism is promoted by Government of India and U.P. State Government with the comfortable Inspection Bungalows built along the trekking route at distances of whole days' journey and night halt. During trekking up to the high mountains, people are advised to stay in each Inspection Bungalow or Rest House for rest and acclimatization and when returning, one or two night halts could be skipped depending upon individual trekking capability. The weather in this zone remains very unpredictable and the best time to visit the Pindari glacier is before the torrential rain starts pouring down in greater Himalayan zone of crystalline rocks with thick vegetation by the last week of June. These higher ranges of greater Himalaya form the barrier for monsoon clouds, which accumulate and condense to rain or snow over southern adjoining zone. The upper wide reaches of the valley of the Pindari river presents a picturesque view of flowers after the melting of snow, and white and purplish bushes of rhododendron blooms all along the slopes. Other plants and birch (*Bhojpatra* tree) start appearing as one crosses the Dwali Rest House.

The traverse from Phurkia to the bank of the glacier should always be taken in the early hours of the morning when the first beams of the sun are on the snow caps of the Nanda Devi group peaks. This should be seen and snapped by the photographers to gain pleasure in and sense of the magnificence of the mountains.

Fig. 71
Fig. 72

5.8 Pithoragarh–Tejam–Munsyari–Milam–Unta Dhura Pass–Topidunga–Khingrudhura Pass–Lapthal Section (Ramganga–Gori Ganga–Gonkha Gad Valleys)

5.8.1 Pithoragarh–Askot–Tejam–Munsyari section (Figure 72a)

Before continuing the section description just over the Main Central Thrust zone from Girgaon near Munsyari, it is thought that a quick acquaintance with the sketch geologic traverse would be beneficial to the readers working in this part of Himalaya. Fig. 72a shows a generalized geological traverse sketch along the motor-road from Pithoragarh to Munsyari incorporating Askot–Klippe crystalline zone.

An expedition in the Himalayan region is always full of adventures and is sometimes quite dangerous. It so happened that our party, consisting of myself, my colleague Dr A. C. Nanda, Veer Pal Singh, Research Scholar from Panjab University, two field attendants Laxman Shrestha and Vijai Arjal, with jeep DHD 3491 driver Shri Puran Singh, was on the way to Pithoragarh from Thal Rest House base camp on 19 July, 1976. It had been raining heavily since the morning but in the second half of the day the weather cleared, although it remained cloudy. About 15 km before reaching Pithoragarh the jeep suddenly turned a sharp bend and we saw a wheel of our jeep running just ahead of us. Being puzzled, we could not momentarily think from where this wheel running before us has come. Watching the rolling wheel going down in the deep ravine 500 m below the road level with gaining momentum, it took us about five to seven minutes to locate the place of its final stop. And then only we could inspect back our vehicle and discovered that the left wheel at it axle base has got broken from the heavy trailer. Thus, we were all saved from the mouth of a fatal accident.

The distance between two camps of Pithoragarh to Girgaon is 128 km. Starting from Pithoragarh, the exposed rocks are massive grey limestone, but half way to Kanarichhina shaly rocks with brownish calcareous shale appears. At Kanarichhina the rock is black graphitic shale in which some sills of basic rocks are also emplaced. At the northeastern flank of the village there is conspicuous zone of crushing marking a tectonic dislocation (see Figure 72a). Going ahead, a thick horizon of whitish, grey and massive quartzites occur with the same 30°–45° dip with inclination towards the northeast. The limestone and quartzites are tectonically juxtaposed to each other. This tectonic zone (*ca*. 200 m) extends NW–SE with the occurrence of mineralization.

Near the village Chama crystalline augen gneiss were observed physically overriding the quartzites (see the detail of the Askot klippe in Fig. 72a). The crystalline rocks, consisting of biotite gneiss, mica schists and feldspathic schists, continue up to the point about 4 km to Thal crossing Didihat and form a synform structure. The northern limb has fresher granitoid and gneissic crystallines. At the lower level layer of klippen synformal structures, the rather massive rich biotitic gneiss appears succeeded by more leucocratic, medium size porphyritic rocks. The augen gneiss is not seen in the southern flank but as we cross Tejam zone the MCT is immediately succeeded by exactly similar rocks.

The quartzites below gneissic rocks of the Askot klippe continues up to Thal. As we go down the road towards Munsyari, near the first bridge, a dark colour rock band (amphibolitic schist) occurs. Further ahead about 4 km from the Thal (Garichan) Rest House, there occurs a thick (50 m) band of talc dipping at an angle of 60° due

(190°N). The talc band is more deformed between competence quartzite and massive limestone in the northern side. The zone is definitely a tectonized one with the emplacement of basics altered to talc.

Onwards a monotonous limestone are exposed with intercalating thick series of blackslate (Fig. 17) crumpled with ferruginous stain with H_2S activity. At Nachni the limestone dip due northeast but further onwards the dip becomes southerly and the northern limb of anticline is truncated out by a normal fault (Fig. 72a) heading northeast. The anticlinal structure continues up to Tejam and further ahead to Girgaon, until the MCT thrusts the crystalline physically over them.

From Munsyari onwards at Girgaon (Figures 71 and 73), calc zone dolomitic limestone assumes more parting and greenish colour, and massive type is succeeded by sericite schist (*ca.* 10 m), white sugary quartzite and quartz-schist. At the contact the limestones are marblized. Several metamorphosed quartzite rocks, called as sericite quartzites, contain small crystals of kyanite as an accessory mineral. This sudden growth of kyanite could be assigned to the dynamothermal metamorphism resulting due to Main Central Thrust movement. This present of sericite schist layers with a maximum thickness of 10 m is marking an important phenomenon all along the MCT zone in the Higher Himalaya with quartzite as allochthon. One such layer is also seen between limestone and quartzite. This persistent horizon is so prominent and exposed in the region that road cuts provided their debris along the slope and escarpment. It remains a great problem to link origin of sericite schist directly from basic or ultrabasic intrusive, since there is no remnant left of these rocks as such in association or in the vicinity. However, their origin from an argillaceous matter rich in alumina giving rise to rich sericite due to dynamothermal metamorphism could not be ruled out. This persistent horizon helps the geologist working in these region to trace the surface of the Main Central Thrust zone all along the high mountain escarpment. Immediately over this surface chloritic mica schist with quartz vein appear all along over the thrust contact dipping of 20° in direction of 320°N (Fig. 73). An amphibolitic sill (20 m) is prominently exposed over this chlorite mica schist zone, which in turn is overlain by porphyritic gneiss. The size of porphyroblasts at the contact layers are rather smaller in size but in the upper layers they assumed bigger size of 5.5 cm × 3 cm. The road section between Girgaon (1930 m) to Munsyari (31 km) provides a new picture which was not available from the previous work of early workers. The road passes here along a hair-pin bend reaching up to the Kalamuni pass (2700 m) where a temple to a Hindu Goddess was constructed in ancient times. This is just half way between Girgaon and Munsyari. Panoramic views of the Panchuli groups from Munsyari could be seen with five magnificent Himalayan peaks with heights from west to east 6304 m, 6312 m, 6334 m, 6437 m (the highest), and 6072 m (Figure 2).

This route particularly at the location of the Kalamuni pass is infested with leeches during the

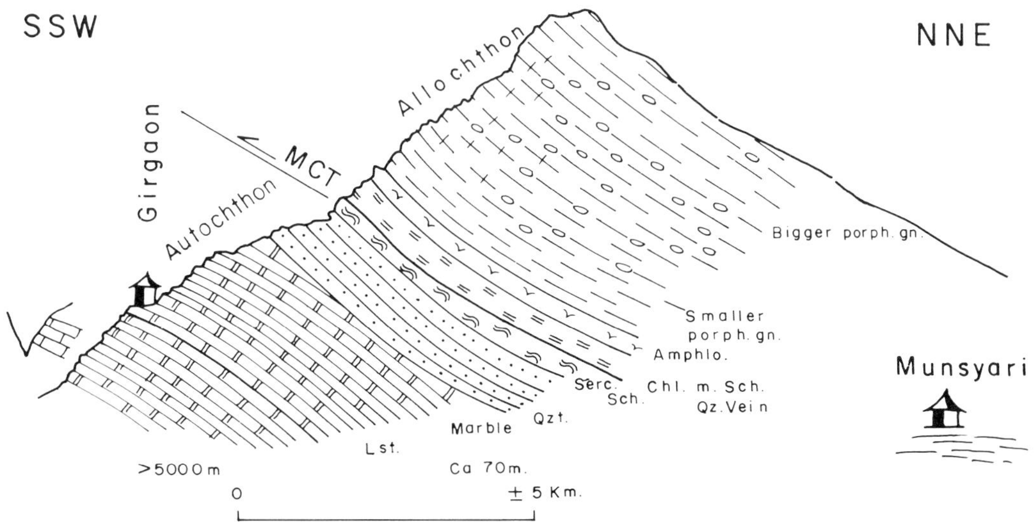

Figure 73. Main Central Thrust (MCT) contact zone at Girgaon–Munsyari Section in Gori Ganga valley

rainy season. The rest half way we drove along tectonic contact zone crystalline sericite schist and quartzite until finally we reach to Munsyari Rest House situated on the flat topography of quartzite sericite schists.

5.8.2 Munsyari–Milam–Unta Dhura–Topidunga–Khingru–Dhura–Kiogad Section (Gori Ganga–Gonkha Gad–Girthi valley) (Figures 71, 72b,c,d) (Photo Panorama No. 5)

The famous section onwards along the Goriganga valley has been sketched by the author on drawing sheets of his diary and reproduced in final form in Figures 72B, 72C, and 72D up to Belcha Dhura Pass through Milam and Unta Dhura Pass. All these meticulous works were accomplished during the 1976 summer expedition. It was tried to give the maximum geologic information along these geotraverses reproduced in the sketches, keeping the scale and general trend of geomorphological features.

The base camp was made in a house of Public Works Department of U.P. Government, to start the trekking onwards, since Munsyari happens to be the end point of the jeepable road.

While camping in Munsyari I had to face the worst attack of malarial fever from 23 July to 30 July, 1976. The fever used to come repeatedly on alternate evenings with strong rigour and temperature up to 105°F (42°C). However, my colleague scientist, Dr A. C. Nanda, could manage to procure quinine tablets which helped to bring down the temperature to normal. I, therefore, strongly recommend that it is most important for the Himalayan geologist to carry with him sufficient quantity of life-saving medicine including quinine. Before departure, on instructions from my doctor-wife, I had all the latest medicines including strong antibiotics and high altitude sickness medicines but the antimalarial drugs never cropped up in my mind since malaria is supposed to be eradicated from India. And these antibiotics not at all help me in this malarial fever.

Regaining physical strength and moral courage, I with my party of ten ponies and sixteen men started trekking onward along the toughest traverse of Central Himalaya, in a fine sunny morning on 3 August 1976. The lowermost unit of Central Crystalline complex represented by chlorite–mica schist (Figure 72b) is intruded by amphibolitic layers (5 m recorded on the junction of the bridle path at village Ranthi towards Jauljibi). Over this sequence the highest of Shankhadhura Chorpatta, and Saring were found to be composed of porphyritic gneiss with the plagioclase crystals of 5 cm × 2.25 cm size (Figure 74). This porphyritic biotite gneiss is also met with in the Girgoan section on the eastern slope of the ridge. The general trend of these rocks with the dip of 30° at a direction 350°N. On isotope analyses the potassium argon absolute age on the porphyritic gneiss collected at Talla Dumar village (Figure 72b) has come to be 217 ± 4 Ma. The approximate thickness of gneiss horizon at Munsyari would be *ca*. 250 m. Below Dharkot village the horizon of quartzite with muscovite–sericite schists (*ca*. 300 m) is exposed dipping very flatly about 20° due 350°N (NNW). This is again followed by porphyritic gneiss of the first type which gradually and transitionally passes into porphyritic grain of lesser dimension. The potassium–argon absolute age determination relates to this horizon of porphyritic gneiss. At Suring ghat, where a temple of a Hindu Goddess is situated at the confluence with Gori Ganga river, we come across paragneiss and biotitic schist with quartzite complex. This whole assemblage or rocks show ptygmatic folding and formation of much less developed type of schistosity. The section about a kilometre further is truncated by a fault having downthrow in the north. Within this complex of gneiss and schist of psammitic type and porphyritic gneisses there are amphibolitic sills, one of which has been shown in the sketch (Figure 72b) near the fault. Further upstream there is a tectonically disturbed zone reading NNW 80°. This fault presumably should not attract much significance, since only the uppermost part of the rocks exposed in the gorge seems to be much disturbed in comparison with the deep cut of the gorge. Further ahead, the streaky gneiss zone is exposed with porphyritic gneiss similar to the type observed near Munsyari. This is followed by gneisses, quartz–biotic schists and finally all the steeply-dipping calc-silicate and gneiss schist complex till we reached the Lilam Dak Bungalow. This rest camp is situated on the pronounced tectonic line with steep-angle dip about 80° due northeast (Figures 71 and 72b). This is the zone of calc-silicate before reaching Lilam and provides the weathering patches of yellow ochre colour. The distance between Munsyari and Lilam is 13 km and such night halts in this are are called '*parao*' by local people. The tectonic line at Lilam assumes a significant character because suddenly, after the calc-silicate zone complex quartzite, quartzose mica schist, garnetiferous porphyritic mica schists and quartzite start appearing with marked different style of deformation (Figures 75 and 76). The general trend of quartzose mica schist is 40° to 35°N. In contradiction to the general monoclinal trend, the dip is almost vertical in the crystalline rocks, before coming to Lilam tectonic line, attract special significant for the root-zone deep-seated tectonics. The quartzite banded with the biotitic mica schists becomes more and more massive and starts showing growth of garnet crystals, especially in the mica schists layers. A such thick zone of (12 km) continue up to Rargari Shiuni through Rupsia Bagar (*ca*. 700 m to Rargari). There appears tourmaline in the quartzite, which contains profuse mica (biotite + muscovites). This zone also shows isoclinal folds and beautiful preservation of cross-bedding structure, confirming its sedimentary origin, as well as a normal dipping sequence. Near the described camp site, Shiuni, there appears blocks of quartzite sericite schist dotted with beautiful cobalt-blue crystals of kyanites. These crystal are preserved in fallen blocks from the upper quartzite horizon.

However, such occurrence *in situ* could not be located along the track since the leeches in this part of the traverse are so awful and menacing that they reach inside the socks and shoes, overcoming all necessary precautions and making men and the animals carrying the load bleed profusely. Particularly, the track between Lilam and Bugdiar is profusely infested with leeches up to the size of 5 or 6 cm. I would suggest geologists take adequate care washing the feet with saturated salt water in the morning before putting on socks and after coming back to camp in the afternoon.

After Shiuni (about a kilometre from the Bugdiar camp site) starts a zone of biotitic mica schist with calc-silicate rocks. Henceforth, the dip direction of schist appreciably changes to just due northeast. Bugdiar (2900 m) is situated at the

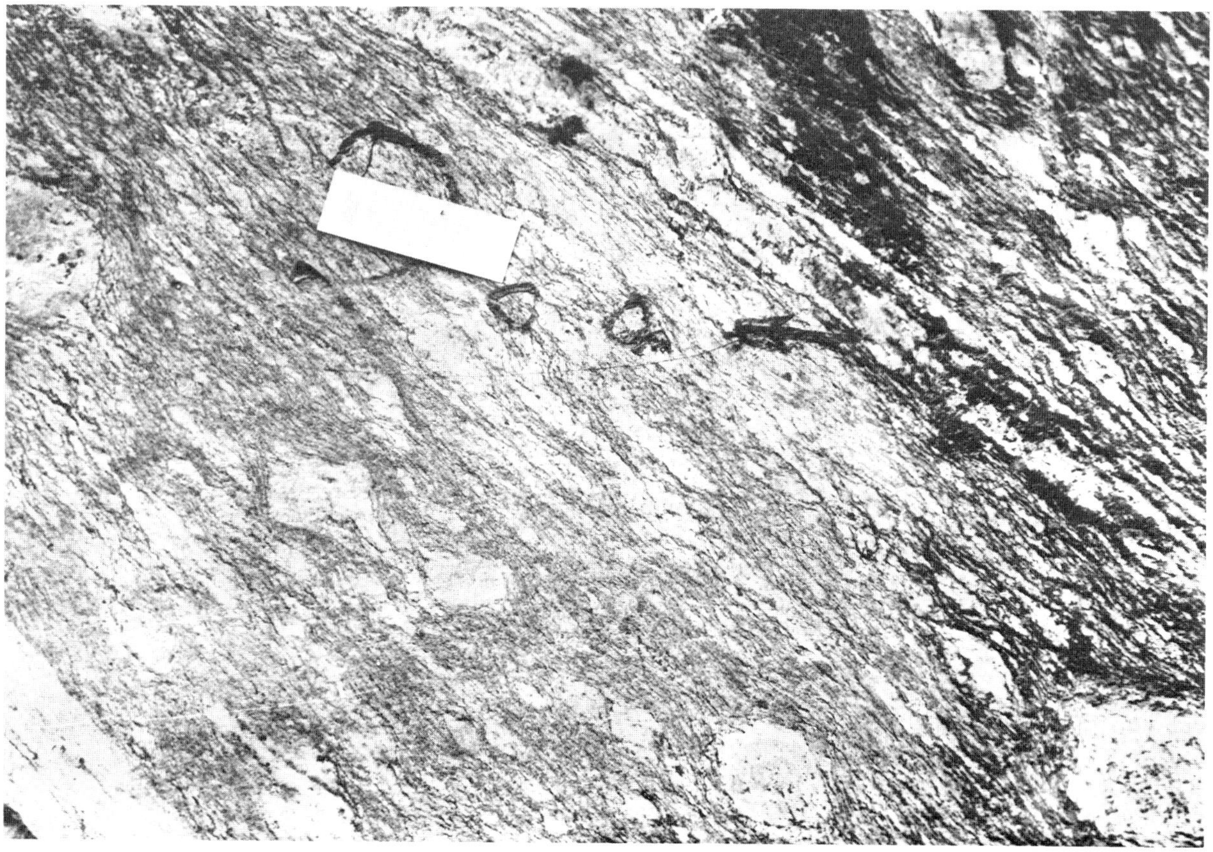

Figure 74. Porphyritic gneiss in Vaikrita near Suring Ghat

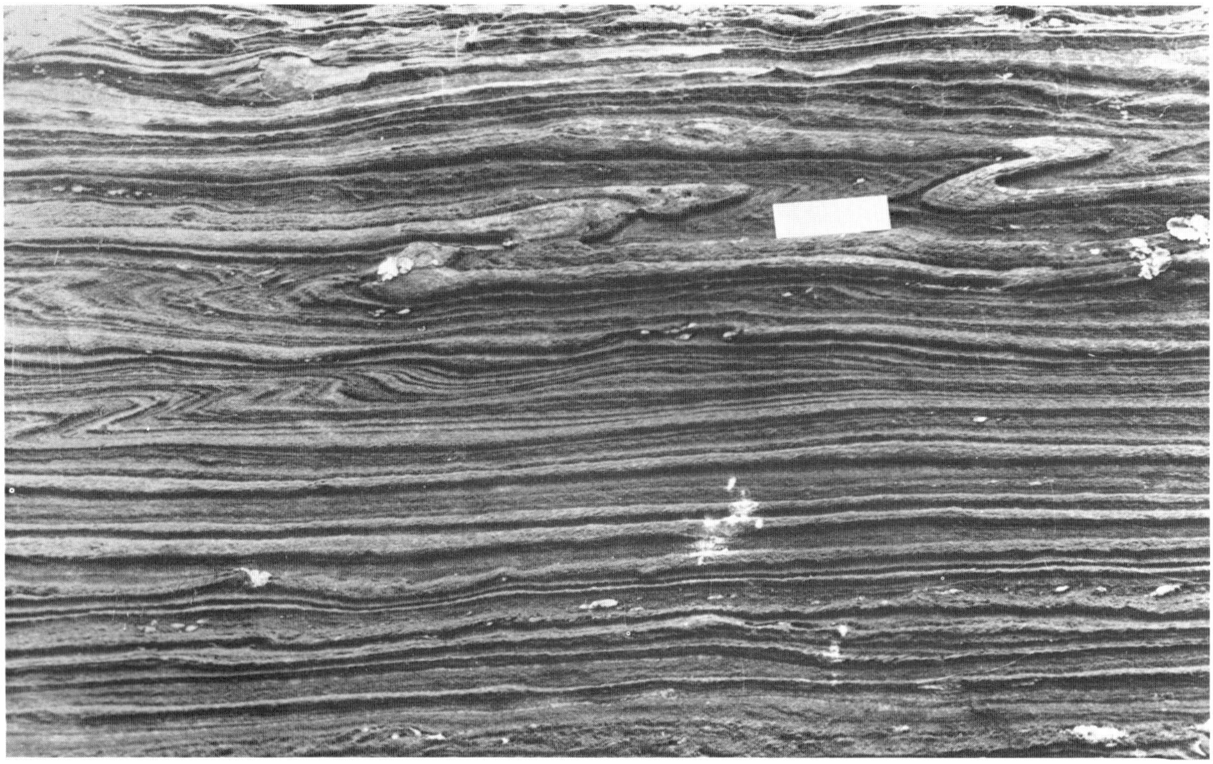

Figure 75. Deformation in Vaikrita crystalline complex near Lilam fault (Vaikrita Thrust) in the Gori valley

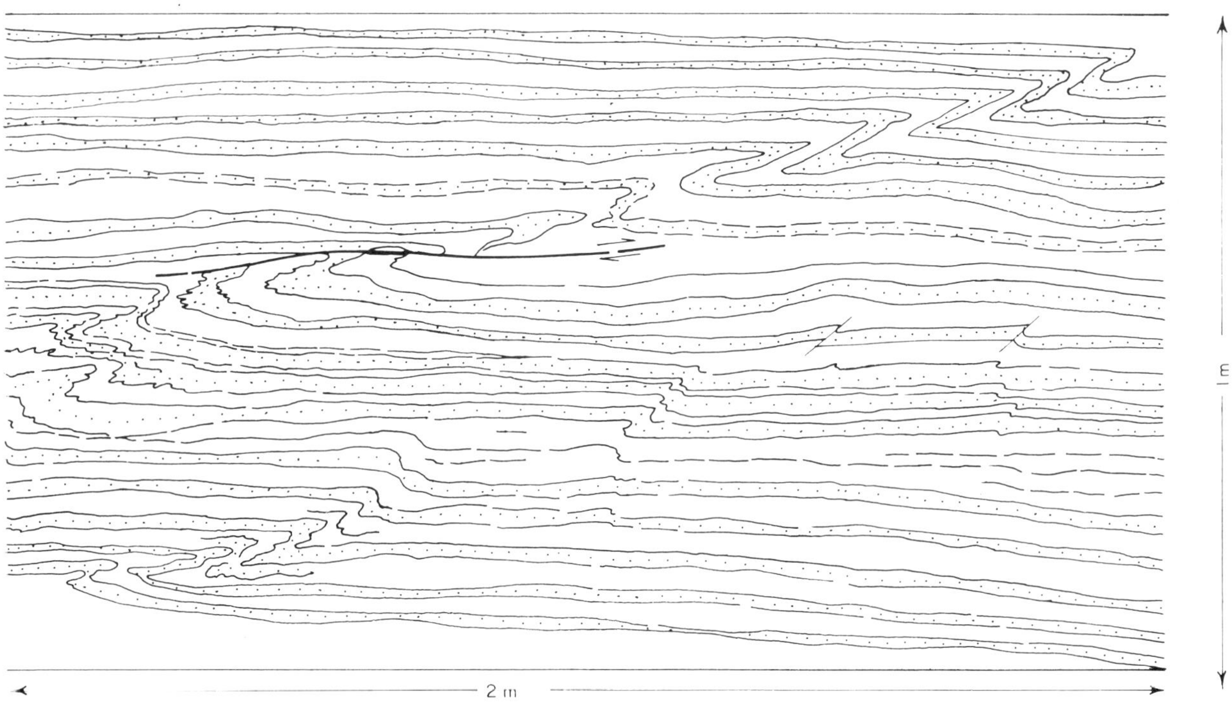

Figure 76. The sketch of the photograph in Figure 75 showing detail

confluence of Gori Ganga and a stream flowing down from the northwest side. The distance between Lilam and Bugdiar is about 15 km.

After the overnight halt on 5 August 1976, our party moved forward for the next camp halt at Rilkot, about 13 km, on the right bank of Gori Ganga. Since the crystalline gorges are very narrow with rapidly flowing rivers cutting valleys with an appreciably steep gradient of about 15° to 20°, the water from rapids and waterfalls very often fall over our heads while walking along tracks on either side of the river, especially during the rainy season. Since there is not much space to avoid these rapids falling down from appreciable heights, the trekkers are left with no alternative but to get soaked in ice-cold water. He cannot go either down to the river or up towards the vertical wall. In the Nala tributary of Gori Ganga flowing from western side, on the left flank of Bugdiar Rest House, there are scattered transported blocks of high specific gravity, dark colour, garnetiferous amphibolitic rocks. Under the microscope these rocks are deformed and amphibolitic with high garnet content. It is obvious that such high specific gravity mineralized garnet amphibolitic rocks constitute the central part of the crystalline complex since other specimens were also collected from a transported boulder along the road to Badrinath near Hanumanchatti. This confirms the presence of basic amphibolitic rocks in the central part of crystalline zone. Another interesting rock transported here are egg-shaped porphyritic gneiss. The section between the Bugdiar and Rilkot concludes the crystalline and sedimentary Tethyan zone starts as sketched in the Figure 72c–d.

Up to the point opposite the waterfall (*ca.* 30 m) on the left bank of Gori Ganga (2552 m), there occur mica schists with calc-silicate rocks, but before reaching Nahar Devi temple the kyanite mica schist zone is followed by granitoid orthogneiss with large augen (10 cm × 2 cm). The granitoids are intruded by pegmatite veins up to 3 m in thickness. The interrelation of granitoid and pegmatite is shown in Figure 72c. The huge batholithic mass of granitic rocks emplaced in the quartzose silicate complex would represent the granitic activity in the Central Crystalline zone prior to the last orogenic cycle or synkinematic phase. The approximate thickness of the orthogneiss body would be *ca.* 1000 m near Tibu. The same calc-silicate group is met near Bugdiar. The frequent aplitic and pegmatic dykes, obviously of the last orogenic phase and having many muscovite and tourmaline crystals of much larger size up to 15 cm, appear criss-crossing between Tibu and Chhirkani (Figure 77). Nearly about 1 km the Gori river is covered by a thick cover of ice (Figure 72, *ca.* 2 to 30 m) on either sides in tributaries. These patches of snow do not melt away at any time throughout the year but the river flows through them. The trails over the snow patches are sometimes very dangerous and any time passing ponies could be trapped inside the dangerous crevasses or in the collapsing ice. We had to walk over such ice patches for about 400 m. In this zone another phase of magmatic activity starts appearing intruding calc-silicate and country rocks. These are the tourmaline-bearing granites which preceded the pegmatitic–aplitic phase. Thus so far this part of Central Crystalline snows three phases of pegmatitic activity represented by (1) Nahar Devi type, (2) Chhirkhani type, and (3) the last phase of aplite–pegmatite (see the section of profile between Nahar Devi and Laspa in Figure 72c). At Bangaela we made a halt for our field-lunch break and sat under the rock cave of calc-silicate gneiss complex intruded by granite and pegmatite. At Chhalobagar the kyanite schist starts appearing and we noted the highest degree of culminating metamorphism in the crystalline zone. These rocks intercalated with calc-silicate rocks continue up to Jolpani. Between the Jolpani and Rilkot (*ca.* 700 m) there seems to be a retrograde type of metamorphism, but the rocks appear to be sieve-like biotite porphyroblastic schists in the ground mass of micaceous materials. This Rilkot horizon typically represents the rock type met with at Pindari glacier zone and the Budhi schist in the Kali river gorge. The gorge at Laspa village is situated about 200 m higher up on the right bank of tributary flowing down on the northern side. At this place highest magmatic activities of leucocratic granite culminates. Considering the overall geological picture with highest degree of metamorphism and strong magmatic activity, it would not be difficult to presume that this tectonic zone should present a deep-seated nature. To my mind biotite porphyroblastic schists of Rilkot overlain by biotitic sericite would represent a passage zone between highly deformed metamorphic crystalline and the unmetamorphosed sedimentary sequence overlying them. But there is all the chances that this passage zone of *ca.* 300 m thickness is produced due to the contact metamorphism of heat gradient ascending along deep seated lineament fracture. Although previous workers regarded these zones as a passage zone without a tectonic break but, however, to my observation there appears to be a pronounced tectonic break below the Rilkot porphyritic biotitic schists marked by kyanite crystalline below it, and profusely intruded by post-tectonic granite.

The Gori river flows in a deep gorge from Bugdiar and the gradient slope reaching to Nahar Devi becomes 20° to 30°. As we go upstream the valley starts widening and near Mapang it assumes a meandering feature within its wide valley. Such geomorphic features continue up to Rilkot. The Rilkot village is situated in a wide flat morainic terrain having three to four houses. The temperature freezes during winter season and the local people go down to the Lesser Himalayan region. The early Rilkot village is completely deserted except for the ruined walls standing on the track side leading towards Martoli village. There is a projected wall cave in the biotitic schist of the Martoli groups of rocks and we spent our night in tents pitched under the cave walls. The weather remained unfavourable and the rain continued to pour down especially during the nights.

After this zone of crystalline we enter the Tethyan sedimentary sequences. The thickness of the biotitic porphyritic Rilkot schist zone should in the order of 250 to 350 m. It is interesting to note that apparently it looks as if the Rilkot horizon transitionally passed into biotitic sericite quartz as we moved ahead and reached opposite Tola village. The hydrothermal quartzite activity in the biotitic schist zone is very intense and it is gradually diminishing as we moved further north towards Martoli.

The sedimentary section over the crystalline has been named the Martoli Group after the Martoli peak (4586 m) (Figure 78) situated in the right flank of the Gori valley. It consists of steep biotitic sericite schist and phyllitic rocks assuming a sheeny lustre. After the gneissic gorge from Rilkot onward the Gori valley flattens and widens. The villages on either side of the river in this wide valley are situated on the old river terrains at about 50 to 200 m above the present current. The Martoli gravel terraces are at the height of 3374 m. As we cross over the Nala coming down from Martoli village from the western side, the phyllitic rocks dip with a steep angle and with a flat cleavage. The dip across the river to Burphu village in the exposed Martoli sequence below the Mapadhura (5206 m) remains steep and the changes in lithology and the rock type are greenish quartzitic sandstone showing

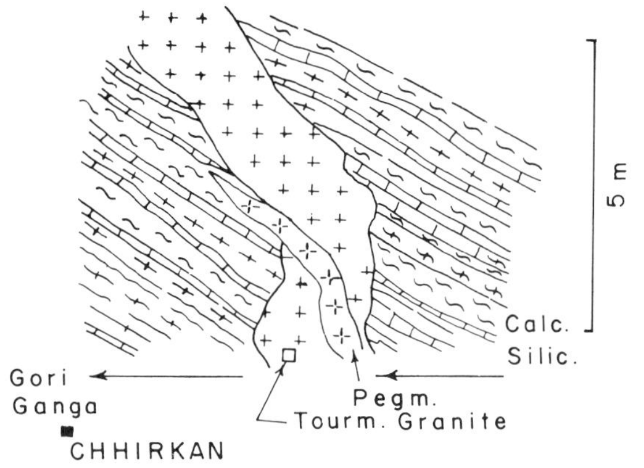

Figure 77. Cross-cutting relationship of two phases of post-Kinematic granites in calc-silicate gneissose rocks of the Vaikrita formation

profuse and well preserved development of sedimentary structure, e.g. graded bedding and cross bedding (Figures 79 and 80). Reaching up to Bilju village in the left bank of Gori (Figure 72D), the entire Martoli series is more quartzitic in its upper part with a monoclinal dip towards NNW or NNE. As in the left bank of Gori between Burphu to Bilju hamlets (Figures 71 and 72), no more fresh exposures could be met with since there is a thick terrace and deluvial morainic mass. Observing with the naked eye and with the help of binoculars, the exposures crop up in the eastern range of the Gori valley showing intensely deformed rocks with folds and faults. These exposures in contrast to their extensions on the western side show rather monoclinal disposition of strata. The dip as a matter of fact on both the sides of valley is very steep, giving an impression of strata standing on their heads. The folds are rather open. All along the track after Burphu village, the transported border of the Ralam conglomerates coming from top are met intermingled with the fossiliferous black slates and carbonates of higher horizons. The quartzite with beautiful cubes of pyrites and beautifully preserved sedimentary features, *viz.* cross bedding, graded bedding and slump structures, are exposed in the section between Bilju and Milam villages (Figures 79 and 80). Approaching Milam about a kilometre over the bridge at the confluence of Gonkha Gad, greenish quartzites are exposed standing on the head (almost vertically dipping with NW–SW strike). This part of the section is a very interesting area for sedimentological studies but the paucity of time did not allow us to go into detail.

Milam village is situated at a very wide valley of the Gori and Gonkha Gad confluence (Figure 72D). This part of Milam between Gori and the Milam glacier, as noted by Heim and Gansser (1939a), is formed by the high left lateral moraine of the Milam glacier, apparently marking a stage of recession after a warm period. Milam village is situated on a levelled gravel terrace of wastage on the side of moraine at a height of 3425 m and about 50 m above the river bed. This village consists of about 100 families and used to be one of the most important trade centres of Indo-Tibet before the conflict with China in 1962. From Tibetan side there used to be free trade on the basis of barter system. The goods like wool, rock salt, borax, and gold were essential items supplied from Tibetan side, whereas cloth, silk, sugar, tea, and various spices were the items exported from Indian side. The trade was in such volume that this region of Himalaya was more flourishing with the riches of gold than the lesser Himalayan region down to the Indian plain where the climate and land both were more favourable for growth of human civilization. In conversation with the local people, I learned that the equivalent exchange rate for gold in those days were Rs. 15/– Rs. 20/– for 1 *tola* (old Indian system of weight) (approx. 12 gm) of gold. The silver from Tibet used to cost 4 to 8 *anna* (old Indian coins making 16 *annas* a *rupee* for a *tola*). Tibetan tea of *ca*. 2 kg packet used to cost Rs. 5/–; ordinary cloth from India would cost 75 *paisa* for a metre, whereas 5 m of silk from Assam was worth of Rs. 50/– to 200/– in those days, 45 years ago. The weight was measured as '*nali*' (one *nali* is equivalent to approximately 1.5 kg. One *nali* of rice or wheat was equivalent to approximately 1.5 kg). One *nali* of rice or wheat was equivalent to five *nali* of salt or borax. Since 1964, the border has been sealed from either side and the trade is completely closed. Because of this historical calamity, the people in the village in the upper reaches of the Gori Ganga valley have all retreated down towards Munsyari in search of their livelihood.

Camping at Milam on the 8 August 1976 we also made a cursory visit to the famous Milam glacier (Figure 72c) which has its gate about 2 km further north from the village. Our field guide, a

Figure 78. Martoli peak (4586 m) in the background of Bilju hamlet in the Tethyan zone of wide valley of Gori Ganga. View from northeast. The deformation in the Martoli Group shows pre-Tertiary Himalayan orogeny

local man (Prem Singh, 60 years old) who lives now in Munsyari but who was born and brought up in Milam, narrated that during his childhood he used to watch the Milam glacier touching right on the northern side of Milam village. But now the glacier has shrunk more than a kilometre in a span of 50 years. As recorded by Heim and Gansser (1939a) the Milam glacier had a length of 18 km (Figures 71 and 72d) stretched in the direction of SSE–NNW and used to be one of the largest glaciers in the Kumaun Himalaya, but perhaps second to the Gangotri in Garhwal Himalaya (27 km situated in the northwest with Greater Himalayan peaks of Badrinath groups). Now if we look at Milam glacier from the southern side, it looks more like a debris of dirty, rusty morainic materials with a lateral morainic ridge on either side. It is extremely slow in movement which almost equals the melting of the ice. The blue ice could only be seen at the gate from where the Gori Ganga starts flowing down (Figure 72d). The snow and ice could only be traced back below the clouded invisible peaks of Hardiol (7150 m) and Tirsuli (7074 m). An assault on these peaks through natural glacier side was made in 1974 by joint Indian and New Zealand women climbers. This unsuccessful expedition cost the life of an Indian girl whose dead body could not be recovered (Figure 13). After making a short acquaintance with glaciers in Himalaya, it is alarming to note that most of the glaciers are shrinking. The time may come when the rivers fed by these glaciers will have much less quantity of water causing an acute problem in the future in northern India, where our agriculture, irrigation, and hydroelectric stations are dependent on these snowbound glaciers of the Greater Himalaya. It is strongly recommended that, in order to know the cause of recession of glacier, a study should be taken up immediately with the help of aerial photographs, Landsat imageries, and field work. The gate of the Milam glacier is rather unusual in comparison with the glacier situated in the shady side of highest of the Central Himalaya with its gate (3550 m) reaching about 300 m further down than other glaciers we come across in the Central Himalayan region. Cotter and Brown (1907) were the pioneers in the study of the snout of the Milam glacier. According to them, the eastern lateral moraines goes 700 m beyond the glacier's gate. Heim and Gansser (1939a) remarked that an old native, Kaisingh Bahadur of Milam, stated that the extension of Milam glacier to Milam village was reached in 1900. Thus high subrecent lateral moraines are found in most Himalayan glaciers corresponding roughly to Alpine glacier earlier in the last century.

Our traverse continued along the Gonkha Gad, a tributary to Gori Ganga (Figs. 71 and 72), in a northeastern direction. The silty greenish grey shale and argillite with their steep dip continued till my observation point OC 22 and 23 were reached with the intense quartzitification in the Martoli series near Lakhuri Bhel. About 2 km from Milam, hydrothermal activity with quartz veins has given rise to the mineralization of copper, pyrite, and polymetallic mineralization (Sinha, 1977). This zone of mineralization occupies an interesting position in all the sections from the crystallines across Tethyan zone. They are found to be confined to the contact junction between the Martoli and Ralam formations. Since the debris, scree and rock waste reach an immense thickness along the bank, no contact

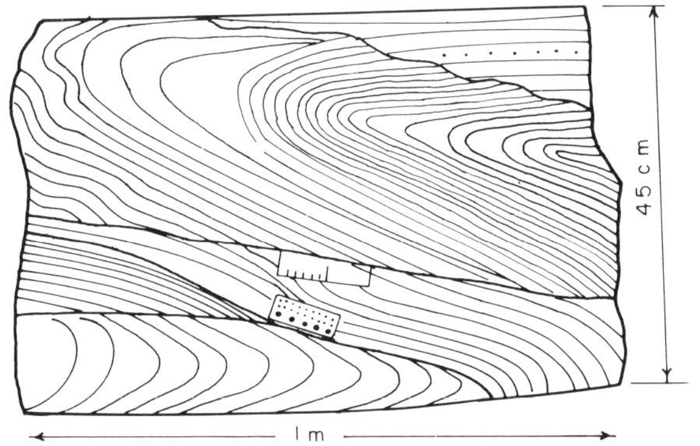

Figure 79. Cross-bedding and graded bedding in the Martoli silty lithology near Milam

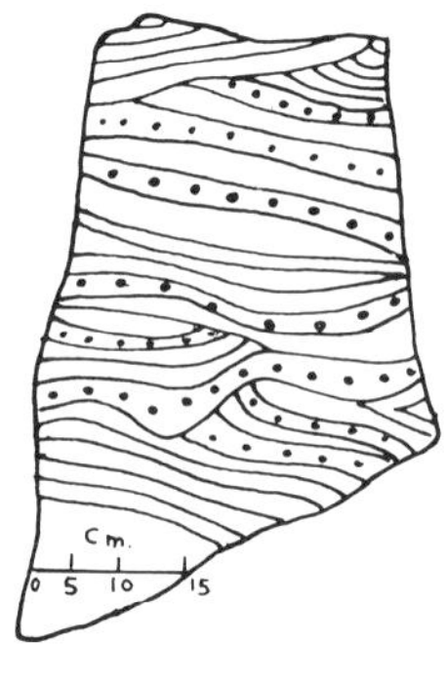

Figure 80. Cross-bedding and other sedimentary structures in the Martoli group

zone between the phyllite and the conglomerate is visible, but on the escarpment, an unconformable contact between Martoli and Ralam is conspicuous (Figures 81 and 82). Such a conspicuous and marked angular unconformity is not seen in other exposed sections but in the Darma valley. This unconformity marks the orogenic hiatus before the deposition of the Ralam conglomerate (Figure 81) in the Tethyan zone. The conglomerates of Ralam assumes the same lithologic character as described in the Dhauli Ganga section and is followed by the quartzite horizon. The calc-phyllitic and dolomitic phyllite zone over the quartzite horizon is conformably overlying the quartzite of Ralam. At Samgaon the dolomitic phyllite horizon is overthrusted by the Ralam Conglomerate and quartzite again. This repeated thrusting of the Ralam over the Garbyang is unusual in comparison to other parts of this area. In other section such remarkably conspicuous thrust plane were never observed. The thrust plane is at an angle of 35 to 45° and it becomes rather low angle and exposures could be observed near the Shillong campsite (Figure 72d). This thrust is being named as Samgaon Thrust. The extension of this thrust could be traced as shown in the geological map of Central Himalaya (Figure 14). The onward sections up to Dunga along the right bank of Gonkhagad valley exhibit an interesting exposure of Palaeozoic rocks complicated by repeated faulting and thrusting. The exposed rock types are greenish calc-shale of the Shiala formation overlain by purple variegated Silurian limestone of Yong followed by the Muth brown dolomitic and white sugary quartzite. The huge cirque of the Haldawa Pinglawa Khola glacier presents the most picturesque geomorphic feature in this height of the Central Himalaya (Figure 72d; Photo Panorama No. 5). Pan 5
On the onward journey, a ground with green grass patches was observed below the lurking glaciers from the western side, and wild goats were found grazing over them, but on the way back this region experienced a very heavy snow fall of more than 2 m of thickness which continued for five days. When we came back to our camp at Dung on the 22 August, 1976, it was under heavy snow. The ponies deserted us, and only with great difficulty, could we trek back to Milam camp, searching for our path over a thick cover of snow.

Figure 81. Ralam conglomerates. Quarzitic pebblen siliceous cement perhaps marking Cambrian–Precambrian boundary

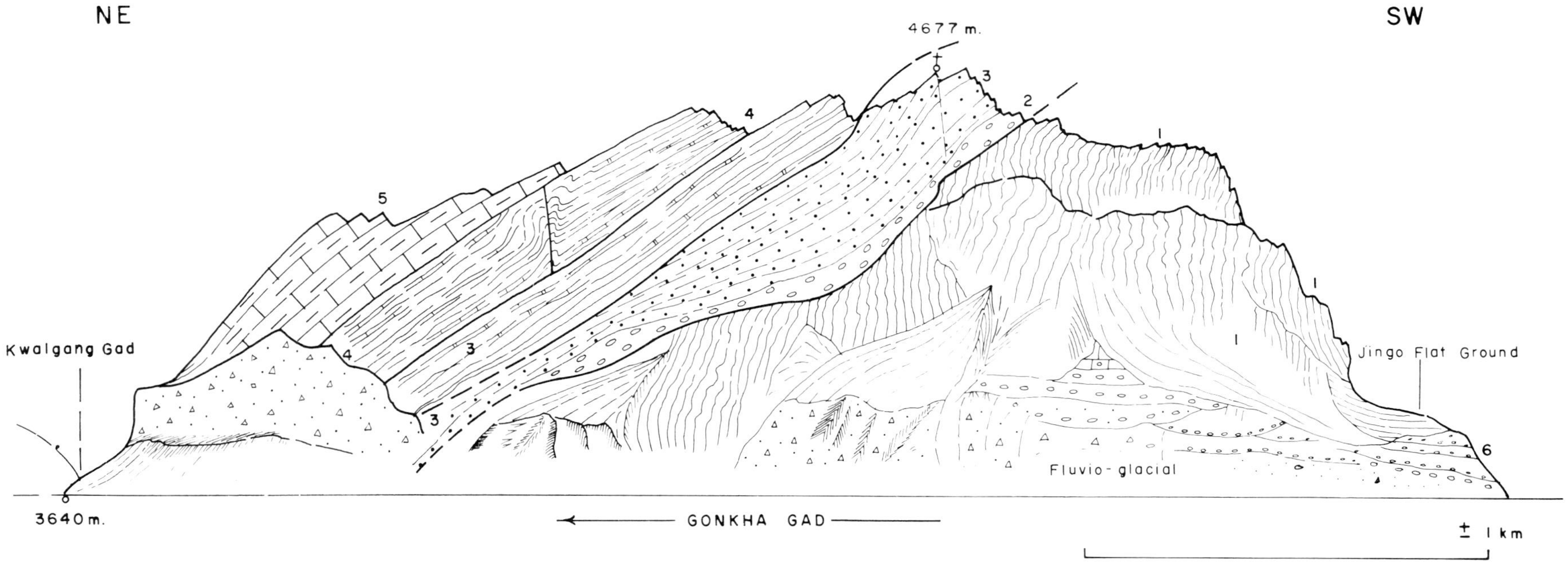

Figure 82. Angular unconformity contact between the Martoli Group and the Ralam conglomerate near Kwalgang Gad (3 km upstream of Milam) in Gonkha Gad valley

After Dunga, we continued our trek towards the Unta Dhura pass (5450 m) and down to Topidunga side in the Girthi valley. The Bamberas glacier (Figure 83) with its gaites, at 4 km to 7 m present a picturesque view with purple colour debris derived from the variegated formation, whereas the Shafed glacier in the eastern flank of the track is white in colour, because of its association with Muth white, sugary quartzites. The Safed glacier ends with a lake '*Pari Tal*' (Figure 84) attaining an altitude of 4813 m. Over the Muth quartzite horizon, the dark carbonaceous shale of Kuling lies unconformably followed by the Kalapani limestone. Disharmonic deformation in the Kuti shale formation down the Topidunga side is clearly visible once the Unta Dhura pass is crossed. We made a rich collection of fauna from the Kuling shale. Crinoidal stems of large size (10 to 12 cm) from the Kalapani limestone were collected near Pari Tal. Near Lauka (4846 m) we got bryozoans and ammonites. Finally the expedition party reached the Topidunga camp, crossing the most treacherous Unta Dhura pass at sunset (Figures 85 and 86). This pass is notorious for its unprecedented weather and snowstorms throughout the year. Negotiation of this part has become so difficult now-a-days that we had to wait for a whole month to strike fair weather with the help of meteorological resources of the Border Police headquarters. Our expedition party consisting of eight ponies and sixteen men were all taken sick due to high altitude sickness and pulmonary oedema problems (Figure 87). One horse broke his hind leg while crossing down the pass along the northern slope. The northern slope is covered with perpetual ice and glacier and half cirque feature (Figure 87).

Down the slope of Unta Dhura Pass up to Topidunga campsites (Figure 87) we passed over Kuti shale. Crossing over the Girthi gorge, another pass, the Khingru Dhura (5244 m), is

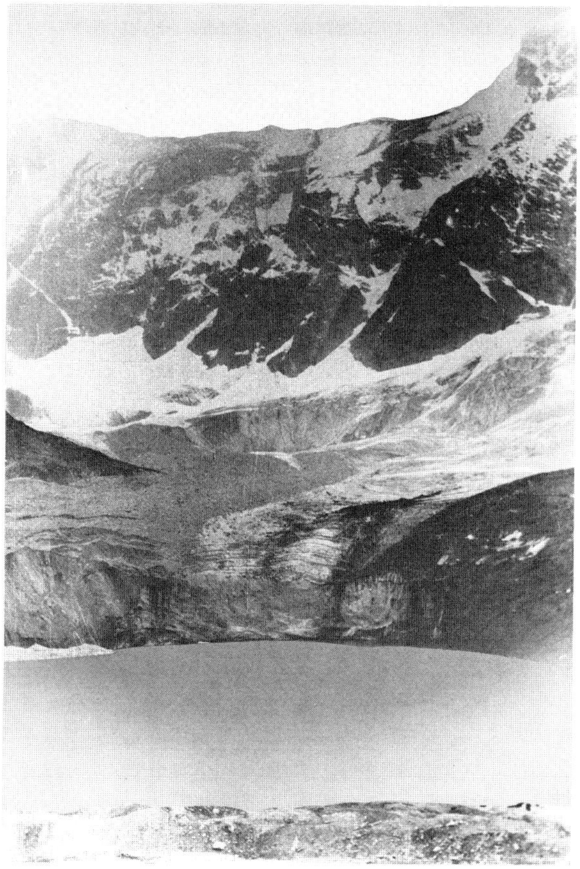

Figure 84. The southern flank of the Unta Dhura Pass with Saphed (white) glacier ending in a lake called Pari Tal

Figure 83. Bamras (red) glacier on the way to Unta Dhura Pass. The hillock is composed of Devonian Quartzite

Figure 85. A breather before crossing Unta Dhura Pass (5450 m) over exposed Triassic Kuti shale

Figure 86. Mesozoic sequence in Tethyan zone. A view looking northwest from the Unta Dhura Pass

climbed up. The tired and exhausted men continued the trek through the Khingru Dhura (Figures 71 and 87) down towards Lapthal base camp through the Chidamu gorge. The Kuti shale formation is described in the Girthi Ganga section is tectonically unconformable and is followed by the oolitic and arenaceous massive limestone of Kioto. Along the track measured column was prepared for tectonostratigraphic purposes (Figure 23). The complicated tectonic picture (Figure 39) has been sketched (Photo Panorama No. 5; Figure 41) from Chidamu campsite. Rich fossil collections of bivalve pelecypods, belemnites, and ammonites were made. We made proper documentation of locations along the section from Khingru Dhura down to the Lapthal base camp (Figure 71). The extended section further north crossing over the Giumal sandstone up to Dobala were taken beside the other described sections all along the Kiogad valley. The tectonic implications with a measured column of this area has been taken up in the previous chapter to avoid repetition.

The Nanda Devi group with their two peaks: the Nanda Devi (30° 22′ 32″: 79° 50′ 22″) 7820 m and Nanda Devi East 7435 m represent the highest peaks in the Central Himalaya in India territory. It was conquered by a British–American expedition in 1936 by N. E. Odell and H. W. Tilman. From India the peak was climbed by Nawong Gomtu and Sherpa K. Norbu (Figure 72D).

The main Nanda Devi peaks (7820 m) and Nanda Devi East (7434 m) can be seen if the weather is clear from Bilju village along the glacier and a ravine coming down towards east in Gori Ganga main stream. But the best observation of these peaks is made through the classical mountaineering route along the Rishi Ganga valley from the Joshimath side. From the Pindari glacier region the proper Nanda Devi peak cannot be seen. Thus it was presumed that the peak regions described by Heim and Gansser (1939) is an open synclinal structure. Maruo (Japan) recently described the structure as an inverted isoclinal fold of calc-silicate quartzite as detached

Figure 87. The Unta Dhura Pass photographed from the north. The exposed rocks are Triassic Kuti shale

klippen lying over crystalline base over a thrust surface. But there is no doubt that the Nanda Devi peaks are composed of rocks of the Martoli Group.

5.9 Dharchula–Tawaghat–Bidang–Dawe (Darma Ganga Valley) (Figures 88, 89 and 90; Photo Panorama No. 2)

Fig. 88
Fig. 89

Dharchula is situated on the right bank of Kali river which marks the Indo-Nepal border. Starting the traverse from Pithoragarh town and after taking a diversion following the motor road up to Ogla at a distance of 35 km, and going ahead further 45 km, the small township of Dharchula is reached. The checkpost to Nepal gives it a special position from the commercial and administrative point of view. The calc-zone of Dharchula with dolomitic limestone, phyllites and basic intrusives is tectonically thrusted by crystallines with a sharp contact called the Main Central Thrust (Figure 90). The orthogneiss thrust sheet starting with the village Ranthi consist in its lower layer large biotite alkali feldspar gneisses (Figure 91). It has been reported (Heim and Gansser, 1939a) that it is a formation of homogeneous orthogneissic rock and orthoclase forms large perthite, some horizons substitute microcline. The quartz in bigger grain size exhibit strong undulatory extinction and mortar structure. It seems to be bordered in parts by grains and groundmass constituting predominantly of quartz-sericite biotite. Muscovites occurs in large scales. Magnetite, apatite, and zircon are accessories. Feldspar and quartz gives an aphinitic structure but on the whole the rocks are gneissose. The basal orthogneiss are succeeded by typical augen gneiss, further up near Ellagad. Closer to Siankur the porphyritic gneiss with blue quartz becomes very prominent and rocks assumes a schistose augen character. The blue quartz in the gneiss is also being encountered in the different Main Central Thrust zone of Sutlej valley, Mandakini valley, as well as from its Nepal counterpart. The blue quartz occurrence is a pointer to quartz porphyry parentage. On the basis of regional correlation and petrological study (Ashgirei *et al.*, 1977a) have described such porphyroid nature of Chails from the root zone of Chail Nappe in the Sutlej valley of Himachal Pradesh. Below the Khola village the orthogneisses are invariably with the basic sills of quartzose biotite amphibolites composition having an average thickness of 2 to 2.5 m. Hornblende gives a strong blue–green pleochroism. The brown flakes of biotite is mixed with hornblende occurring in the large amphibolitic crystalline. Since quartz is in abundance with biotite with less mafics the basic sills may be of dioritic composition towards Tawaghat. The drivable road along the Kali river valley has been constructed but there is a zone of constant landslides about a kilometre from Tawaghat. During the heavy rains a traveller with a vehicle cannot pass along this unstable road. One has to take the arduous long climb up to Khela to come down to Tawaghat. Tawaghat is

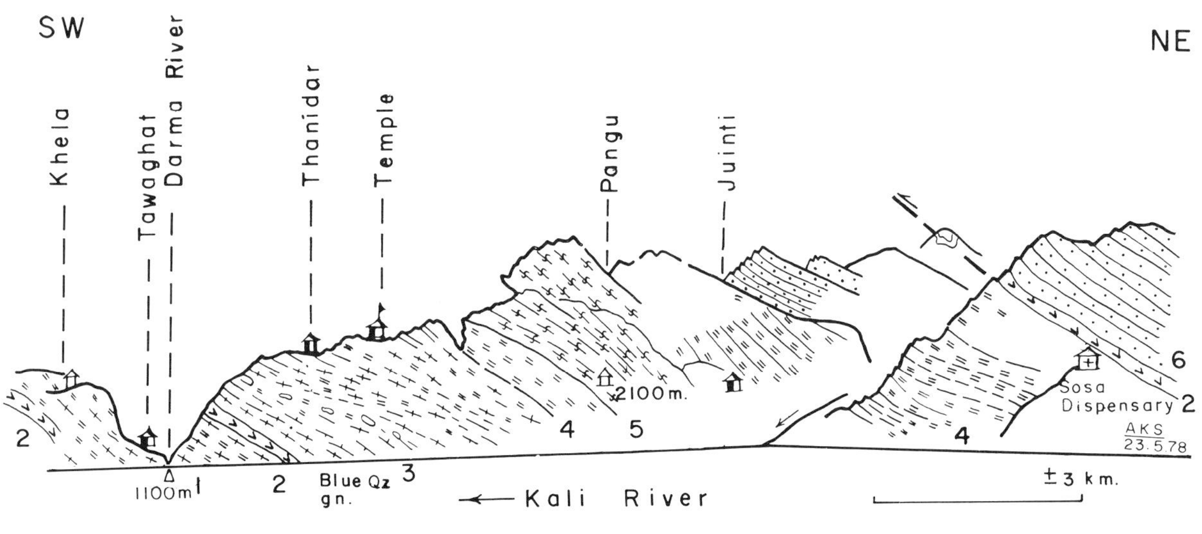

Figure 90. Sketch geological section along the Kali river between Tawaghat and Sosa. 1. Quartz–mica schists with blue quartz; 2. Basics or amphibolites; 3. Mica schists with relict structure (10 cm × 3 cm) of biotitic composition; 4. Mica schists; 5. Porphyritic gneiss; 6. Quartzite

situated at the confluence of Darma Ganga river and Kali river (Figure 88).

The present traverse will be described along the Darma Ganga valley. First few kilometres with river flows along the strike direction of the crystallines, but after reaching Chirkila village the direction turns across the strike. The schistosity trend of the crystalline is 60° northeast. Reaching Khet village rocks becomes highly porphyritic (Figure 91). The porphyroblasts of plagioclase feldspar are as big as 5 cm × 3 cm. This zone of *ca*. 800 m thickness is intruded by amphibolite sills. Porphyritic gneiss zone with the amphibolitic sills with a tectonic contact passes over to sedimentary quartzite, phyllite and limestone zones of Jhamri Goan-Nyu. These sedimentaries within the Sirdang–Gam Malla zone also exposed in the Kali valley assume a special character.

Here from the porphyritic gneiss exposed at Nyu onwards we continue our traverse description along the Darma Ganga valley. Up to Dhar village these porphyritic gneisses are rather monotonous in nature but two amphibolitic sills of about 5 m thickness are met after Sobala village. At Dhar village the zone of migmatization starts. The young granites are seen to be intruding at Sobala and Dhar. There also appears strong migmatization. Between Dhar and Nogling we had our night halt at Bauling and Sela. The structure just after the Dar faulted zone exhibits an anticlinorium, and until Sela there is no change in the monotonous northerly dip. The monoclinal dip continues due northeast. Between Dhar and Bauling on the northern wing of these anticlinal structures and quartzite occur within the gneiss and schist, and the younger granites are found to be intruding at the confluence of Nandarma Nadi. Such granitic bodies also occur in batholithic size between Urthing to Sela and continue at Baling till the whole crystalline complex is attenuated by the deep-seated lineament zone (TT) with overriding Martoli group of sedimentary rocks. The crystalline complex in this section is mainly mica schists and gneisses. Calc-silicate rocks also occupy a large volume and are mixed up with biotite gneiss and intruded by a younger granite containing garnet, kyanite, and sillimanite (fabrolite). The potassium–argon radiometric age of these granitic bodies is 37 ± 4 Ma. Another pronounced magmatic phase is represented by the pegmatitic rocks of a distinct post-tectonic phase, which occur throughout the section between Bauling and Baling covering a distance *ca*. 80 km. At Bauling the grade of metamorphism looks to be visibly increased to kyanite grade and garnet–mica schist appears with kyanite. Between this kyanite garnetiferous zone near Urthing village the traverse of 5 km predominantly rocks of calc-silicate complex, intruded by granitic pegmatite assume almost vertical dip and the steep zone

Figure 91. Porphyritic gneiss. Porphyroblasts (10 cm × 5 cm) of plagioclase. Locality Khet, Darma valley

culminate with the high angle fault with gradual flattening of dip to 40 to 50° due northeast on the hanging wall side.

This observed fault zone is clearly seen as a geomorphologic feature along a deep gorge in the right bank of Darma valley followed by a similar deep-cut valley from the left bank flowing from height 4965 m (Figure 89). The height point in the western side is 5772 m from where down to deep-cut valley stream flows along the fault. The section following the Urthing to Sela dominated by granitic emplacement and the country rock near Sela is of schistose type with quartzite. To my mind, the zone of the Urthing fault should be correlated with the Badrinath and Pindari Glacier divergent zone. The post-tectonic aplitic intrusive rocks give an absolute age of 20.7 ± 0.1 Ma by the potassium–argon method. A batholithic mass of granitic composition with mica schists dominate the section with pegmatite (Figure 92). At Nagling quartzite bands reappears again dipping at an angle of 50° NNE. The zones of the ptygmatic folding and migmatization are very prominent near Nagling (Figures 93 and 94) and immediately after that the exposed wall rock dipping at 35° or less northeast are intersected with snake-like dykes and sills of aplitic granitic veins up to the Daling village. Kyanite–sericite schist zone with calc-silicate rocks and quartzose schists are overlain by biotite porphyroblastic mica schist of Pindari–Rilkot–Budhi type. This marks the end of crystalline zone over which the pyritiferous carbonaceous schistose, rock suite is overthrusted. Thus we come across major tectonic break before reaching Baling (Figure 88).

The pyritiferous carbonaceous slate belonging to the basal part of the Martoli sedimentary sequence is systematically folded in minor folds with their axial planes parallel to the axial plane of major thrust plane dipping at 30° with N20°E inclination. They are highly cleaved and their major cleavage direction is also towards W20°E but with 50° inclination. From the section Figure 89, at Baling a simplified picture of this transition has been attempted. Calc-silicate mica schist, quartz schist and garnetiferous mica schist repeat due to faulting at near morainic debris material over the Baling Gal Fault. Pyritiferous carbonaceous schists are repeatedly thrusted parallel to the main deep lineament of Baling which has assumed the character of a duplex thrust system during the final phase of thrusting movement. Calc-silicate rocks in the adjoining crystalline suite weathered differentially, with the eroded portion due to chemical weathering of the calcareous material in them. The trend of this rock is 45°NNW inclined to N10°E as we traverse from Duktu to Baling. The rock sequence changes in the following descending order: biotite calcareous garnetiferous mica schist, quartzitic garnetiferous sericite schist. Due to the presence of pyrite crystals the carbonaceous slate have a small amount of H_2S and profuse limonitic leaching along the cleavage parting. The carbonaceous slates also have sufficient quartzitic material at the core of small fold. The phyllitic rock follows the black slate with sericitic and biotitic needle flakes. They serve as roofing slate for the local Bhotia people. This roofing slate has typical resemblance with the slates of Chail and Simla in the Lesser Himachal Himalaya.

Figure 92. Younger aplitic–granite intruding the biotite schists and quartzite with calc-silicate in the upper reaches of the Darma valley

Our next night halt camp along Dharma Ganga valley was made Rama campsite beside the Public Works Department gang hut. The slate in its upper part are interbedded with quartzite and profusely deformed synsedimentationally and due to the direct folds development parallel with the major thrust plane. The traverse between Baling and Dugtu exposes mica schist is to be composed of carbonaceous material and possibly the calc-silicate. The quartz component also increases and it becomes typical quartzite about 5 m ahead. There appears again garnetiferous mica schist and gradually rocks change into quartzite schist and carbonaceous schists also start appearing. Further on the rock type show typical intense tight, recumbent and tight folds with the development of cleavages. The fold trend shows southwest movement and reaching Dugtu it becomes a typical pyritiferous view Panchuli groups of peaks (6904 m) are visible with wide view morainic valley over the slate sequence of Martoli (Figure 2).

Figure 93. Ptygmatic folding and deformation in the migmatized zone in the Darma valley

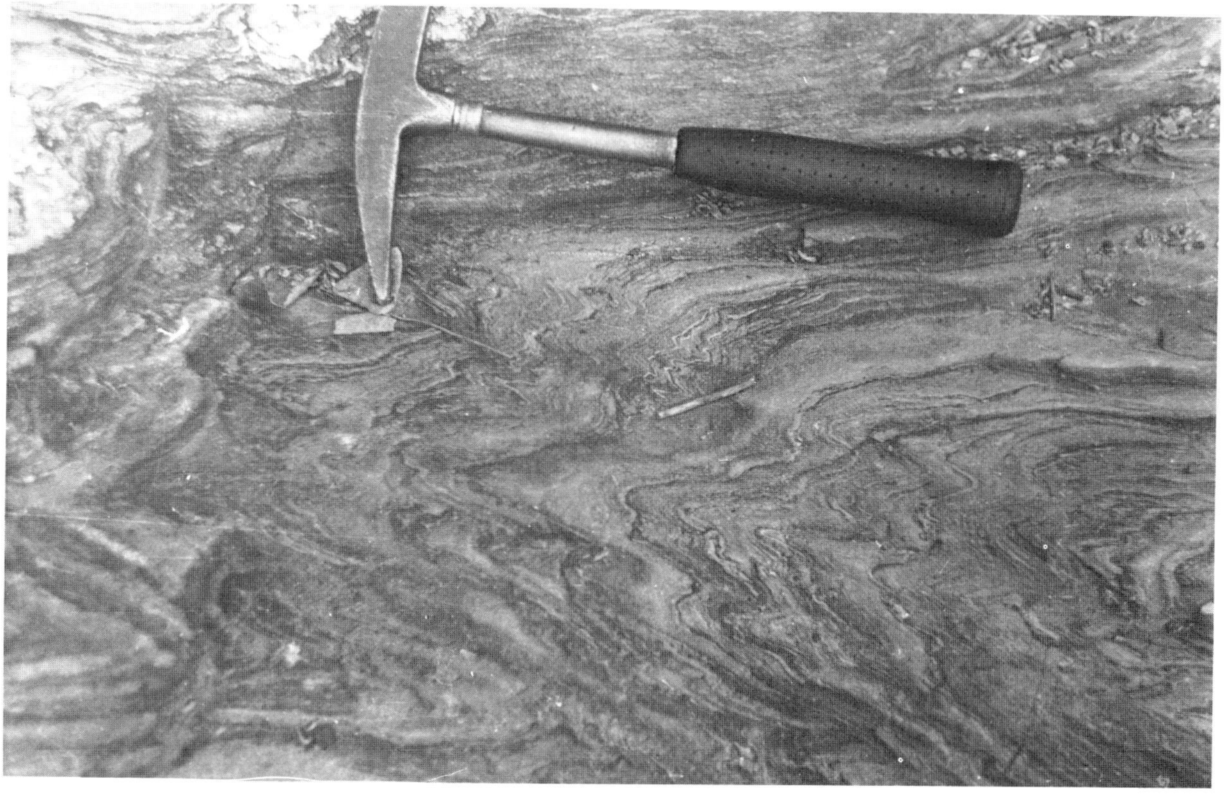

Figure 94. Deformation in the migmatized zone in the Darma valley

It seems that the sharp top of the Panchuli peaks are made of Martoli group of peaks rather than crystalline as thought earlier. Since the peaks are still unconquered, the geological rock composition of the Panchuli peaks has remained a conjecture in the absence of a natural rock specimen from the top. The unconformable contact of Ralam conglomerate cross before reaching the village of Go. As observed in the Dhauli section in the Gori Ganga valley, the basal part of the Ralam comprises red–purple conglomerate with quartzite pebbles and siliceous matrix, overlain by quartzite with sedimentary feature of cross bedding, current bedding, graded bedding structure (Figure 95). At the turning point towards Rama camp in the left bank of Dhauli about 300 m above the river bed at the contact Ralam with the Garbyang (Figure 96) a rich mineralized zone of copper polymetal and pyrite and barite–quartz association appears (Sinha, 1977a). This zone has been found in wide strike area all along the tectonic contact zone of the Martoli with the Ralam or Garbyang. The sediments with polymetal copper–barite–quartzite association remain occupying the same geological position. The traverse sequence from this point up to the Rama camp and further 1 km ahead is the exposure of Garbyang formation comprising dolomitic limestone with the phyllitic layers become more arenaceous in the higher horizon (Figure 97). This dolomitic shaly and arenaceous strata of Garbyang are also tightly folded (Figure 96). This is a pointer of shallowing of basin and deposition of strata with cross bedding and pelmel manner. Similar to other sections Garbyang are followed by Shiala calc-argillitic Yong Reefal limestone followed by dark purple argillitic, variegated and white sugary quartzite of Muth until we reached Bedang campsite at the height of 3850 m. As shown in this section towards the southeast along the escarpment, following the trail towards Joling Kong Pass (6010 m), the Muth is unconformably overlain by the Permo-Trias Kuling and Kuti horizons. Since the carbonaceous Kuling horizon is mixed and intercalated with limestone bands and shales, it is difficult to differentiate this horizon, therefore, they are mapped together (Figures 13 and 89). This in turn is thrusted by an inverted syncline with Muth quartzite at the core and variegated Silurian and Yong in the limbs. The lower and upper limbs are repeated by two more thrusts. They are observed in this section with their base of Muth quartzites on the hanging wall side (Figure 88). These thrusts have been beautifully cut

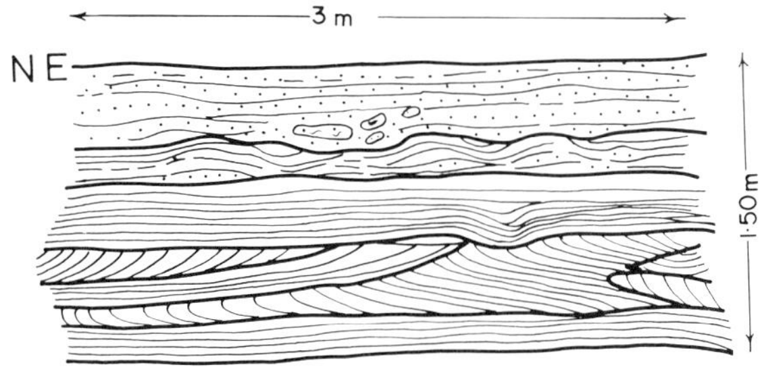

Figure 95. Sedimentary structures in the Ralam quartzite in the Darma Valley

by a fault with a vivid displacement towards the western side of the thrust units. The panoramic photographs show (Photo Panorama No. 2) the deposition of Tethyan zone strata between Bidang and the Jolinkong pass region. Exposure of an anticlinal axis in NW–SE direction and minor axes of similar pattern at Piungrung with steeply dipping strata, complicated by repeated folding and faulting and thrusting (Figures 50 and 89) could be observed in this region. A fish fossil remain was collected from the Permo-Trias horizon by the expedition palaeontologist, Dr S. P. Jain (Figure 97). Our expedition party moved northwest along the left bank of Dharma Ganga. This leads to the Darma pass to cross to Tibet. At Piungrung camp, it could be established that whole region is folded and thrusted into a huge anticlinorium structure, the northern limb of which has been completed by three repeated

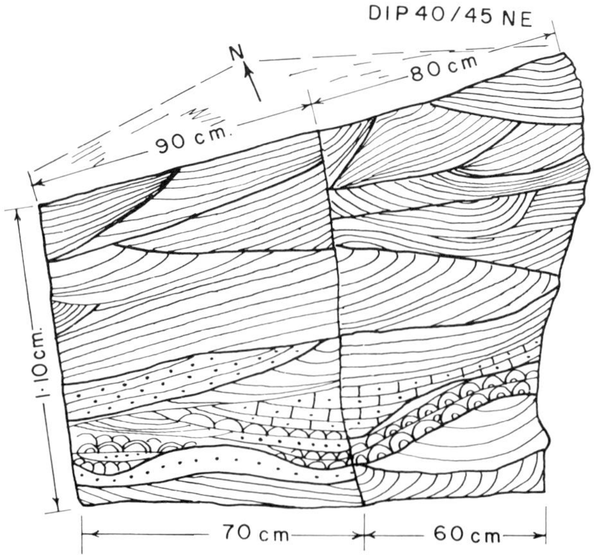

Figure 96. Cross-bedding and biostromal growth in the Middle Garbyang sequence of the Darma valley near the Rama campsite

thrusts and successive faulting (Figure 89). There is a copious amount of debris of different rock types brought by the glaciers from cliff of buff brown quartzite which look to be fossiliferous, but to identify the fossils is not possible. Other components of the debris are white quartzite with the friable black shale and limestone from the Permo-Triassic horizon. The Muth quartzite on the way to Dawe assumes a greenish tinge which is easily differentiated and which represents their lowermost horizon constituting the core of anticlinal structure. It is coloured in the middle part and snow-white at the top. The purple variegated Silurian cliffs folded with the Muth quartzite at the source of Darma river represents a very wide U-shaped geomorphic valley. The right flank is covered with black carbonaceous Kuling shales unconformably overlain by the Muth quartzite. On the left bank of cliff one comes across Dawe campsite. Over the thrust, the Yong algal limestone is followed by variegated Silurian repeated in isoclinal fold with the core of Muth quartzite and repeating strata. The porphyritic gneiss from the exposures near Khet, the absolute age by potassium–argon method has been determined to be 740 ± 20 Ma (Figure 89). It would not be surprising to have such older Precambrian dates of crystalline rocks, but it would be interesting to note that the porphyroblastic gneiss rock could have also given the last phase of metamorphism.

5.10 Kali–Kuti Valleys Geotraverse

5.10.1 Tawaghat–Garbyang–Gunji–Nihal section (Kali–Kuti valley)

From the previous section we have had a general view of the geological set-up from Dharchula up to Tawaghat from where the traverse route along the Kali gorge strata takes a side western track along the Darma valley. Since Tawaghat is the confluence of the Darma and Kali rivers bordering Nepal, the track along the Kali river from Tawaghat onwards is not followed along the river bed but across the ridges over the right bank. The ridges have general NW–SE strike trends. Thus this traverse is a rather difficult one in comparison to the other traverse along the river valley. This route has been traversed not only by the local inhabitants but for the trade between India and Tibet. This has also been the pilgrimage route to Holy Mount Kailash, the seat of Lord *Shiva*, and Manasarovar lake.

To recapitulate the figures described in last section (Figures 89 and 90) we switch over to the section from Tawaghat up to the tectonical contact of Sirdang–Jhamrigaon sedimentary zone and let us refer to the Figure 104 which sketches the geology along the track. Just near the bridge at Tawaghat over Darma river the exposed rock

Figure 97. Rama camp site with the lower Palaeozoic rock groups in the background. W.I.H.G. Expedition party. L to R, Drs S. K. Tandon, S. P. Jain, V. C. Thakur and A. K. Sinha

are biotite porphyroblastic biotite schist. Biotite occurs in lensoidal form of dimensions 15 cm × 10 cm. Climbing up along the bridle path towards Thanidhar, there is blue quartzite bearing augen gneiss. Further along the route 1 km ahead occurs amphibolitic sills 5 m thick. To summarize, the rock sequence over the Tawaghat area in ascending order are: porphyritic biotite schist, quartzose mica schist, gneisses with an amphibolitic sill and with quartzitic bands in its upper parts, porphyritic blue quartz, porphyritic gneiss dotted with big xenoliths and scyaliths (10 cm × 3 cm) of biotite with schistose structure up to 100 m below the Thanidar village 930 m (Figure 89 sketch section between Khala and Sosa in 5.1). This zone represented by chloritic mica schist with quartz veins continues up to the Pangu village (2100 m) from where big porphyroblastic gneiss of Khet type of Darma valley appears (Figure 91) the size of the porphyroblasts of feldspar are as big as 15 cm × 7 cm. This zone is the eastern extension of similar zone as we come across near Khet in Darma valley. From Pangu onwards we track downwards and reach Juinti. The porphyritic gneiss is overlain by mica schist with multilayered weathered amphibolitic sills further up till we reach the dispensary at Sosa (2500 m). At this contact, the well-marked tectonic zone is exposed with amphibolitic sills physically overthrusted by an unmetamorphic quartzite–limestone phyllites sequence of the Sirdang–Gam Malla sedimentary zone.

This Sirdang–Gam Malla sedimentary zone present an interesting tectonic phenomena within the highly metamorphosed crystalline. Above Samari (2533 mm) the inverted anticlinal zone is truncated by the crystalline thrust sheet overlying the inverted northern limb of Sirdang–Gammalla anticlinorium structure. The track after going down to Samari goes up and cross through the Rungling pass (3000 m) composed of the porphyritic gneiss and at the low level it is succeeded by biotite schist. Till we come down across the pass in the Shinkhla valley, the coarse porphyritic gneiss is the dominating rock in the central part of Rungling pass, where a mica schist appears again in the northern part near Shinkhla with schistosity plane of 45° due 30° NE. This mica schist zone with minor folds having large garnets (3 cm × 2 cm) in them continues to Galagad.

I would like to mention purposefully here that the section being described, henceforth, now has been so well and illustratively described with sketches (Heim and Gansser, 1939a; Valdiya and Gupta, 1972) that I prefer to concentrate on some new aspects. Hence I have not sketched a systematic profile along this valley as in the case of other traverse in the higher central Himalayan region. I would like to refer to sketch IV B, plate II of Heim and Gansser (1939a). I would also like to recall that the metamorphic zonation zone by Heim and Gansser (1939a, Fig. 62) holds good until we reach Dudhi, except for some minor correction and editing in some places. Hence, I would not dwell on the aspect already well illustrated and described. The section between Galagad and the Jipti rest house is composed of mica schist quartzite granitoid with hornblende overlain by rich garnetiferous mica schist with garnet of euhedral shape. This is followed by injunction gneiss with fine grain mica schist. A detailed observation of about a kilometre of section before reaching the Jipti rest house (2700 m) is represented by the following section from bottom to top. The quartz mica schist with quartzite, arranged along the schistosity plane (30 m thick), continue to become porphyritic garnetiferous mica schist again (about 50 m) followed by quartzite (7 m) and overlain by aplitic garnetic intrusive injuction with big crystal of hornblende (thickness 15 cm) within the porphyritic granodiorite–gneiss.

The traverse between Jipti rest house and Malpa has been nicely sketched by Heim in 1936 and a zone of tight isoclinal folds has been described before reaching Malipa rest house (2707 m). Garnetiferous mica schist zone continues down to the track, until we cross a tectonic contact starting with the calc-silicate interrelated with quartz mica schist and micaceous quartz schists assuming (Figure 88) tectonic deformation character with repeated isoclinal folds. The quartzite assuming a schistose character are interlayered in them schist showing tight isoclinal character on the way to Malipa in the right bank of the gorge, *ca*. 3 km from the Jipti rest house. The garnetiferous quartzite mica schists and the quartzite are intruded by a dyke of pegmatite of *ca*. 5 m about 50 m above the track. The pegmatite dyke is situated in the right bank of 50 m of water fall which has to be crossed by a temporary bridge. The zone of garnetiferous mica schist intercalated with quartz schist and quartzite continue up to about 1.5 km until the kyanite bearing mica schist appear in fallen boulders intruded by younger aplitic granites. This zone is followed by pure quartzite horizon up to the Malipa Rest House.

The weather in this part of Himalaya in the month of July is very unpredictable, and has virtually all the time to face clouds and rains hampering and disturbing the mapping work. It has reported by a local man that the temporary cross bridge of about 1.5 km onward from Malipa has been completely washed away due to heavy downpour, and so were forced to retire in rest house. Next morning with the help of porters and the gangmen belonging to Public Works Department, somehow we managed to cut big wood logs and erect a temporary bridge to help cross pony without supply load.

The kyanite and garnet bearing quartzite assume a schistose character, continuing for a thickness *ca.* 1000 m before the mylonitized garnetiferous mica schist zone appear (*ca.* 100 m) followed by a typical porphyroblastic gneiss in the middle of this rock type. We had to cross over the damaged temporary bridge and about a kilometre further ahead a dyke of aplite–pegmatite start intruding kyanite-bearing porphyritic gneiss succeeding a zone of migmatization. Before reaching Lamari village porphyritic gneisses start appearing with abundant of calc-silicate complex. The relationship of these calc-silicate rocks and gneiss intruded by tourmaline granite and pegmatite has been sketched in Figure 98. The quartzite opposite the Lamari tea shop exhibits conspicuous interrelation with calc granulite and mica schist which in turn are intruded by young granite (Figure 98). It would be interesting to recall here that such marvellous fold (Figure 38) with an axial plane dipping 60° NE and plunging towards the southwest has been recorded by Arnold Heim (Heim and Gansser, 1939a, plate IX, photo 12). The hornblende bearing intrusive pegmatite appears in profuse volume hence onwards in the country rock of quartz fluorite mica schist, biotite schist, and calc-silicate rock, as described in the last paragraph the earlier phase of granitic intrusions as dyke and sills with its typical light colour whitish character, looks to be hanging in the dark biotite schist and calc-silicate crystalline complex before we reach Budhi. This reminded me with the granitic rocks type with which we come across at Lapsa in the Gori Ganga valley. Budhi is situated on the steep tectonic zone marked by a pronounced geomorphic feature of deep-cut valley as a tributary to camp from north NNW flank (Figure 88). In the traverse geological map of Kali river valley it has been clearly shown that this deep seated tectonic lineament, to be called as *Budhi Malari Lineament* which is now at an angle of 35° parallel to the schistosity plane towards the northeast. This as a whole could be called the Tethyan Thrust (TT) to be traced all along the crystalline Tethyan sediments boundary. It represents the similar zone which we have been observing in the Pindari Gori Ganga, Darma valleys as Pindari schist, Rilkot and Baling schist respectively. This has been a crucial junction to ascertain whether this zone is to be regarded as passage zone between highly metamorphosed Central Crystalline with Tethyan sediment, or a product of dynamothermal metamorphism of basal sediments of Tethyan sequence. If we look at Figure 88, it is clear that the entire Martoli horizon is missing above the tectonic contact over the Chhiyalekh climb and suddenly the dolomitic–quartzitic horizon of Garbyang appears physically overlying garnetiferous mica schist overlain by porphyritic biotite schist in the northeast of Budhi village (Figures 13 and 88).

Crossing down the Chhiyaklekh (*lekh* words in this part of Himalaya is used for the high pass *viz.* Lipu lekh etc.) we come down in a wide broad valley of dolomitic rock with phyllitic partings. After seeing higher contortion and higher degree of deformation imprinted in minor folds we see now major fold of larger dimension with a general trend of axial planes parallel to the major thrusting movement plane along the Budhi tectonic lineament of Tethyan Thrust. The Garbyang village is situated on the fine fluviolacustrine deposit of Pleistocene and Recent periods. In the western part of the village and fold and monoclinally dipping blue dolomitic limestone assumes steeper dip. With flexure and parting as shown by the typical character of limestone under stress and strain condition. The dolomitic limestone is in general bluish but after the Garbyang village but it assumes a rosy colour. Garbyang village is mainly habited by the *Bhotias* tribes of this region. It is situated at the height of 3800 m is facing a serious problem of slumping and cracking of the houses. The roads all showing the cracks of fresh landslides. The whole region is slumping down towards Kali river due southeast. It is worth to point out that Chhiyalekh height in the post Pleistocene time must worked as a high dam to block the flow of Kali river, forming a huge lekh in this Garbyang rock horizon of Tethyan zone. The varvite of fine silt and clay deposit in the lake subsequently formed as a flat plain after the breaking of high Chhiyalekh dam it become a fertile plain for habitation and cultivation. Now the *Bhotia* tribe inhabiting this village grow barley, wheat and other crops. Due to the seepage of water from the northwestern side below the varvite accompanied by the constant cut by Kali river current the whole region of fluviolacustrine deposits is slumping fast destroying the house, crop fields, and roads. The Government has asked the villagers to vacate their houses and go down to stable land area but the people are still clinging to this Garbyang village.

The remaining part of the traverse continued up to the Gunji village (Figure 99) characterized by more arenaceous horizons, showing no inversion, and sedimentary features of beautiful graded cross bedding and ripple marks (Figure 79). Before reaching Gunji, the phenomena of recent neotectonic movements were recorded in the left bank of Kali river in the traverse section sketched 4 km to Gunji village. This is to be marked that unsorted gravel and fine silty clay show synsedimentational slumping and minor folding in the lower level with the flow of water where on the top a successive the sequence of coarse and fine grain unsorted sediment are deposited. Studying the section up to the village Nepal chu on the right bank of Kuti river (this called *Kuthi Yankti* by local people) and comparing with other stratigraphic column in this part of central Himalaya, I am convinced that standard work and observation of Arnold Heim remains unaltered so far as the stratigraphic classification is concerned. There is

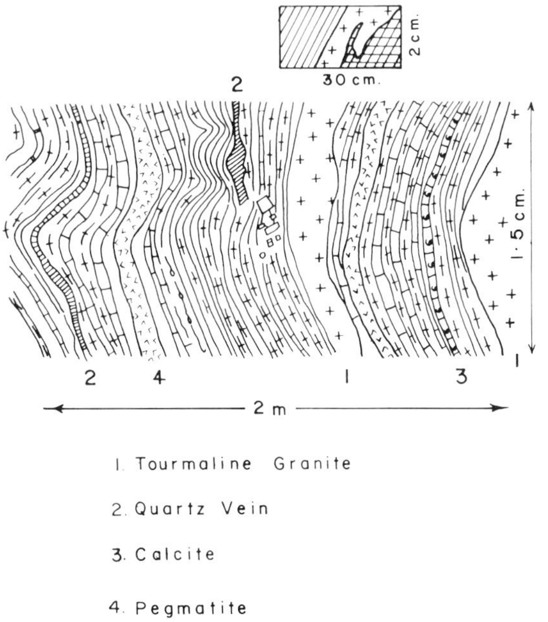

Figure 98. Intrusive country rock relationship in Vaikrita. Locality—Lamari Vill, Kali valley

only possibility to divide the whole Garbyang in three parts upper, lower and middle rather than creating new names within this popularly worldwide known horizon of Himalaya. This Garbyang sequence in the Girthi Ganga valley should be taken as a classical column for the division of Garbyang for the whole basin. Village Nepal Chu in the right bank and Gunji is in the left bank across Kuti serves as a base camp with its beautiful Gunji rest house. The confluence of Kuti with the Kali river in a wide valley with quaternary sediments cut by Kuti river show at least three distinct phases of uplift marking the new tectonic movements.

Here we shall first take up our towards Kuti village (3758 m) the highest village known from Himalaya and later follow the Kali river towards Kalapani. I recall that we could not cross the Jolingkong pass (5494 m) to come down to Kuti valley from Bedang side (Figures 88 and 89) which was successfully negotiated in 1937 by Augusto Gansser. This traverse is not negotiable now-a-days for the explorer of this inaccessible zone of Himalaya. At Gunji we collected fossil bryozoans and bivalves belonging to Shiala affinity, and hence there is a little revision is required in the original diagram drawn by Late Professor Heim.

Traverse upwards towards Nihar (distance from Gunji (3433 m) to Kuti (18.7 km): Gunji to Garbyang (10 km): Gunji to Kalapani (9 km):

We come across at Ronkong village the folded strata of Shiala, Yong limestone, Variegated series, Muth quartzite, Kuling shale and Permo-Trias limestone as shown in Figures 99 and 100, further about 3 km upstream along the left bank of Kuti river at the Nihal the geologic exposures were sketched (Figure 48) depicting the position of Nihal thrust identified by Heim and Gansser (1939a). The rocks in the northern slope are not shown in Figure 100. This is very well documented whereas Shiala Gad Variegated capped by Muth quartzite in the SSW flanks assumes rather steep folds and flexures. Onward without dwelling in detail on the description of this section: the classic work of Arnold Heim and Gansser (1939a) should be referred to as the standard work. At the most of the horizon the Shiala has yielded varieties of Silurian fossils and it would be incorporated in the chapter of biostratigraphy. Now coming back to the Gunji base camp at the confluence of Kali and Kuti rivers, we resume our traverse along Kali river following the right bank.

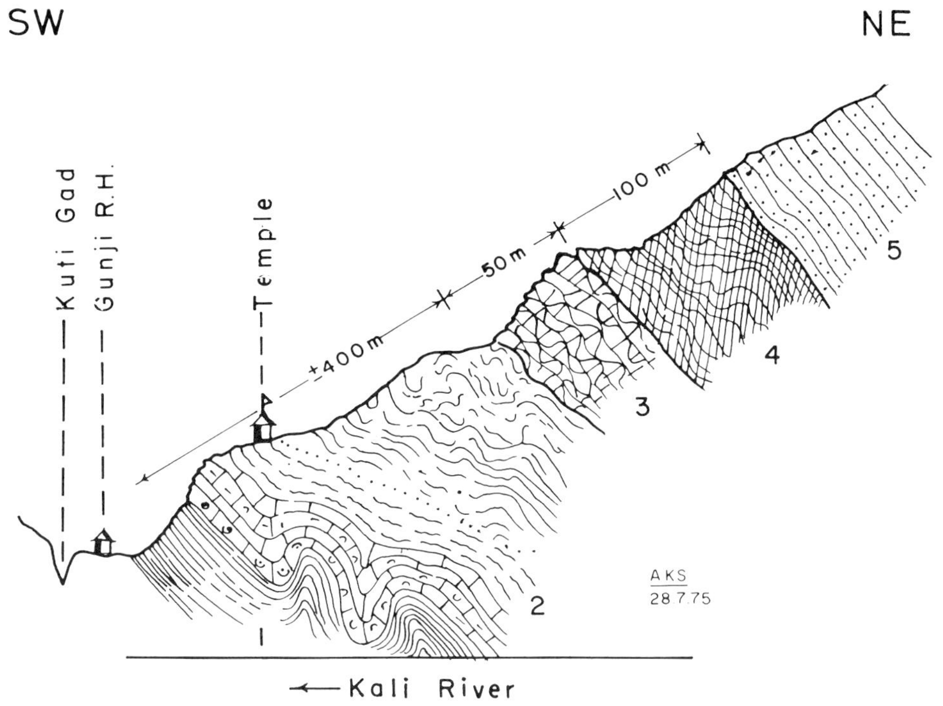

Figure 99. Geologic profile in the right bank of Kali: 1. Upper Garbyang greenish shale; 2. Biostromal limestone and splintery shaly siltstone—Shiala; 3. Biostromal Yong limestone; 4. Variegated calcareous multicoloured purple shale

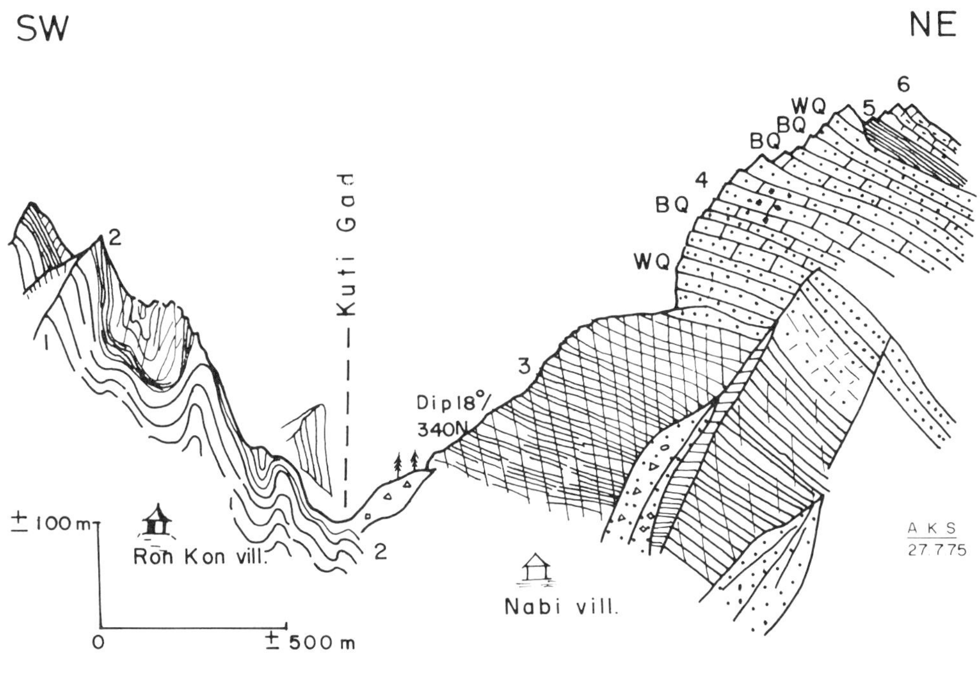

Figure 100. Disposition of folded and monoclinally dipping Shiala near Ron–Kon village in Kuti valley, 1. Shiala formation. 2. Yong formation. 3. Variegated formation. 4. Muth quartzite; WQ—White quartzite; BQ—Brown quartzite; 5. Kuling shale with Productus and Spirifer; 6. Permo-Triassic limestone

As I have mentioned earlier this river marks the boundary of Nepal. The southwest slope in Nepal could be easier to map with our available data to extend simply the necessary bedding plane feature. Here I would like to note that although the Kuti is subsidiary in its role as tributary to Kali river but the total volume of water of Kuti is more than the quantity flowing in the Kali river. Since Kuti flows through the exposed clay rock with more carbonaceous material the colour of the water is dirty and could be easily differentiated at the Gunji confluence.

5.10.2 Gunji–Kalapani Section (Kali Valley)

Fig. 102 Fig. 103

Along the Gunji rest house slope from the northeast, a major section (Figures 21 and 99) was prepared and fossils are collected from them. Here it could be clearly seen that upper Garbyang shale is clearly followed by a biostromal limestone, splintery shale and silty stone with limestone band of Shiala formation. The average thickness of exposed Garbyang with Upper Shiala along the slope would be *ca*. 400 m whereas the biostromal limestone belonging to Yong reefal limestone is 50 m thick underlain by 100 m quartzite purple colour carbonaceous shale with oblique to bedding plane cleavages at the top with brown and white Muth quartzite. Further upstream about 2 km in the right bank of the Kali, the section has been magnificently sketched by Arnold Heim (Figure 47). I took up the study of the section while traverse along the Kali gorge towards Kalapani. This section beautifully demonstrates the back folds in the Muth quartzite at the core and the successive strata of Kuling shale, older horizons in isoclinal folds riding over younger mostly Kalapani limestone. Here I would like to emphasize that we have collected no *Fenestella* fossils from the horizon of arenaceous shale of Kuti which is also having similar lithology as the *Fenestella*-bearing silty micaceous carbonaceous shale. This Triassic horizon bears typical fossils of ammonite but *Fenestella* was nowhere found. Previously Valdiya and Gupta (1972) might have collected *Fenestella* shale not belonging to this horizon. I emphasize that Heim and Gansser (1939a) would have also liked to accept this horizon belonging to Triassic, overthrusted by isoclinal fold of Muth quartzite with variegated in the core as depicted in Figures 13 and 88. The outcrop in the Pankha Gad Nala in the left bank of Kali near the campsite has been sketched and document as shown in Figure 101. The deposition of strata and folding pattern with their excess inclined to the northeast is to be noted here. Crossing further towards Lipu Lekh following the pilgrim route to Mount Kailash and Manasarover lake could not be possible, because the border has been sealed from Tibetan side by the Chinese border guards.

5.11 Sirdang–Gam Malla Sedimentary Zones (Kali–Darma Valleys) (Figures 13, 88, 102, 103, 104, 105, 106 and 107)

Earlier the description of traverse covering the locations between Dharchula and Tawaghat further up the boundary of Sirdang–Gam Malla sedimentary tectonic window zone at Sosa and Jhamrigaon respectively in the Kali and Darma valleys have been taken up. Tawaghat (1008.9 m) Rest House, situated at the confluence of Kali Darma, was made the base camp for a detailed geologic and tectonic study of Gam Malla–Sirdang Sedimentary Zone confined within the high grade crystalline rocks. This particular zone with its enigma of tectonic setting was first described by Heim and Gansser (1939a) working in the Central Himalayan Crystalline Zone.

The author started his research in this zone with the Wadia Institute of Himalayan Geology (WIHG) Expeditions since 1975. The detailed field work to map this window zone was taken up during the WIHG expedition in 1978 summer again. Especially in the last field season author could spent time to prepare a measured litholog of Darma section (Figure 107).

In the Kali valley (Figures 102 and 103) above the porphyritic gneiss (*ca*. 300 m) zone of Pangu (inclination of foliation 35–40° NE) at Jiunti an intermediary zone of quartz biotite chlorite sericite schist intruded with series of amphibolitic sills, is followed by a thick (*ca*. 500 m) white quartzite zone at village Sosa just over the hospital building. The tectonic contact zone (see insert detail in SW end of Figure 104) is marked by weathered amphibolite sill and sericite quartzite zone (dip 35° at 45° NE) overlying them. A zone of deformed mica schist (15 m) is observed intervening with the thick massive white sugary quartzite zone which follow subsequently over them. Another thick green body of amphibolite in quartzite is just met about 200 from Sosa village along the western flank of the track to Sirdang. This body has been petrologically described by

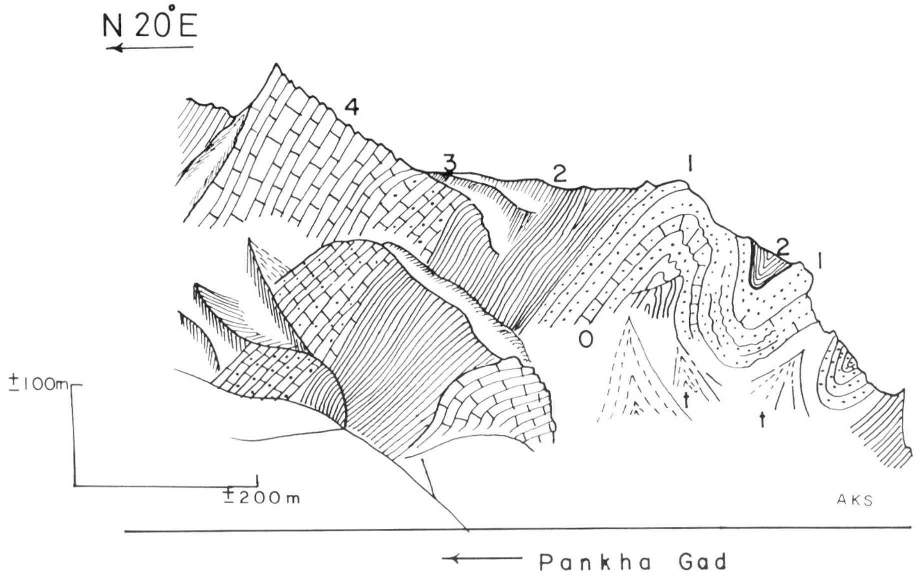

Figure 101. Sketch section near Kalapani camp site. 0. Black limestone; 1. Muth quartzite; 2. Black shale; 3. Chocolate buff quartzite; 4. Kalapani limestone (100–200 m)

Heim and Gansser (1939a) and an attempt has been made to date these rocks and their geochronology has been taken up. Under the microscope this quartz amphibolite–hornblende is highly pleochroic: X = light yellowish green; Y = olive-green; Z = blue green; with extinction along Z/c is 17°. Quartz inclusions are sieve like, the lath of andesine in full in inclusions of hornblende, biotite and epidote. Biotite is abundant. It has an ophitic type texture and points towards its dioritic composition of ortho amphibolitic origin. Such two amphibolitic sills are again observed about 400 m ahead (Figure 103). The quartzite continues further till a tectonic zone of thrust we encounter before the thinly parted limestone with slate intercalation and calc slate zone comes up at the turning point of traverse along N60°E. These calcareous rocks stands steeply vertical at their base over the trail but stays down towards the southwest and further steepen up on the top of the ridge before crossing through the Tithal pass (2764 m) one has to take a short pause holding the breadth and enjoy the fantastic panoramic view of Greater Kumaun and Nepal Himalayan snow covered peaks. The phyllitic rocks folded in open folds and flexures are interspersed with cleavage and parting parallel to the general thrust plane direction dipping at 45°NE. As we descend down towards the Sirdang village pass through white quartzite zone showing inverted graded bedding structural (Figures 102, 103, and 104; Outcrop 6 at Tithal Pass location). Just behind the Sirdang village (3100 m) an inverted anticlinal fold is observed with axial plane dipping 35° towards the northeast having a deformed limestone core. This nucleus of inverted anticline provides us with the data to profound an inverted antiformal structure of the whole of sedimentary Sirdang window zone (Figures 103 and 104). Calc-phyllites also occur in association with the limestone. In these limestones, I had made thorough search on the exposed surfaces along the trail but no stromatolitic structure could be seen. The northern limb of Sirdang inverted antiform is composed of white sugary and greenish quartzites intercalated with chlorite schists and amphibolitic sills, followed by thick (*ca*. 500 m) micaceous graphitic slate to be precisely called as graphitic–sericite biotite–phyllite. Small scales of biotite together with sericite form minutely folded layers near Sirkha village, and sericite calc-slate are also observed (*ca*. 5 m) further towards Sirkha Rest House (R.H.). The section after Sirkha R.H. is represented by Pinkish marble (*ca*. 200 m) dipping 60° due 45° NE, overlain by white quartzites emplaced with two thicker amphibolites sills (*ca* 15–20 m) as we turn to climb Rungling top (3646 m) from Samari village. Again a prominent thrust zone (MCT) appears inclined to the northeast at flatter angle of 30–35° with muscovite–biotite gneiss and paraschists as allochthon.

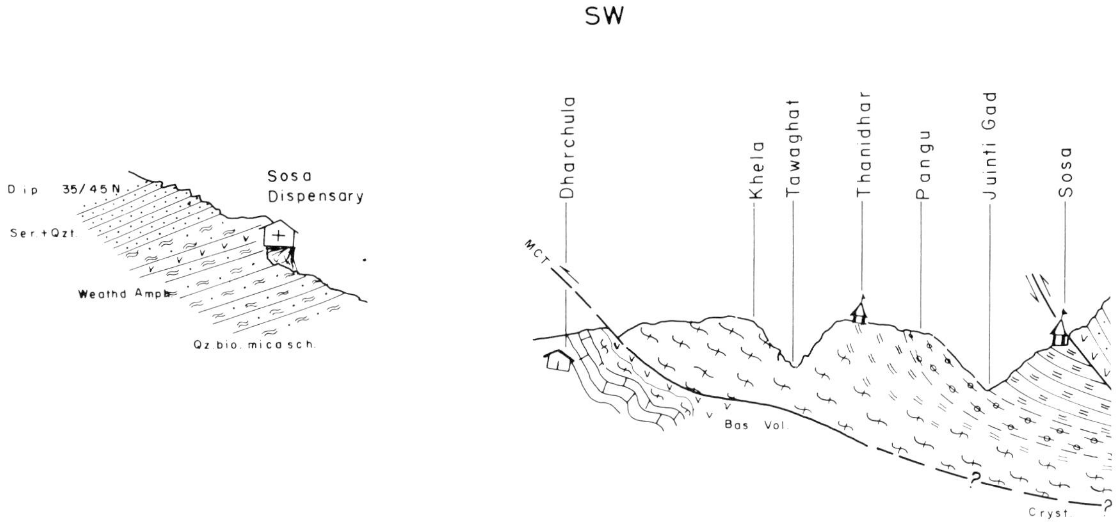

Thus to summarize this section it becomes apparently clear that we are dealing with an inverted antiformal tectonic window with their core at Sirdang village. The southeast and northeast wings of this structure is systematically followed, but the setting of lithounits have been subsequently deformed, refolded, and faulted within the framework of major structure. The pinkish and red marble zone of northern limb is caused due to dynamo- and contact-metabolism of the MCT and the thermal effect of emplacement of amphibolitic sills along the thrust zone. On close examination of this traverse, an important conclusion emerged that the tectonic contact zone at Sosa should be envisaged as a reverse medium angle fault rather than the tightly folded MCT as thought earlier (Fuchs and Sinha, 1978; Valdiya and Gupta, 1972). A geologic sketch profile (Figure 104) has been prepared to illustrate this new interpretation. The fact is substantiated by presence of a brecciated zone near Jhamirgaon in the right hand wall of Darma Ganga (Figure 105).

After completion of this traverse, the expedition party followed the Darma valley from Tawaghat to make a closer study of the similar zone exposed in the gorge with particular attention of the southern and northern contacts and whether its eastern extension joint with the main calc zone of Goriganga valley is exposed over the Belchidhura pass (3850 m), the water divide between Darma and Gori rivers. Our aim was to cross this pass and go down to Munsyari via Bona (Figure 106), tracing this zone across 4324 m ridge striking N–S, but it remained unaccomplished due to inaccessibility.

Traversing along the left bank of Dhauli river about 10–50 m above the bed we passed through similar biotitic gneiss with quartz–feldspathic augens of rather smaller size with minor percentage of blue quartz. The schistosity plane is dipping at 30° to 45° NE. After the Chirkila village at a point opposite PWD Gang Hurt (6.2 km from Tawaghat milestone) is situated the contact of biotite–gneiss with the mica schists of Thanidhar type in Kali valley. Their foliation is dipping 40° due 45°E. The mica schists zone continuous up to 4.5 km before reaching Khet. And henceforth the porphyritic gneiss of Pangu type starts having more thickness in comparison to Pangu exposure (*ca*. 1000 m to 3000 m). At Jhamrigaon village the contact zone of porphyritic gneiss overlain by quartz feldspathic gneiss (50–100 m) and at contact quartz mica schist is observed (Figures 107, 108 and 109). As sketched in documented Figure 107 and measured lithology on millimetre graph paper this tectonic contact bears fact that there is the presence of marked tectonized mylonite zone (2 m). Quartz-gneiss rock with radiating sillimanite crystals is followed by intervening 12 m thick mica schist zone. White sugary quartzite with basic sills physically overlie them constituting the southern inverted limb of Sirdang–Gam Malla antiformal structures. Apparently it seems that the thickness of mica schist zone as observed in the slope between Juinti and Sosa in the Kali

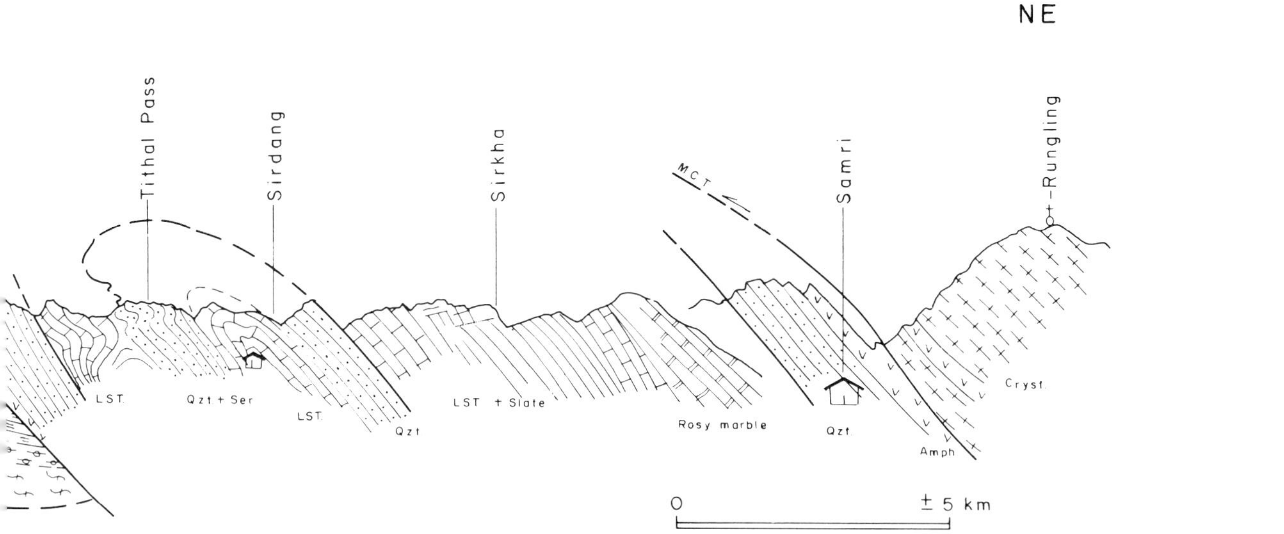

Figure 104. Cross-section along the Kali valley between Dharchula and Rungling showing interpretation of reverse fault at Sosa and inverted anticline of Sirdang

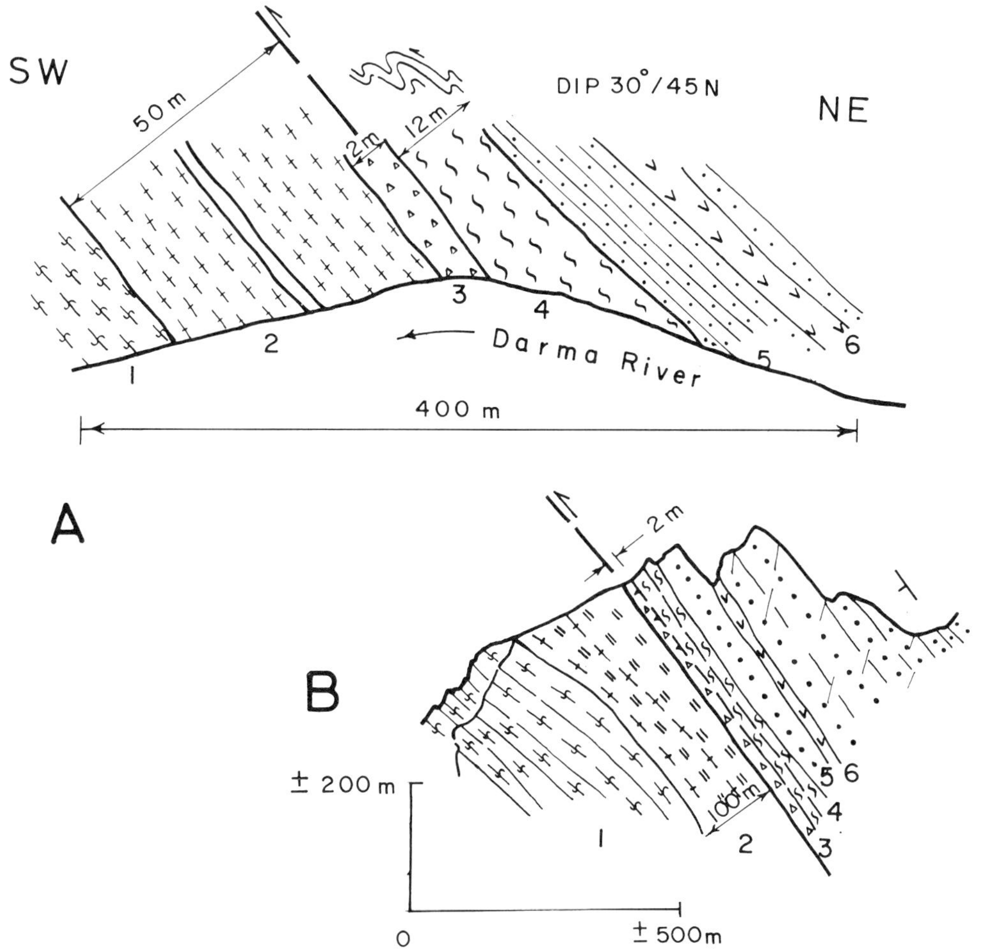

Figure 105. Detailed sketch showing the tectonic contact near Jhamrigaon in the right bank of Darma A and B; 1. Porphyritic gneiss; 2. Quartz–feldspathic gneiss; 3. Tectonitized quartz–gneissose rock with sillimanite; 4. Mica schists; 5. Quartzite showing inverted sedimentary structure; 6. Amphibolite

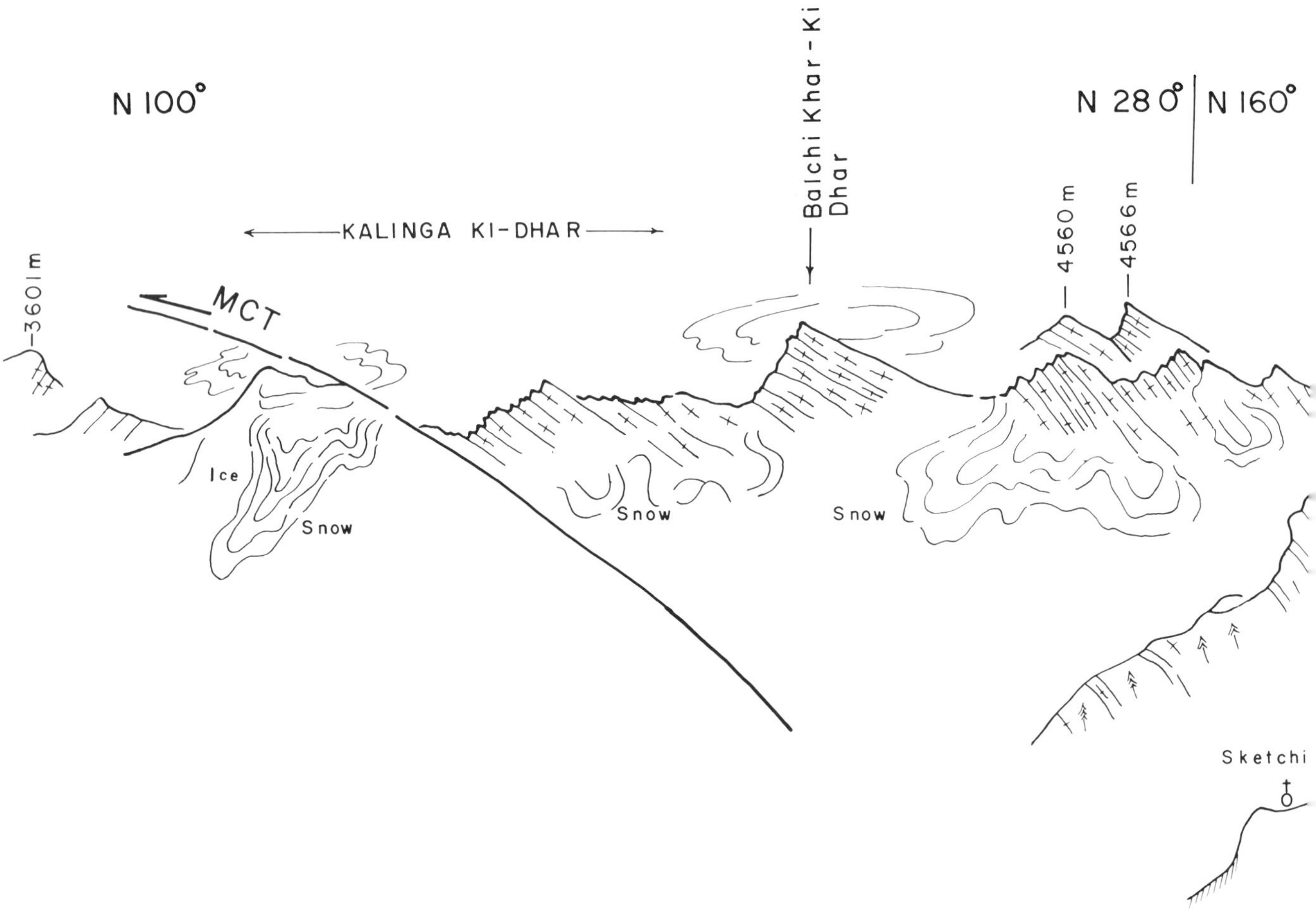
N 100°
Balchi Khar-Ki Dhar
N 280°
N 160°
KALINGA KI-DHAR
4560 m
4566 m
3601 m
MCT
Ice
Snow
Snow
Snow
Sketchi

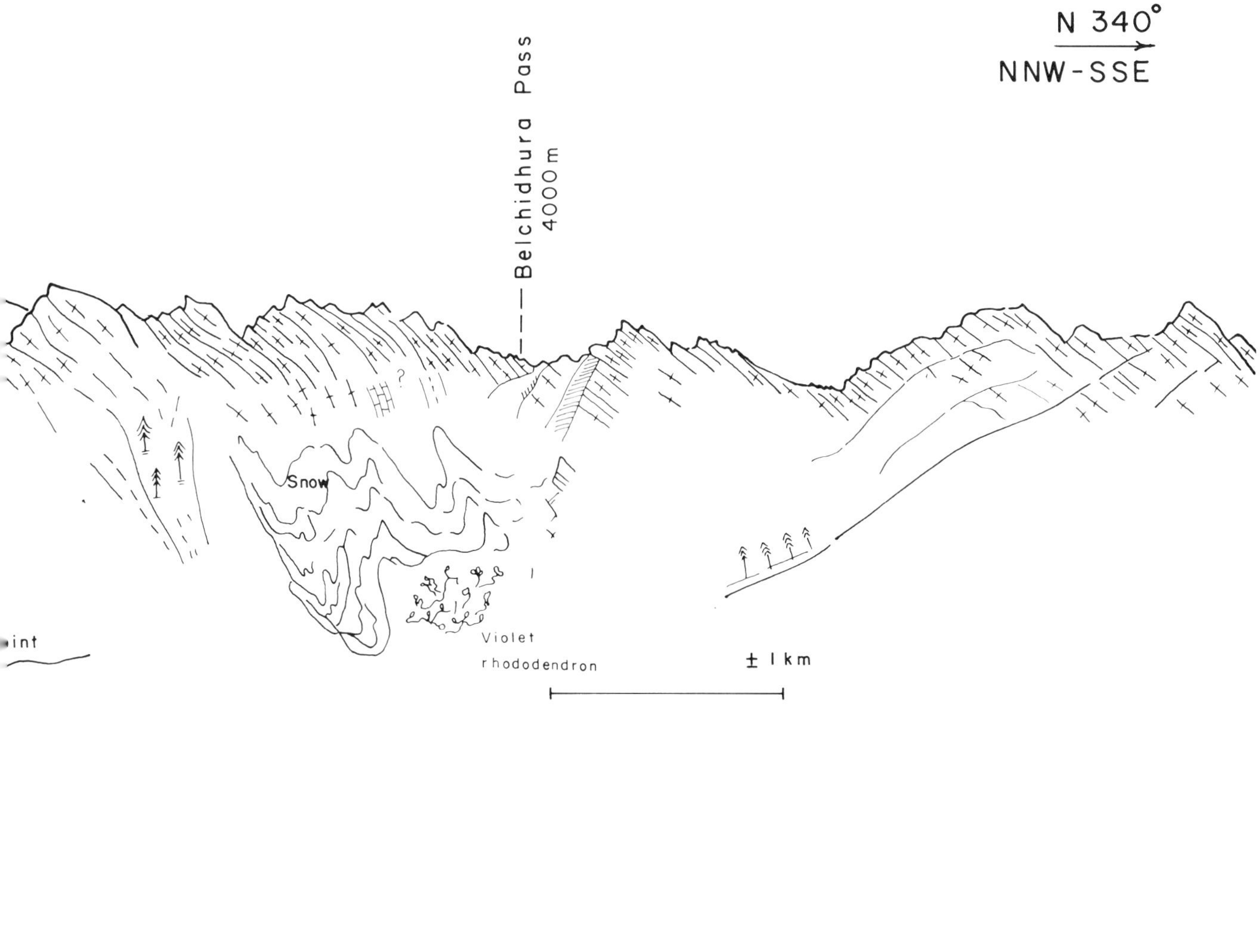

Figure 106. Sketched geological cross-section showing the flat dipping continuous crystalline sheet overlapping the sedimentary zone of Sirdang–Gam Malla

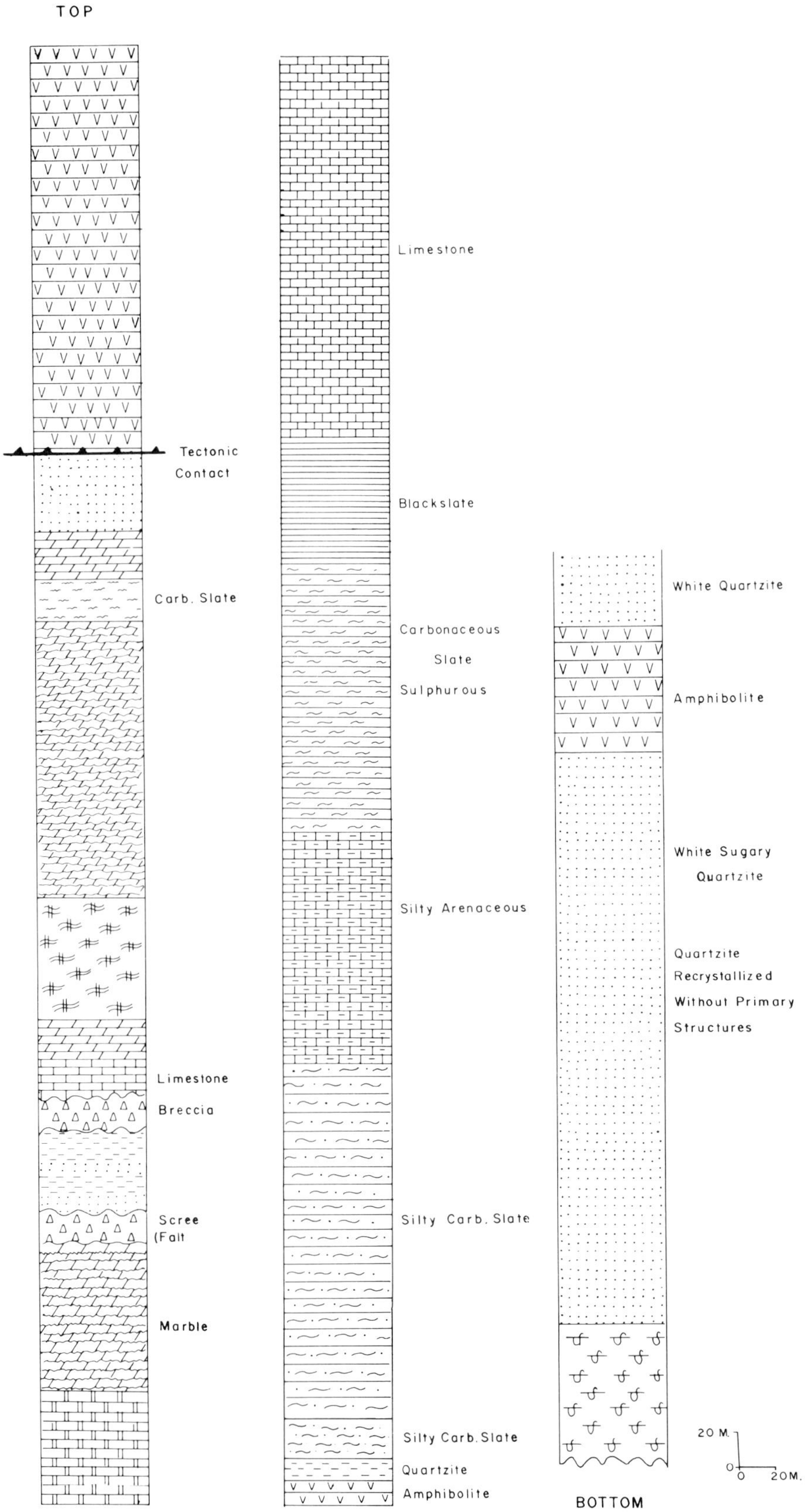

Figure 107. Litholog measured along the Darma valley of sedimentary Gam Malla–Sirdang zone

valley (Figure 102) is appreciably squeezed along the Sosa–Jhamrigaon reverse fault movement of Sirdang–Gam Malla zone against competent crystallines.

The inverted anticlinal structure is carefully documented in the map, section and measured litholog section in Figure 102 and further description is uncalled for. Before reaching Nyu a reverse fault has been observed along the right bank wall of gorge cutting the dolomitic limestone against phyllitic rocks. At Nyu a prominent thrust zone with 40–50° inclination due northeast is observed marked with emplacement of amphibolite over quartzite of northern limb, and henceforth the gneissose rocks follow.

Now to ascertain the western extension of this sedimentary zone traverse was continued along the Sumdum tributary from Sobola upstream towards Sumdum (Figure 102). Traverse was extended up to the Belchidhura pass (4324 m) water divide and it was confirmed that the MCT takes a sweeping curve along the spur of Balchikhar *Ki dhar* and its enclosure could be seen near Hartola Tal, the lekh in Kalinga *Ki dhar* abutting against the Sosa–Jhamrigaon reverse fault (Figure 106). I could document the features with the traverse to Belchidhura pass (4000 m), but could not venture to climb the Hartola Tal area due to extremely inclement weather of snowfall and hailstorm. The sketching point (♁) has been shown in Figure 106 and sketch is self explanatory of extension of crystalline rocks at a very low angle covering the window like a 'roofing thrust sheet'. Thus it could be established that this is closed tectonic window rather not an open zone joining with calc-quartzite of lesser Himalaya. There is much scope for further research in the tectonic window. To my mind it seems that the roof-like crystalline thrust-sheet must be eroded in the northwestern flank of Belchidhura exposing the part of elongated extension of sedimentary zone near Bona village. However, on the basis of observations in the Gori Ganga valley it could be concluded that the sedimentary zone gets tectonically squeezed and definitely in the left bank of Gori Ganga, since there is no sign of such zone in the right bank. Looking over the Belchidhura pass towards the Gori Ganga valley one may have an overview of another pass of Giundhura after crossing the steep down slope of Bhyu Dhunga. This traverse taken on the 30 May 1978 would remain memorable day in my lifetime, not only for the discovery of geologic fact but how much difficulty I had to face with my colleague, late Dr A. K. Khanna, in climbing through the thick bushes of white and purple rhododendron during hailstorm.

CHAPTER 6

Synthesis of the Problem of Structural–Formational Zonation, Thrust Structures, Crustal Evolution, Compression and History of Geological Development of Central Himalaya

6.1 Tethys Himalaya

The Higher Central Himalaya, where the author accomplished his major research, has been recognized as typical structural facies zone since the later part of last century. It has been named as Tethyan Himalayan Zone constituting pretty thick 6–13 km mainly marine sediments and partly of continental type. The sequence spans almost entire Phanerozoic time. In this sequence the earlier investigators had found a lot of similarity in their group and formational affinity with their counter part in Alps. In the process of author's research there has been established new biostratigraphic zonations and new formational members with characteristic lithology and faunal assemblages have been established.

Perhaps, Ashgirei was the pioneer geologist to point out that considering all other aspects of the rock complex in the Tethyan zone of Himalaya, it should represent the typical geoanticlinal zone. He takes the opportunity to refer the unanimously agreed thinking of alpine geologists that usually there is either absence of or badly preserved biostratigraphic zonation amongst the sequence, occupying central, axial, eugeosynclinal zones in geosynclines, and even in the adjoining to eugeosyncline the miogeosynclinal depressions (for example Penninian and Peumotian sources in western Alps). Such picture is exhibited by the major part of tectonic zones of Southern Himalaya.

In contrast to the Southern Himalaya, in the Tethys Himalaya we come across a large number of different groups of rocks with typical marine neritic and sublittoral fauna, having often much less thickness. Amongst them there are frequent gaps and breaks in the deposition sequence of entire rock group. For example, many stages of Carboniferous and Triassic are absent, although in the beginning it starts with a complete conformable sequence. Such features supports to consider the Tethys Himalaya as fossil geoanticlinal built up, and rather not as peripheral miogeosyncline. However, the Tethys Himalaya from both the north and south sides is bound by real geosynclines and zones of deep-seated faults.

It is being elaborated that the concept of a Precambrian Tibetan Slab (*Dalle du Tibet*) underlying the Tethys Palaeozoic sequence (Lombard, 1958; Gansser, 1964) is untenable (Shackleton, 1981). The structures and metamorphism now seen are Tertiary, whatever the age of the sediments themselves.

The northern periphery of Tethys Himalaya is limited by so called Indus–Brahmaputra (Tsangpo) 'Suture' (Gansser, 1964). This structure represents a colossal deep-seated fault named as Gilgit–Dras–Darchen deep-seated lineament zone by author (in Sinha and Jhingran, 1977) on the base of which 'at least from Permian to Palaeogene' developed comparatively a narrow zone of eugeosyncline. According to classification of Schuchert (1931) this may be called mono-geosyncline. However, the terminology 'mono-geosyncline' is unwarranted in this case.

The second structural–tectonic feature of Tethys Himalayan belt is its *en echelon* framework. W. H. Bucher (1957) observed that *echelon* structures are common distributed in the mobile belts of the earth. Irrespective of the discontinuity and isolated tectonic zones with relation to *echelonement* distribution along the strike, the Tethyan Himalaya as a whole extends as one single huge tectonic strip or belt of more than 2000 km length all along the Himalayan mountain range. This tectonic zone in the System of Northern Himalaya is separately well defined and comparative narrow in comparison to Southern Himalaya which possesses completely a different structural facies character.

Many tectonists in recent years consider that Himalayan mountain system has been formed due to collision of Asian continental plate with the Indian subcontinental plate (Argand, 1924; Gansser, 1966; Dewey and Bird, 1970; Dewey and Burke, 1973; Le Fort, 1975, and others). Taking into account the palaeomagnetic data and oceanographic research in the Indian Ocean a fast northward drift of large magnitude of the Indian plate is deduced for the late Mesozoic and early Cenozoic (Klootwijk *et al.*, 1979; and others). During the Cretaceous the Tethys ocean between India and Asia started to close, and heralded the deposition of flysch accompanied by the *emplacement of ophiolitic lava* (Sinha and Bagdasarian, 1977; Sinha and Jhingran, 1977; Fuchs, 1981) some 38 Ma age (Molnar and Tapponier, 1975). There exists the view that the subduction of oceanic crust led to anatexis producing the intrusion of the Ladakh—or Tanshimalayan plutons as well as intermediate to acid volcanism in the north (Powell and Conaghan, 1973b, Powell, 1975; Frank *et al.*, 1977; Andrews-Speed and Brookfield, 1981). Initial collision between India and the magmatic arc zone of N–W Himalaya occurred during Late Palaeocene–Early Eocene (Thakur, 1983; Fig. 108). It has been suggested by palaeomagnetic data that between 55 and 38 Ma the above mentioned plates moved together with about 10° of latitude northwards (Klootwijk, 1979). The final intimate collision between Asia and India occurred about 38 Ma age (Molnar and Tapponier, 1975). It has been envisaged that post-collisional compression sheared off a northern marginal part of the Indian Continent and as such the continental Slab was underthrust along the MCT (Powell and Conaghan, 1973b; Le Fort, 1975). The Lesser Himalayan nappe movements would have taken place in post-Lower Miocene. In the Indus Suture Zone (Gansser, 1964; 1980), this and other compression produced the counter thrust in Himalaya. Seismic and gravimetric studies (Figures 51 and 52) affirms that in the Indian plate the continental crust has an average thickness of 30–40 km, whereas the thickness below Himalaya goes up to 80 km with a mean value of 60 km below Tibetan plateau (Qureshy, 1969). So far there is no consensus of opinion regarding the crustal thickness below Tibet. In the contrary view of doubled thickness expressed by Powell and Conaghan (1973b) and Powell (1975), Gansser (1980) explains the enormous concentration of young volcanism in Central Tibet as a 'hot spot', which implies an exceptionally thin lithosphere.

The crustal shortening according to Molnar and Tapponier has been compensated by thickening of crustal layer below Tibet, movements along the transcurrent faults, and the rest of the shortening was absorbed by thrusting in the Himalayan region, especially from north to south on a large scale.

There are indications that facies and faunal affinities cross the 'Suture' to the north. Himalaya and Southern Tibet gave the basis to Crawford (1974) of the idea that the Tibet plateau remained still a fragment of Gondwanaland, and thus Kaila and Hari Narain (1976) propounded the concept of a plate boundary passing through the Tien–Shan zone through China. But the contestant to regard Indus–Tsangpo as plate margin are more in number (Shackleton, 1981; and others).

The Chinese geologists Mu *et al.* (1973) also support the view that the facies and faunal affinities actually cross the Suture zone of Himalaya. They think that there was a close interrelation between the Jolmo Lungma (Everest) region south of the suture line and north, and Central China, especially in the Palaeozoic. This is not in conformity with the plate boundary along Himalayan Suture. There has been advanced some explanation like assuming a series of smaller plates getting broken off from an older larger India and drifted northward to collide with ancient Eurasia (Chang and Cheng, 1973). In accordance with these authors, each of these plates shows a migmatized southern margin, and are bordered by a flysch through along the northern margin associated with an ophiolitic melange.

They think that successive accretion of these plates to Asia occurred in Caledonian, Hercynian, Indo-Sinian (Early Mesozoic), Yenshanian (Late Mesozoic) and Himalayan cycles (Fuchs, 1981, p. 121). Stöcklin (1980, p. 30) states, 'The Central Domain', comprising Tibet north of the Indus Suture 'can be conceived as a mosaic of Gondwana continental fragments'. However, this does not explain the close affinities between the Palaeozoics of the Himalayan region and the northern and central China (Mu *et al*., 1973).

So far the question of width of ocean separating the Asia and India is concerned Gansser (1976, p. 189) estimated it to be 5000–6000 km at the most. The suggestion from Rickard and Belbin (1980) envisaged a revised assembly of continents in Pangaea during the mid-Palaeozoic featuring no large Tethyan ocean and consequently more coherent pattern of early Palaeozoic orogens. The model is closer to the conformity of palaeomagnetic data and includes the distribution of facial and faunal similarities across the Himalayan Suture.

The geology on both sides of the 'Suture' is very complicated. The author had an opportunity to traverse the Karakoram Highway across Himalaya from the Karakoram region up to the Hunza valley during the International Geodynamics Conference, hosted by Peshawar University in November, 1979. The field excursion was attended by a large number of active Himalayan geologists including Professors Gansser, Crawford, Stöcklin, Valdiya, Marussi, and others and there was lively discussion on the spot. After working for more than a decade in the southern zone adjoining to Himalayan Suture and crossing this suture along the Karakoram traverse, the author experienced a quite uneasy feeling (also felt by G. Fuchs, 1981, p. 122) while standing with one leg on the Indian plate and the other on the Asian plate, imagining the welding plane which would have been an oceanic gap of thousand kilometres until the Jurassic, as suggested by palaeomagnetic data. I frankly share the view with Dr G. Fuchs (1981, p. 122) that a continental boundary cannot be drawn within its gigantic structural formational units with a sharp line.

6.2 Aspects of Crustal Evolution and Compression in the Central Himalaya and adjoining Tibet

It has been contended that Himalayan mountain building is the product of collision between Indian and Eurasian plates which began during Eocene time (Figure 108). The collision between the Indian and Eurasian continents is considered to be one of the major tectonic events in Cenozoic

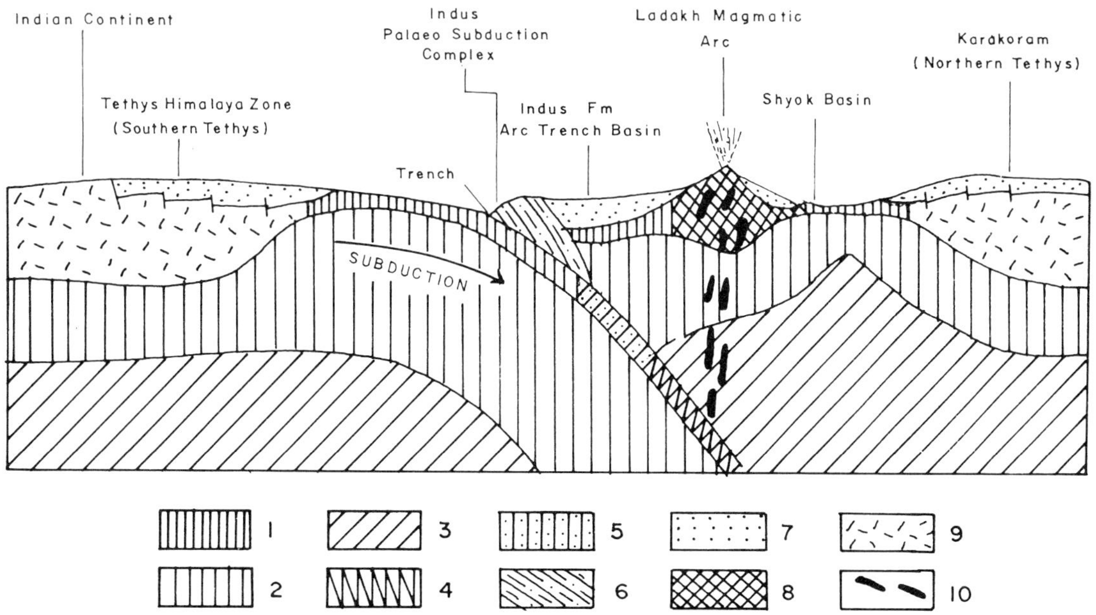

Figure 108. Plate tectonic model of the evolution of Himalaya (after Thakur, 1983) 1. Basaltic oceanic crust; 2. Bottom of the lithosphere; 3. Asthenosphere; 4. Eclogite layer; 5. Transitional zone of eclogite to basalt; 6. Deformed crust; 7. Marine sediments; 8. Crust of the arc; 9. Continental crust; 10. Magma

time. Comparison with kinematic models derived from magnetic anomalies constraints, the age of the collision is estimated to be about 50 Ma (Besse *et al*., 1984). An older age for the initial collision has been proposed from geological evidence and sea floor spreading and it is suggested that it began before the middle Eocene (*ca*. 53 Ma) at least in the Ladakh area. On the basis of palaeomagnetic data after considering the motion of the Indian plate, it has been calculated that the total amount of north–south crustal shortening which followed the collision is in order of 2600 ± 900 km and it has occurred in three episodes. Firstly from initial collision say from 50–44 Ma due to subduction of continental crust (approx. 300 km) and internal deformation (44–37 Ma), and then lateral extrusion (Patriat *et al*., 1984), along large left-lateral strike-slip faults as proposed by Tapponier *et al*. (1982). Crustal shortening in major orogenic belts involves, in addition to folding, one or several modes of deformation: subduction of continental crust in the wake of pre-existing oceanic subduction, and sideways escape of crustal blocks along major strike-slip faults. A direct measurement of the amount of shortening is often not possible and each method has remained controversial in the case of Himalayan orogen and associated Tibetan plateau. So far palaeomagnetism has remained the only indirect means to obtain the quantitative estimates. Large-scale continental shortening followed firstly along the original GHS (Indus–Sangpo Suture) Zone and then to the south of the suture along newly developed major continental thrusts such as the MCT, the MBF in the south of High Himalayan mountain chain, and the TT in the north (Besse *et al*., 1984).

It has rightly been observed that comparatively small amount of shortening and deformation of the surface sediments has taken place north of the Indus Ophiolite. This is in sharp contrast to the huge thrusting movement and crustal deformation in the southern Himalaya (Tapponier *et al*., 1981). A 300 km long north–south seismic section has been obtained by Hirn *et al*. in 1983 (Allegre *et al*., 1984) using critical reflection profiling on the Mohorovicic discontinuity complemented by fan profiles. Such magnetolluric and seismic explosion data indicate that like some orogenic belts in other parts of the world the Tibetan plateau is also characterized by obvious layered structure, and the depth and geometry of the Moho discontinuity are distinctly varied along the N–S direction. The section reveals a thick crust (between 50–70 km) with a complex structure. An interesting feature is the 20 km step in the Moho precisely below the surface trace of the GHS which is consistent with strike-slip reactivation of a major thrust contact. Crustal repetition occurs along parts of the section, particularly north of Yangbajain; in Tibet where two superimposed Mohos are observed with the lower crustal layer being approximately 20 km thick. Another 20 km step in the Moho, very similar to that has been observed below the GHS near Indus. Thus, the Tibetan Moho displays a complex topography with slopes and faults and in places implies in the Gangetic plain of India. 300 km width of crustal repetition by continental thrusting (Allegre *et al*., 1984). In the southern region of Tibet the crust gets thinner and gradually, towards south, and reduces to about 38 km and region between the GHS and the Gangetic plain represent the transitional collision and compressional belt.

Three structural layers have been identified with different thickness and composition along the aforesaid profile (Li Tingdong *et al*., 1984).

The uppermost layer represent 30 km thick sialic crust, including covering strata and basement. At the base of it there commonly occurs a 5 km thick low-velocity and low-resistivity sublayer, which is gradually filled up to the south and named as 'intracrustal asthenospheric layer'. The middle layer represent a lower-crust simatic structural layer with an average thickness of 35–40 km in Himalayan region. The lowermost layer is a mixed crust mantle layer occurring north of the GHS with an average thickness of 25 km showing the property of mixed crust and mantle material. It has been observed that the crustal shortening and thickening in the Tibetan region have been brought by motions of these three crustal layers in different manner. The sialic layer of the upper crust is thickened by a series of imbricated nappes along with brittle listric thrusts. The increase in the thickness of the lower simatic crust is mainly caused by plastic folding or ductile shearing. The intracrustal asthenospheric layer constitutes a décollément surface between the upper and lower crusts, through a layer by layer deformation and by converting the upper crust into imbricated nappes (Li Tingdong *et al*., 1984). It has been now widely accepted that the structure of High Central Himalayan region is associated with a major intracrustal thrust, the MCT with the Northern Himalayan crust thrust southwards. It has been suggested that there may be as many as two continental collisions and six continental thrusts between 26 and 33°N longitude. A crude estimate of the surface and volume loss in Tibet over say the past 150 Ma can be made over a 2500 km long front from Pamir to Assam syntaxis. It seems clear that since the cessation of subduction 45 Ma ago along the GHS there has been no oceanic lithosphere left between India and Asia. Yet intracontinental motion continues on the MCT and the TT around 20 Ma and now on the MBT, at a rate of about 2 cm yr^{-1}. And the average total motion between India and Asia is 5 cm yr^{-1} (Tapponier *et al.*, 1981; Lyon-Caen and Molnar, 1983; Molnar, 1984). Northward drift of India continued since Jurassic and the subcontinent was subjected to considerable tensile stress since 110–60 Ma with the flow of Deccan trap (Allegre *et al.*, 1980). Thus it could be summarized that there seems to be a steady southward progression of thrust faulting from GHS since 50 Ma through the MCT, MBT to MBF.

6.3 Structural Facies in Southern (Higher and Lesser) Himalaya

It is traditionally written in the geological literature that the major part of the Southern Himalaya is occupied by the Lesser or Lower Himalaya. Only the northernmost part of Higher Himalaya, mostly constituted by central crystalline geomorphotectonic unit, is included in the Southern Himalaya. As mentioned earlier this division of Himalaya into Southern and Northern zones is adopted in this work on the basis of the Main Axial Deep Fault zone, coinciding with the tectonic contact between Vaikrita and Tethyan zone sediments.

The geologic framework of the Southern Himalaya remained obscure until today because of poorly studied palaeogeologic setting due to barrenness of fossils for dating the sequence. Moreover, the whole zone has undergone repeated thrusting obliterating the original imprint of geological history. Thus the series of basic geologic questions, related to the actual dividing geological boundary between southern and northern part, are only now getting clearer with the finding of new fossils from different horizons, as well as the geochronological dating by different methods.

Although the main theme of this work is about the Higher Central Himalaya, without touching with the problem of structural facies zones of lesser part their continuation with higher zones would not be possibly understood.

Keeping in mind the restriction of area to be described in this volume of monograph, the area adjacently south and north of the Main Boundary Fault (MBF) known in Himalayan geological literatures as Siwalik Molasse zone and Krol belt would not be discussed. Particularly, the Simla Krol belt geology has been brilliantly described in the works of Auden (1934, 1937, 1970), Bhargava (1972, 1976); Fuchs and Sinha (1974, 1978); Sinha (1980a); and others. It is to be borne in mind that only this belt is a typical miogeosynclinal comparatively narrow tectonic zone (50–150 km wide, length about 350 km; area—40,000 km^2). The autochthonous base is general in the Simla Krol belt is constituted by flyschoidal black Simla plates. Although nowhere their base is exposed, but it is estimated to be about 2–3 km thick. The age of the autochthon Simla on the basis of stromatolites (Sinha, 1977b; Sinha and Raaban, 1979) and acritarchs (personal communication Dr Mme Nina Volkova) is inferred to be late Precambrian to lower Cambrian (*Ilicta deobanica* f. nov., Sinha and Raaben, 1979). In this very belt localized the peripheral graben of great geosynclinal system of Southern Himalaya. This geosynclinal regime got its inception perhaps in the Lower/Middle Palaeozoic with the deposition of terrigenous, possibly marine sediments with phosphorite horizons and conodonts in the basal layers (Azmi *et al.*, 1981). Carbonate layers are intercalated with shale of red and green colour enriched in coccolith of Oxfordian to Maestrichtian and younger age in the Krol B horizon (Sinha, 1975b) identified by Dr (Mrs) Olga Dmitrenko. The thickness (5–8 km) of the Simla–Krol belt sequence is partly beheaded by not large thrust sheets, and in the Dehra Dun area there has a widespread covering klippen of Chail (Chandpur) thrust sheets from north exposing the Simla slate basement with unconformable capping of nummulities (Figures 7, 8, and 10).

The zone of the Simla–Krol peripheral geosyncline is separated from main geosyncline of Southern Himalaya by *en echelon* positioned deep-seated lineament faults of second order. Autochthon of Southern Himalayan main geosyncline is mainly constituted by thick predominantly marine carbonate–terrigenous complex having different local names from northwest to southeast. In the northwestern part the autochthon is constituted by thick (2000 m) carbonate complex of Shali–Deoban intercalated with slates and quartzites. Thickness as a tendency increases in the southeastern part. The age of this complex exposed in the Larji, Shali and Deoban tectonic windows (Figure 10) has been established by author with the help of Dr (Mrs) M. Raaben to be not younger than 1000 Ma (although *Ilicta deobanica* f. nov. appears in lower Cambrian, Sinha and Raaben, 1979), but in general the typical *Colonella–Conophyton* assemblage and other identified forms are the pointer of entire Riphean up to 1700 Ma (Sinha, 1977b; Ashgirei *et al*., 1975). In the northern and eastern direction in the volume of autochthon increases the amount of metabasites with vesicular amygdaloidal texture showing effusive character (Sinha, 1977a). Such metabasites are known from west to east in Mandi, Darla, Simla, Rampur, Yamuna valley, Chamoli–Karnaprayag and Naini Tal–Bhowali area.

In the Pithoragarh area of Tajam belt stromatolites have been studied by Valdiya (1980b), Bhattacharya (1981), Kumar and Tewari (1978) and others, and the forms studied from the carbonate layers intercalated with thick quartzites of Berinag and slates, are identified to belong to Middle and Late but on the whole a general Riphean period.

Miogeosynclinal character of deeper depression filled up with Riphean sequence in the Southern Himalaya does not warrant any doubt, since there are less metabasites. But in the Northern zone where the volume of basic rock suites considerably increases, and there are grounds for proposing an eugeosynclinal regime and the corresponding conditions. A detailed documentation ofthe section along the contact zone of Deoban limestone and flyschoidal Simla (Damta) formation along Bini Gad representing a rift–graben system of geosynclines in the Southern Himalaya (Sinha, pl. 3 in Fuchs and Sinha, 1978). A study by Valdiya (1980) and the author (Sinha, 1972b) of the Pithoragarh and Larji windows in Kulu Valley respectively shows the inversion of beds associated with isoclinal anticlinal structures, adding to the conclusion that in general of the presence of inverted Riphean structures. This is also the pointer to possible Precambrian or Baikalian phase of orogeny in this part of Himalaya (Sinha in Ashgirei *et al*., 1975).

A considerable two to three fold increase in thickness in the Riphean complex of Southern Himalaya in Chamoli–Tejam area takes place in the direction of adjunction between Southern and Northern Himalayan zones. This vital point has been observed by Auden and Gansser, and it should not be underestimated. There has been a general northward deepening of geosynclinal basin of Southern Himalaya towards its proposed axial position quite within the existing mountain ranges.

Therefore, the lithological composition of the Riphean rock complex represented by the marine, pelitic, and fine-grained psammitic terrigenous rocks, and significant role of thick carbonate formations, are not supporting a platform type of deposition. Moreover, the increased role of metabasite with typical geosynclinal effusive character (not the traps) of considerable thickness, and significant dislocations not related with younger data dislocations (Sinha and Jhingran, 1977; Rupke, 1974), combined with all other related features do not provide basis in support of calling the Southern Himalaya, as is sometimes done, reactivated part of Indian platform.

In the following paragraphs there is a discussion of the arguments against the support of unacceptable hypothesis that the Lesser Himalaya is a reactivated part of Indian platform.

6.4 The System of Thrust Sheets and Lineament Structures in the Southern Himalaya

The first interpretation of thrust structures in Himalaya dates back to the Darjeeling area as early as 1907 by Hungarian geologist L. V. Loczy. In recent years it has been established that in southern territory of Himalaya exists the system of thrust-sheets (Pilgrim and West, 1928; Auden, 1934, 1937; West, 1939; Heim and Gansser, 1939a; Gansser, 1964; Fuchs and Sinha, 1978; Valdiya, 1980b; Sinha, 1972–1981; and others), covering a large surface area more than hundreds of thousand square kilometres, constituted by autochthonous-base units. Now the question arises from where these thrust sheets originate, or in other words where lies the 'root-zones' of above-mentioned thrust system. It has been made clear from the available tectonic interpretation that the Southern Himalayan thrust-system in no way is related to the deep-seated lineament of the Indus Suture (Gansser, 1974b, 1976, 1980a,b; Sinha, 1972–1981; Sinha and Jhingran, 1977), since there is no similarity between the lithological and petrological composition of rock suites in these two regions. Considering all the factors we are coming closer to an important inference which has remained the matter of discussion for all these years of geological studies of Himalaya—that there must be in existence a rather major system of steeply dipping deep-seated fault zone incorporating the 'root-zones' of the thrust-sheet system in Southern Himalaya. Principally this root-zone deployment must be south of Tethyan sedimentary basin.

Based on research experiences in the other mobile belts of Tien–Shan, Greater Caucacus and Altai, it is felt that there is a possibility of the existence of such a deep-seated fault system related with a root-zone also in the Himalaya. In this connection the data collected by 1969 Centre of Advanced Study in Geology, Punjab University, Manali–Leh expedition were taken into account to explain the display of divergence character of disposition of sequence near the contact zone of crystallines with Tethyan sediments, and young intrusive granitic suites of a Jaspa granite-like body. In subsequent years, the traversing by Powell and Conaghan (1973b) was published with a detailed section and elaboration of the issue of the root-zone problem (refer to Fig. 3, Sinha and Jhingran, 1977, p. 53; Figure 37, for comparative picture of the Main Central Axial zone of Himalaya). Thus, it was understood that in this part of Higher Himalaya there existed the root-zone of Southern Himalayan thrust-sheets with a steep-angle fault system. As shown in the comparative Figure 54a, such zone was observed by Heim and Gansser (1939), in Badrinath area and beautifully plotted by the genius geologist of our era, the late Professor Arnold Heim, as early as 1936.

The author, while preparing the documented sketch of different cross-sections through the Central Higher Himalayan Crystalline zone observed another zone of divergence in the upper reaches of Pindar river close to Pindari glacier, 150 km southeast (Figure 70b). In other sections through Goriganga, Darma and Kali rivers this zone is either coinciding with the Tethyan Thrust or other fault system which has been superimposed by late Himalayan orogenic events. Frankly speaking, more close observation and documentation would be necessary to elaborate this problem further.

The concept about the existence of deep-seated fault zone in Himalaya was propounded by author at the Indo-Soviet Symposium on the comparative Geology of Himalaya and Caucasus in the summer of 1975. On the basis of the available data on stratigraphy, tectonics, magmatism and seismicity (Figures 51 and 52) of the region, it was proposed that the deep-seated lineament structures, parallel to the general strike of Himalaya, should be grouped as follows (Sinha and Jhingran, 1977; Sinha, 1977a).

1. Gilgit–Dras–Darchan deep-seated fault (Indus–Brahmputra or Himalayan Suture of Gansser, 1976, 1980);
2. Kylong–Badrinath–Malari Main Central Axial Structural line of Himalaya (Main Axial Zone (MAZ) of Himalaya (Figure 53) (Ashgirei, 1977; Sinha and Jhingran, 1977);
3. Pir Panjal deep-seated lineament with associated Besites (Sinha, 1977a);
4. Main Boundary lineament (Main Boundary Fault).

It is essential to emphasize that the deep-seated fault of the Indus–Brahmputra incorporates the root-zone of thrust-sheets translated to the autochthon towards south over 30–35 km (Fuchs, 1981; Tewari, 1981; and others). Often the thrust-sheet is made up of Cretaceous and Palaeogene flysch. There are 'exotic blocks' of limestone with Liassic fauna of Adnetian facies typical for the Eastern Austrian Alps. Comparatively the nappe of moderate but effective type with 30–50 km amplitude is known the south or northern structural formational zones in Tethys Himalaya.

In the northern boundary of research area von Kraft (1904) investigated in detail the 'exotic blocks' of Malla Johar situated along the Kiogarh water divide. Furtherance of this investigation was carried out by Gansser in 1936 who discovered the Jungwa thrust of ultrabasic–serpentine ophiolite complex of rocks (Gansser, 1964, 1979) and the existence of the Indus–Tsangpo (Brahmputra) Suture extending over 1500 km and displacing the flysch of Cretaceous and Palaeogene age associated with ophiolite, greenstone and amygdaloidal porphyry, spilite, fresh peridotite, diabase and varieties of basic tuffs. Now the author, in his collection from limestone lenses in the flysch of Malla Johar, discovered with Dr N. C. Mehrotra dinoflagellates of Palaeogene age (Mehrotra and Sinha, 1978, 1981). It is a significant discovery to fix the lowermost age limit of northern thrusting, and interestingly it is same as in the zone of Southern Himalaya, where nummulites marks the age limit of thrusting found exposed in tectonic windows (Auden, 1934). The age of the dinoflagellates are also typical Eocene in general. The essence of the concept leads to the inference that it is not the peripheral 'Himalayan Suture' along the Indus–Brahmputra, but the Main Axial Zone in the centre of Himalayan width playing a more significant role in the history of geological development of present Himalayan mountain range, dividing the whole Himalayan mobile zone width into two prominent structural formational (facies) zones: the Tethyan Zone in north and the Southern Zone in south, having an *en echelon* disposition along the entire strike length.

It has been clearly established that there exist four stages of successive and different tectonic units in the southern zone of Himalaya, with evident upward increase in the metamorphism. From the bottom to top these thrust-sheets are named as follows: 1. Krol thrust-sheet; 2. Chail thrust-sheet; 3. Jutogh thrust-sheet; and 4. Crystalline or Garhwal thrust-sheets of Auden (1934). In places the Jutogh and Crystalline sheets are inseparable.

The Crystalline and Jutogh, being the topmost, are characterized by their higher degree of metamorphism and association with a late granitic body in them. They mostly occur as detached klippen showing a displacement amplitude of more than 100 km from north to south. Such klippen (2000–2500 m) are widespread from the Beas valleys in the northwest to Almora crystalline klippe in the southeast and further east in the Nepal territory of Central Himalaya (Figure 10). The roof-zone location of this thrust-sheet is discussed in an earlier paragraph under the Main Axial Lineament zone of the higher central Himalaya.

The next thrust-sheet below the Crystalline is known as the Chail thrust-sheet. It is a thrust tectonic unit is the most widely spread one in entire span of southern lesser zone from Beas in the west (Figures 7 and 8) to Kali valley in the east (Figure 10). It is estimated that the Chail in the Beas valley is as thick as 2000 to 4000 m. This thickness holds good also in Sutlej to Alaknanda valleys. The general area occupied by Chail thrust-sheet spans about 300 km long strike and 100 km across, i.e. 30,000 km^2 in the investigation area.

An exhaustive description of the Chail thrust-sheet geology has been presented by the author and Dr G. Fuchs in their monograph on the Southern Central Himalaya of Garhwal and Kumaun with coloured geological and tectonic maps (scale 1:500,000) and sections (Fuchs and Sinha, 1978). In this work there have been reviewed the existing geological ideas pertaining to this zone, and a new interpretation in correlation with the adjacent region of Nepal and N–W Himalayan zone. Thus, including the entire zone the occupied area by Chail thrust-sheet goes upto 40,000 km^2. Thus, the colossal dimension of Chail thrust-sheet and its early geosynclinal facies characteristic does not warrant any doubt. In the monograph, Dr Fuchs covered the Chail under three units: C_1, C_2, and C_3; whereas the author regarded the C_3 as the true Chail nappe of his description (Figure 10); moreover, the existence of *eugeosynclinal* regime in Southern Himalaya remained substantiated. However, the question is raised—in what period were the Chails deposited? This question has been touched upon earlier in the stratigraphy chapter 3 that it was in the Middle Palaeozoic; the typical uppermost Palaeozoics are completely absent (Ashgirei *et al*., 1981, Sinha *et al*., 1983). This is also supported by Fuchs (1975, 1981, and other works) and author's view is also supported by his concept of the existence of geosynclinal regime in the Southern Himalaya. The miogeosynclinal Simla–Krol belt is clearly defined in the author's work with Fuchs (1978). The description follows in next paragraphs.

There have been geological works (Gansser, 1964) in which the importance of a separate Chail thrust was not recognized and it was mixed up with either the Simla black slate or the Jutogh sheet. In our recent work (Ashgirei *et al*., 1977), the detailed lithoformation character of Chail was analysed and it was established that in Chail there existed tuffaceous acid volcanogenic ingredients like greywackes with albitic fragments and even the grains of albitized volcanic rocks. This character was similarly noticed in Chandpur of Alaknanda valley and Lansdowne area. Thus, it is the clean lithofacial feature of a separate tectonic zone. The black slates of Simla and partly of Chandpur were found to be distributed in a limited area bound by south and north faults of the Simla Krol belts. And beyond the limit of above-mentioned zone which is closer to miogeosynclinal type in its character, there are widespread exposure of black–green and grey–greenish Chail rock-suites composed of, besides the terrigenous material, well distributed lenses of volcanic greywackes with albitized grains of rock fragments plagioclase. So it is leading to the conclusion that the Chail complex including the Chandpur represents a eugeosynclinal formation. This regime persisted further north closer to the Main Central Thrust of root-zone of deep fault belt separating north Tethys and South Himalaya.

This conclusion carries rather important significance, since after the research during last decade, it is firmly established that over the large thrust-sheet of Chail in the northern zone there is persistently present the higher more metamorphosed cover of Jutogh. In the Simla area the contact zone of Jutogh thrust over Chail is always marked by persistently thick (approximately 100–200 m) horizon of graphitic schist showing strong crumpling and crushing effect (Sinha, 1980a). Physically over this zone lies the quartzite showing tight recumbent folding which in its turn is physically overlain by marblized limestone and garnetiferous mica schists in association with other high grade metamorphics.

In the Kulu Valley, as we go up from Naggar (1717 m) upwards to Rohtang Pass (3930 m), there are exposed metamorphics of different grades. In general, the rock suites belong mainly of amphibolitic facies constituting the crystalline axis of Himalaya (50–80 km), as termed by Pande and Saxena (1968).

Gansser (1964) proposed that the crystalline axis of Himalaya is bound in south by the Main Central Thrust (MCT), and attached a very significant role to this zone in the genetic evolution of present Himalaya. He went further in his assumption to propose that, except in the Kashmir area, the sub-Himalaya, Lower and Higher Himalaya are made up of elements formerly connected to the marginal part of the Indian shield. Only in the Tibetan Himalaya, according to him, is there geosynclinal influence with sinking Tethys Sea. But it is to be noted with caution that even in the world of Gansser (1964, p. 252)—'In Sikkim and Bhutan, where the Lower Himalayas consists predominantly of crystallines a clear division between Lower and Higher Himalayas is somewhat arbitrarily traced'. Such is the picture in the west in Beas valley and further northwest. Hence, with a cautious note, the role of the MCT is to be thought in a rather limited sense, considering the overall picture of Himalayan tectonic framework. However, it has been argued by Gansser that along the MCT the thickness of continental (granitic) crust is doubled and hence below the Higher Himalaya the continental crust is 60–70 km thick.

In the Mandi–Kulu–Rohtang section (Pande and Kumar, 1974; Mehta, 1980) the weaker side of Gansser's idea is substantiated by the very thick (4000–6000 m) Chail thrust-sheet. Firstly, such enormous thickness cannot be completely ignored; secondly, the Jutogh thrust-sheets 2000–4000 m position is clearly explained; and thirdly, what is represented by the massif of crystalline granitized rock exposed in the Chandra section as a gigantic wall of a thickness not less than 3000–5000 m? Moreover, there is no substantiative data to support the idea that this crystalline complex continues below the Tethyan unmetamorphosed Precambrian layers, either in Spiti or the Kumaun–Garhwal section, which are represented by the Haimanta or Martoli series. They attain thickness up to 10,000 m (Sinha, 1982), and are succeeded by immense thickness 5000 m of Phanererozoic–Palaeogene flysch.

6.5 A Synoptic Overview of the History of Geological Development of the Central Himalaya

Although there are still only scanty data, with that available and the data pouring in every year in different journals in India and abroad, we are coming closer to a consensus that there is little doubt about the existence of Phanerozoic and Riphean geosynclines (with average thickness of 4–6 km) in the Southern Himalaya. Keeping in mind that the Aravalli orogeny was the last major tectonic event in peninsula India followed by fairly uniform thick sedimentation during the Vindhyan period, in the southernmost peripheral zone of Central Himalaya developed the narrow tectonic basin (20–30 km) of Simla–Krol, (Auden, 1934) bound by graben type structure mainly controlled by the Main Boundary Fault (MBF) in the south (Sinha and Jhingran, 1977; Ashgirei, 1977) between the Indian platform and the Himalayan Geosyncline. This narrow belt started receiving carbonate–terrigenous sediments with profuse development of Riphean stromatolitic biohermal growth and early Precambrian form of acritarchs. The exposures of Shali and Simla type flyschoidal sediments along the truncating border zones as seen near Sataun, Thundapathar, near Bilaspur etc. (Figure 7) confirms this view of a Riphean basement. There has been basic extrusion during this period (Sinha, 1977a). It has to be borne in mind that the depth increase of Southern Himalaya continued towards north, perhaps due to activity along Main Axial Zone of general transition between southern to northern Tethyan zone.

The activity of the Main Axial Zone extends as early as Precambrian since on the both sides of it we come across Riphean period sediments, *viz*. Shali in the south and Martoli in the north. In the southern zone where the autochthonous–parautochthonous tectonic units are exposed average width ranges between 50–100 km at the length of about 350 km. Further west the width increases up to 75–150 km and length further up to 100 km (average area 40,000 km^2). Over this area there is a physical cover of Chail and Jutogh thrust-sheets, but in the window zone the autochthons are exposed with a Palaeogene capping. On both the sides in north and south of Main Axial Zone (MAZ) (Figure 53) there is evidently one Precambrian phase of deformation (Baikalian?). Evidently this is confirmed by deformation and steep angular unconformity above the Martoli group by the Ralam conglomerate in the north. In the southern structural facies zone the isoclinal and inverted fold pattern in Riphean Carbonate sequences (Valdiya, 1980b; Sinha, 1972) are the corresponding Baikalian phase of deformation. The Caledonian deformation record is not evidently imprinted in the southern zone except the age of granitic body closer to 500 ± Ma of Mandi, Jaspa, Naushera and other granitic bodies (Jaeger *et al*., 1971; Bhanot *et al*., 1980, Mehta, 1980; Le Fort *et al*., 1980). But in the northern part there is an apparent disconformity with marked erosion surface over variegated Silurian at the base of the Devonian Muth quartzite sequence.

The Hercynian gap is prominent in the northern zone with the absence of large part of the Carboniferous except some patchy occurrence of *Fenestella* shell in the eastern part of the basin. Interestingly it has to be noticed that the thickness of sedimentary column of Tethyan basin of Central Himalaya increases towards eastern part of basin. Comparing the two columns in Figure 15 it would be clearly seen that besides the Martoli with the thickest column in the Gori valley, the Muth quartzite itself, multiplies its thickness many times. Calculation is obviously done avoiding the duplicating of strata by repeated thrusting and folding.

After the Hercynian gap the Mesozoic sequence covering the Triassic–Jurassic–Cretaceous up to the Palaeogene (Eocene) represent uninterrupted sedimentation with rich faunal assemblage of each biozone. The incompetency of the Triassic Kuti shale between the competent Devonian Muth and the Rhaetic Kioto limestone horizons led to experience disharmonic style in deformation. Evidently the Lapthal series of Lias with an overload of Spiti shale (Portlandian–Tithonian) also experienced a Jura-type open fold pattern. Markedly the appearance of glauconite and radiolarians in the Guimal sandstone and Flysch of the Sangchamalla formation is the pointer to deepening geosynclinal regime towards north nearing the Himalayan Suture where flourished another basin from the Permian to the Trias, giving way to deposition of 'exotic-block' carbonate facies of Dachstein type. But now we are unable to map *in situ* this squeezed in (or subducted) basin of deposition. As described in the stratigraphy chapter, the blocks of these exotics are found as olistostrome within the flysch and some of them like the Kiogarh peaks were also thrust over flysch along with ophiolitic melange. We have as yet scanty data from the adjacent northern area of Tibet across the Indian border but obviously no major tectonic events took place in this zone until the outbursts of ophiolite suites in the Upper Cretaceous followed by northbound large scale thrusting movement of flysch and Jungwa nappe (Gansser, 1964) preceded by folding and faulting. It should be borne in mind (on the geologic map of Higher Central Himalaya; Figures 13 and 88), that recumbent folding in the Bedang–Jolingkong area between the Darma and Kuti valleys were in their second phase were thrusted in schuppen fashion presenting the repetition of strata as sketched in extreme end of Figure 89 along the Bedang–Jolingkong pass section, and in Figure 47, on the right bank of the Kali, folding in a reverse direction with inclination of the axial plane towards the south. This is further complicated by the Nihal thrust bringing the Silurian over the Permian. If we look further north, on the right of the map, we also find that there is a set of parallel thrust zones along the water divide border land of India and Tibet (Heim and Gansser, 1939a; mapped by the author during 1975–1976). Finally, the movement from the north after the Palaeogene (Eocene) flysch caused the oblique and transcurrent faults displacing the squeezing sequence from north to south to compensate the thrust movement causing an arcuate disposition of Central Himalaya. This has also caused the isoclinal fold pattern within the substantially squeezed crystalline zone as seen in the tectonostratigraphic columns (Figure 14). This zone has been reduced to mere 10 km thickness giving rise to widespread nappe like the Almora–Askot thrust-sheet over the autochthons of the Southern Central Himalaya (Figures 8 and 10).

6.6 Geological Framework and History of Development of Main Axial Zone (MAZ) of Deep-seated Fault System in Himalaya (Figure 53)

The deep-seated, very large in dimension, regional faults in the earth's crust, and on the whole encompassing the entire tectonosphere and mantle of the earth, has always attracted the untiring attention of the world's geologists. Especially, after the development of the plate tectonics concept, the zone of deeper tectonic boundaries on our earth is being looked at with the greater attention, though the interpretation, if not based on actual field data, leads to rather preconceived prejudice. It would be in right occasion to recapitulate about the 'tectonic line of Logan', the famous Canadian geologist who established the colossal fault zone in the north of the Canadian Appalachians, sometime in the second half of 19th Century. Similarly many geologists' attention was attracted by gigantic faults in island arcs (William Hobbs, 1950) and in the Mexican cordilleras (Spurr, 1916).

After the classical work of Nikolaev (1959) and others in middle Soviet Asia and the discovery of the prominent structural line in the Tien-Shan mountains, the problem of a deep-seated fault started to gather momentum. This followed the understanding about the relationship of first-order Recent regional fault with deeper foci zone (Zavaritski, 1946). The works of Hugo Benioff (1958) and many others are known in the world of geological science for studies around the peripheral boundary of Pacific Ocean related with deep earthquake foci. Many research papers have been published by Peive (1945, 1950, 1956) and others on this problem of deep-seated faults.

The concept of the presence of a deep-seated fault zone with the modern viewpoint was first advocated by Gansser (1964) in recognition of his 'Indus Suture' line, on the basis of his observation in the Kailas–Manasarovar area during 1936 Swiss Expedition with the late Professor A. Heim. Successively he elaborated the idea with an ophiolite association, a plate tectonics model and the extension of 'Himalayan Suture' along entire Indus–Tsangpo (Brahmputra zone) from west to east (Gansser, 1964, 1966, 1974a,b, 1976, 1980). This first order 'Great Himalayan Suture' extends for more than 1500–2000 km all along the Himalayan mountain range from Gilgit in the east to the turn of Brahmputra in the Nagaland syntaxial arc of the Eastern Indian territory.

Comparing the deep-seated lineament structures of Himalaya and the Caucasus, the author presented his concept about the presence of a deep-seated fault zone in the Himalayan region, and named it the Great Himalayan Suture (GHS), a collision zone with ophiolite. In the same paper the author also propounded the idea of the Main Central Axial Structural Line of Himalaya on the basis of deep-seated divergence in regional structure of the Higher Himalayan rampart (Sinha and Jhingran, 1977). Ashgirei's paper (1975) published by the Slovak Academy of Sciences, Bratislava first envisaged this dimension in a broad preliminary framework.

Now after more than a decade of work by the authors since 1973, in a series of eight expeditions conducted in the inaccessible higher zone of Himalaya, it is becoming clear that such a zone might exist along the Axial Central part of Himalaya dividing the entire Himalayan width into two distinct structural facies or formational zones in the south and north. If there are associated with the 'Great Himalayan Suture' a comparatively modest thrust-sheet and nappe, then within the zone of the MAZ there exists large-scale amplitude (say 100 km), each thrust-sheet on average 2 km thick, covering more than 40,000 km^2 in the Southern Indian territory of Central Himalaya, as well as extending further east to Nepal. The scale of deformation confined to the zone of MAZ is more than that in the Himalayan Suture (Figure 53).

The MAZ by its genetic disposition is also the home or 'root-zone' or 'würzel-zone' of highest crystalline thrust-sheet of central Himalaya, *viz.* Chor, Dudatoli–Almora, Baijnath, Askot, Mandakini klippen, and those klippen and nappe in the Mandakini–Okhimath zone. Its width is not less than 30–50 km in average and an area of not less than 20,000–30,000 km^2. This is approximately the surface area occupied in the southern zone coverage over autochthon (parautochthon) by thrust-sheets of the Chail and Jutogh crystallines. These speculative figures and facts illustrate the magnitude of geological processes taken place in the southern territory of Central Himalaya resulting from lateral translation of thrust or nappe. Southernmost outer tonguing part of the Chail and Jutogh-thrusting is of post *Eocene age*, utmost in there major part *up to Miocene (10–15 Ma)*. This figure is derived from the age of young post-tectonic leucocratic tourmaline granite (see Chapter 6 for details) defining the uppermost age limit of thrusting.

Thrusting due to gravitational activity might be limited phenomenon only in the southernmost periphery of Himalaya, only in the molassic unit of Siwalik. Whatever is concerned with the prominent translation movement in the southern peripheral zone of Himalayan mountain system, here it is adequately shown that the dip of northern limb of thrust-sheet is towards the north, hence the general lateral movement of the thrust-sheets do not have any relation with gravity phenomenon. The gravity slumping near Barmatiya in northern zone of Garbyang rocks is perhaps to be attributed to the slumping after thrusting related with fault zone near Malari.

According to the latest data Himalaya is rising at the rate of 0.55 to 0.8 mm per year (Sinha, 1981).

In a recent synthesis (Sinha, 1987), the author has attempted to redefine the collision phenomenon and compression in the central segment of Himalaya in the light of various aspects of plate tectonics models and their constraints.

Chapter 7

Geochronology of Magmatic and Metamorphic Episodes

7.1 Introduction

Geochronological studies of the Central Himalayan Zone was taken up in late sixties, and publications on K–Ar dates started since 1972 (Sinha, 1972b; Ashgirei *et al.*, 1975; 1977a; Sinha, 1977a; Sinha and Bagdasarian, 1977). Besides the works on the Lesser zone of Central Himalaya, the collections of expeditions in Higher zone of Central Himalaya, along with systematic profiling of sections, provided valuable material for geochronological studies. All these results are tabulated in Tables 7.1 to 7.8. The earlier published results are mainly from the northwestern part of Himachal Himalaya, but in successive years the northeast zone of Garhwal and Kumaun from Lesser to Higher zones have also been covered. Most of the dates from this zone are being published first time.

In synthesis, the other published dates by K–Ar and Rb–Sr methods have also been discussed. Varieties of rocks ranging in composition from ophiolite–basite to diorite, granitoids, gneisses, and younger granites, and amphibolites have been geochronologically analysed. As far as possible the location of samples taken up for geochronological studies have been marked in the respective profiles and geologic maps.

Analysis of dates obtained from Southern Himalayan zones are taken up first in synthesis followed by the higher northern Central Crystalline and ophiolite zones.

7.2 A Discussion on the Mode of Decay of Radiogenic Potassium

An excellent treaty on the mode of decay and mathematical relationship of potassium–argon geochronology has been presented in many monographs and papers. The work of Gerling (1961), Hamilton (1965), and Schaeffer and Zahringer (1966) are the most informative amongst them.

Potassium has three isotopes ^{39}K, ^{40}K, and ^{41}K. ^{40}K is the least abundant of these and is unstable with respect to both beta emission and electron capture. The eventual fate of each atom of potassium of atomic weight 40 is to be transmuted either by emission of a beta particle to become an atom of calcium (^{40}Ca) or by electron capture to become an atom of argon (^{40}Ar) (Damon in Hamilton and Farquhar, 1968). Due to this transmutation the relative abundance of ^{40}K is decreasing and ^{40}Ca and ^{40}Ar is continuously increasing simultaneously with the time. Thus in 1290 Ma, the half-life ($t_{\frac{1}{2}}$) of ^{40}K, half of the atoms of ^{40}K in existence today would have decayed to either ^{40}Ca or ^{40}Ar and ^{40}Ca to ^{40}K would increase with time and be a measure of the time elapsed. Hence K–Ar method is applied to the dating of geologic events involving magmatic activity and thermal metamorphism.

Thus a mineral or rock containing a radioactive isotope such as ^{40}K becomes, in a very real sense, a radioactive clock or geochronometer that can be used to date the geologic events consisting of the crystallization of a mineral or solidification or a rock or more vigorously, the cooling of the rock unit to a temperature at which loss by diffusion of the daughter decay product is negligible. And so if the clock is thermally, mechanically or chemically perturbed, the geologic event may correspond more or less closely to the perturbation rather than solidification. According to Nier (1950) the relative distribution of K is as follow:

$^{39}K - 93.08 \pm 0.4\%$
$^{40}K - 0.0119 \pm 0.001\%$
$^{41}K - 6.91 \pm 0.04\%$

The following equation in relation to weight is used for calculation of age:

$$\frac{^{40}K}{^{39}K + {}^{40}K} = 0.0222.10^{-2}$$

7.2.1 Analytical procedure in the laboratory

The age determination of the samples discussed in this chapter has been carried out in Geochronological Laboratories of Armenian Academy of Science, Yerevan; Institute of Geology of Ore Deposits, Petrology–Mineralogy and Geochemistry, Moscow (IGEM); and Institute of Geology and Geophysics, Novosibirsk, USSR Academy of Sciences.

At the volumetric method, the 25 gm dose of the sample after preparatory treatment (warming to 150–200°C, in the course of an hour) was melted in quartz pipes at the temperature of 1250°C within four hours. The amount of argon, purified in the usual way (in furnaces with CuO and Ca) was measured, argon (percentage) was determined on the MS 2 M mass-spectrometer by the two ray method of measuring of the isotopic ratio $^{36}Ar/^{40}Ar$ of the sample and in standard technical argon. Isotopic composition of the latter is periodically controlled in the laboratory according to argon captured from the air. Using the method of isotopic dilution, 10–12 gm was melted in a metallic reactor made of anticorrosive steel and filled to the equipment for the volumetric determination of argon.

In the function of the indicator mono-isotope ^{38}Ar diluted with CO_2 was used. Coefficient of dilution of the mono-isotope (^{38}Ar in the mixture of $^{38}Ar + CO_2$) was determined also by the method of isotopic dilution, following a special series of experiments. In this case the air argon was used in the function of dilutor, obtained from the air on the same apparatus.

For the dosing of the admixture of ^{38}Ar and CO_2, there is an additional dosing device fitted to the apparatus. Measuring of the precise quantities of the dosed indicators were done on the normal MacLeod manometer. The error of dosing represented $+0.5-1$ per cent. The indicator was added after melting the sample and leading the exsoluted gap into the measure part of the apparatus. Analytical error in the determination of argon in both methods was found to be $+5$ per cent.

The potassium content in each sample was determined by three different methods; the perchlorate, dipicrilaminate and flame-photometric. The medium difference among the determinations of potassium by these methods is less than 5 per cent. In the calculations of age the mean values of chemical analysis and the following constants of disintegration of ^{40}K recommended by the Commission for Absolute Geochronology at the Soviet Academy of Science, were used:

$K = 5.57 \,.\, 10^{-11}\, yr^{-1}$
$P = 4.74 \,.\, 10^{-10}\, yr^{-1}$
$^{40}K = 0.000122$ K (by weight)
$1\ ng = 10^{-9}\ gm$

Chemical analysis of potassium by means of perchlorate method were realized by dipicrilaminate and flame-photometric methods. Argon was exsoluted on apparatus constructed by Chopin–Gerling. The summary error in determination of the values of absolute age for the samples was estimated 6–8 per cent against 10 per cent admitted for the potassium–argon method.

It is necessary to mention that in order to get the most precise absolute age, the whole series of investigations of each sample was carried through twice or thrice, independently of each other. In case of different results the experiments are repeated until sufficient agreement is reached.

In addition to this, as a control of the radiologic research of sample, the controlling analysis of the standard samples are carried through by the Commission on Absolute Geochronology at the ONZ by Academy of Sciences. This would increase the reliability of the age values of the samples under study. Determination of argon is done by two methods for each sample by isotopic dilution and volumetric methods.

7.2.2 Constraints of K–Ar method

Some caution is required in the interpretation of K–Ar 'ages' measured on whole-rock samples, unless the samples can be shown to be completely fresh. This is because alteration usually appears to facilitate loss of radiogenic ^{40}Ar leading to erroneously low apparent ages (MacIntyre *et al*., 1981). However, alternation can sometimes be tolerated. For all metamorphic rocks, the dates obtained on micas by the argon method points to the time elapsed since the moment of metamorphic event.

To evaluate the data obtained by K–Ar method it is very important to have a clear picture in mind of possible deviations of these figures from real age. So it is always desirable to check by either the Rb/Sr or the Sm/Nd method the authenticity of obtained data. Similar figures present a more reliable data to be adhered to, and to explain the incongruities caused due to superimposition of other processes.

It has been argued (Moorbath, 1967) that in the majority of minerals in which potassium is homogeneously distributed as a major constituent, practically all the argon can escape by volume (lattice) diffusion only, and it has been argued that diffusion at the condition of atmospheric temperature is so small that there is no significant argon loss.

It is felt that the actual temperature at which different minerals become effectively closed systems to argon in nature is not known with any accuracy, but threshold temperature probably does not exceed about 150–300°C from biotite to muscovite (Moorbath, 1967, p. 115).

The results obtained from the area where there might have occurred two or more major events, such as metamorphism and/or igneous intrusion, are much more complicated due to 'overprinting'. This is more prevalent in the case of Himalayan dates. Considering sufficient radiometric date from variety of minerals and rocks limiting age values for two or more major thermal events, i.e. regional metamorphism and/or igneous or orogenic activity could be ascertained with utmost accuracy.

It is worth quoting the following passage from Armstrong and Hansen (1966):

> 'The differences between expected and observed results and between different areas can not be explained by experimental error and should give pause to these geologists working in older metamorphic terrains who are using K/Ar dates to establish 'absolute' ages of metamorphic episodes'.

Armstrong (1966) while proposing the 'model orogen' comes out with the idea that no method of radiometric dating can give the exact age of development of structures and fabrics in metamorphic rocks. The dates most commonly measured are always less than the time where the metamorphic culmination has reached, probably by millions to hundreds of millions of years. Hence, *'metamorphism' has no exact age—it spans a period of time*, a period which may be difficult to interpret and define but nontheless a span of time for which no single age applies. Thus we are determining an approximate time when the rock system cooled through some critical isotherms. He further developes the idea of *'metamorphic veil'* defining it as a conceptual surface in space–time that synthesizes the dating results on metamorphic rocks. And when enough information is available, a contour map for the orogen can be made which represents the time when the various portions of the orogen now observed at the surface passed through the critical isotherms. The contour lines may be termed 'chrontours'.

7.3 Geochronological Data Analysis

7.3.1 Lesser Himalayan Zone

The geochronological picture in the southern zone of central Himalaya is quite divergent and interesting. Some typical dates from different rocks from the Garhwal–Kumaun Zone have been tabulated in Table 7.1. Sheared feldspathic gneiss (6–6/76) which is confined in the Chail and quartzite zone near Dharasu in the Bhagirathi valley yielded K–Ar age of 180.8 Ma. This is quite close to 191.5 Ma age of gneissose acid metavolcanics (1–1c/74) from Chandpur Chail formation from the section between Pauri and Lansdowne (Sinha and Bagdasarian, 1977). These late Triassic–early Jurassic ages from Himalayan region are quite uncommon, and may support a phase of late Triassic–early Jurassic metamorphism. The feldspathic augen gneiss (15–7/76 exp) collected from the Askot Klippe zone of crystalline rocks yielded K/Ar (whole-rock) 41.1 ± 1.8 Ma (Upper Eocene) value. Earlier Rb/Sr dates from the same suite of rocks have yielded 1620 ± 90 Ma (Powell *et al*., 1979) and 1983 ± 80 Ma (Bhanot *et al*., 1981) (Figure

72A). If the samples collected are also from the same suite of metamorphic rock, it is to be concluded that these allochthonous crystalline mass are of Precambrian age but they have suffered Upper Eocene phase of metamorphic episodes. Two figures of 325 and 414 Ma of Lower Carboniferous and Pre-Devonian significance are obtained from the mylonitized granitic and quartz dioritic intrusive of Yamuna Valley area confined in the Chail tectonic unit. It proves that the age of Chail rocks are close to Carboniferous which is being proved with the finding of palynomorphs of the same age (Ashgirei *et al.*, 1981; Sinha *et al.*, 1983). However, the 414± Ma age is obtained from the rocks which is altered variety and may have a disturbed radiogenic argon concentration.

The dates of basic rock suites (Table 7.2) (Sinha and Bist, 1986) from lesser zone of central Himalaya broadly fall in three main categories: (1) 42 ± 2; 51±; 69 ± 5 Ma; (2) 228 ± 10; 338 ± 12; 410 ± 10 Ma; and (3) 710; 1190 ± 35 Ma. These dates are made available from the entire zone of Himachal, Garhwal and Kumaun Himalaya, proving that three main phases of basic magmatism are encountered, a major part of which is confined along a deep-seated fault zone (Sinha, 1977; Sinha and Jhingran, 1977). We are actually dealing with the magmatic activities of three different epochs in this zone. First is the Precambrian activity, probably contemporaneous with the deposition of Simla flyschoidal and *Shali* terrigenous–carbonate formations. On the basis of stromatolites they have been dated as Riphean to lower Cambrian (Sinha, 1972b, 1977a,b; Sinha and Raaben, 1979). The second episode starting from 410 ± 10 Ma to 228 ± 10 Ma through 338 ± 12 Ma gives a broader range of Upper Palaeozoic. These data are also supported by the associated *Glossopteris* plant fossils in the Panjal range of Kashmir. The recently discovered palynomorphs of Upper Palaeozoic age (Ashgirei *et al.*, 1981; Sinha *et al.*, 1983) also supports the view of presence of geosynclinal regime, and hence the associated magmatic activities during Phanerozoic, especially during Upper Palaeozoic. The third and the final basic magmatic activity is related with the last phase of Himalayan orogeny marking the syn- and post-tectonic Tertiary episode.

It has been recapitulated that in the Southern zone of Central Himalaya two thrusts known as the Chail and Jutogh sheets have drawn the attention of geologists since Pilgrim and West (1928) published their work from the Simla area. The same units have been extended in eastern part to represent a more homogenized picture of regional geologic setting (Fuchs and Sinha, 1978). Chail tectonic units consists of different types of black carbonaceous (graphitic) dark grey and greenish grey shales converted into phyllites, sericite mica schists, quartz chlorite schists and muscovite–chlorite schists. The most typical of this complex is the porphyroids derived from sedimentary volcanogenic origin or acid metavolcanics. The volcano sedimentary origin of the above described rocks is proved by their extreme irregular granularity, the presence of well rounded, angular grains of quartz and plagioclase: fragments of microgranular rocks of volcanic appearance; and widely distributed common appearance of process of albitization, characteristics for eugeosynclinal volcanic origin. Within the Chail complex we also come across granitoids and gneiss which has been also dated by Rb/Sr methods. The K–Ar date have been tabulated in Table 7.3. The figure varies between 28 ± 2 to 124 ± 9 Ma, i.e., between Upper Oligocene to Lower Cretaceous. Most of these are the mineral dates on muscovite and sericite, and hence it shows the acme of metamorphism of regional polycyclic type. All those mentioned samples were collected from the Simla and Kulu region of Himachal but from the same tectonic unit C_3 per existing geological maps.

The gneissic rock from the migmatized Chail yield Rb/Sr date of 1430 ± 150 Ma from Baragaon locality in Sutlej Valley (Bhanot *et al.*, 1980).

The most baffling is the results of gneissic rock dating from Garhwal region in the Mandakini Valley. The gneisses at Chandrapuri with six samples isochron of $^{87}Rb/^{86}Sr$ yielded 1546 ± 12 Ma, and those gneisses from Kharbatia Khal with four sample isochron yield 1582 ± 57 Ma, and Toneta Granite at locality of Mayali with four sample isochron yielded 1454 ± 62 Ma (personal communication Prof. V. B. Bhanot) Geology of these granitic body should be referred from Fuchs and Sinha (1978) and Negi *et al.* (1983). These gneisses and granite bodies are located in the roof part of Chail nappe and yield all Precambrian ages, whereas those K/Ar figures could be the metamorphic episodes, but the Upper Palaeozoic palynomorphs from lesser zone of Chail and Chandpur are not consistent with the Precambrian age, if we assume that this is a continuous one single thrust-sheet. There are only two alter-

Table 7.1. K–Ar isotopic data on absolute age determination of Kumaun–Garhwal Himalayan zone crystalline rocks

S. No.	Lab. No.	Name of the rock Locality of the sample	K(%)	^{40}Ar (g/g $. 10^{-6}$)	^{40}Ar (% rad.)	^{40}Ar (cm^3 g/g $. 10^{-6}$)	^{40}K (g/g $. 10^{-9}$)	$^{40}Ar/^{40}K$ (10^{-3})	Age (Ma)	Mean value (Ma)	Remarks
1.	6–6/76	Sheared feldspathic gneiss,	4.22	5.15	92.7	30.46	54.37	10.56	180.8	180.7	Lower Jurassic
		Dharasu, Bhagirathi valley	4.22	5.15	94.2	30.45	54.35	10.55	180.6		(whole-rock)
2.	1–1C/74	Feldspathic gneiss, acid metavolcanics	3.02	3.68	93.1	22.46	40.2	10.9	186	191.5	End of Triassic, beginning of
		in Chandpur Fm, between Pauri and Lansdowne	3.02	3.68	74.0	23.80	42.6	11.55	197		Jurassic (whole-rock) (Sinha and Bagdasarian, 1977)
3.	15–7/76Exp	Feldspathic augen gneiss	1.32	1.61	44.1	2.00	3.56	2.21	39.4	41.1 ± 1.8	Upper Eocene (whole-rock)
		Askot tectonic klippe, Didihat	1.32	1.61	41.9	2.18	3.89	2.42	42.8		Rb/Sr: (1) 1983 ± 80 Ma (Bhanot *et al.*, 1978, 1981) (2) 1960 ± 90 Ma (Powell *et al.*, 1979)
4.	41–13/74	Mylonitized granitoid	2.25	2.74	100.0	31.3	56.05	20.04	334	325	Lower Carbo-
		Barkot, Yamuna valley	2.25	2.74	89.4	29.4					niferous (Sinha and Bagdasarian, 1977)
5.	12–1/74	Quartz-Diorite Kuwa, Yamuna	2.24	2.73	100.0	40.07	71.7	26.2	421	414	End of Silurian
		valley	2.24	2.73	97.2	38.55	69.0	25.2	407		Pre-Devonian (Sinha and Bagdasarian, 1977)

Petrographic Descriptions

1. 6–6/76 ‘Sheared feldspathic gneiss. Augen of K-feldspar, quartz and plagioclase, strongly foliated in fine sericitic and quartzo-feldspathic material.

2. 1–1C/74 It occurs as concordant bands (20–30 m) in phyllites of Chandpur between Pauri and Satpuli bridge. In this section it shows high alteration and porphyroblasts of fresh quartz crystals surrounded by fine medium grained hydromicas. Highly altered relict plagioclase crystals in the matrix (Extn. angle: 12–15). The arrangement of minerals around quartz porphyroblasts are so fluidal that it is tempting to call these rocks metamorphosed acid volcanics.

3. 15–7/76Exp Feldspathic augen gneiss. Augens of altered K-feldspar and plagioclase (1 cm × 1 cm in hand specimen and 2 mm × 1 mm in thin section).

4. 41–13/14 Mylonitized granitoids intrusive in the Chail group of rocks near Barkot.

5. 12–1/74 Quartz diorite with altered plagioclase (35%), biotite (25%), hornblende in relict and quartz (30%) with accessory minerals: rutile, zircon and opaques. Hypidiomorphic texture.

Table 7.2. K–Ar: geochronology of basic rocks from the Lesser Central Himalaya

S. No.	Sp. No.	Rock type	Dating of	Locality	Age (Ma)	Remarks
1.	296–113	Amphibolitized gabbro–diorite	Whole-rock	Naogaon Giri River valley, Simla Himalaya	42	Upper Eocene
2.	29–9/73	Diabase	Whole-rock	Naini Tal, Tiffin Top, Kumaun Himalaya	51	Lower Eocene
3.	A	Basite (fresher)	Whole-rock	Bhowali, Naini-Tal, Kumaun	69	Maestrichtian (Varadarajan, 1977)
4.	B	Basite (altered)	Whole-rock	Bhowali, Naini-Tal, Kumaun	228 ± 10	Lower Triassic Permian boundary (Varandarajan, 1974)
5.	C	Diabase	Whole-rock	Tattapani, Simla Himalaya	338 ± 12	Carboniferous
6.	175–61	Amygdaloidal diabase	Whole-rock	Darlaghat, Simla Himalaya	410 ± 10	Lower Devonian
7.	10–1/74	Dioritic basite	Whole-rock	Kuwa, Yamuna valley, Garhwal Himalaya	710	Upper Riphean
8.	186–65	Amygdaloidal diabase	Whole-rock	Darlaghat, Simla Himalaya	1190 ± 35	Middle Riphean

For detailed information regarding petrography and petrochemistry of specimens 1 to 8 (except 7), refer to paper Sinha (1977). For specimen 7, refer to Sinha and Bagdasarian (1977).

Table 7.3. K–Ar: geochronology of the Chail rocks

S. No.	Sp. No.	Rock type	Dating of	Locality	Age (Ma)
1.	145/53	Quartz chlorite schist	Muscovite	Pandoh, Kulu valley	28 ± 2 (Upper Oligocene)
2.	144/62	Quartz chlorite schist	Muscovite	Pandoh, Kulu valley	
3.	33/14	Quartz sericite schist	Sericite	Simla	45 ± 2
4.	24/10	Porphyroid (tuffo-argillite)	Whole-rock	Simla	63 ± 2
5.	20–64	Muscovite–chlorite schist	Whole-rock	Simla	63 ± 2
6.	70/28	Sericite schist	Sericite	Pandoh, Kulu valley	89 ± 6
7.	126/66	Quartz sericite schist	Sericite	Kathlighat, Simla	124 ± 9 (Lower Cretaceous)

natives: either those phyllitic zones of Chail and Chandpur are not related with those of northern root zones but rather they are the separate structural facial zones of deposition. The northern part is of different set up; or there is something wrong with the basic palynological or geochronological data. However, the first alternative is more rational to accept.

Geochronological figures obtained by K/Ar method from the more metamorphosed thrust sheet of Jutogh lying over the Chail tectonic unit is tabulated in Table 7.4. The metamorphic ages are ranging between 23 (Lower Miocene) to 123 Ma (Lower Cretaceous). This picture is very similar to the Chail geochronology., Thus the regional metamorphism of the Jutogh and Chail are contemporary.

Table 7.5 presents the geochronological data from the highest sheet consisting crystalline and migmatized complex of Central Crystalline. The cited figures could be divided into four groups: age corresponding to the age of thrusting in Himalaya (13 Ma) in general the paroxysmal phase of orogeny; secondly, 37 to 50 Ma, metamorphism corresponding to Chail and Jutogh; thirdly, 148 to 179 Ma, the episode corresponding to acid metavolcanic metamorphism in Chail and Chandpur; and fourthly, the higher value of 250 to 338 ± 5, 160 ± 14, 250± 13 & 259 ± 13 Ma are the K/Ar value on higher side, and two values of 179 ± 6 and 338 ± Ma from Mandi and Pandoh rocks are pointing towards the earlier phase of regional metamorphism.

7.3.2 Higher Central Himalayan Zone (Tables 7.6, 7.7 and 7.8)

This section includes the dates obtained from the rocks collected from the Central Crystalline zones north of Main Central Thrust, and the adjoining Tethyan zone with ophiolitic melange. The Central Crystalline complex in the Himalaya occupies the major part of the Higher Himalayan rampart with lofty peaks, which are usually capped by sediments of Tethyan zone lying in adjacent northern part. These highly deformed and metamorphosed group of rocks known as 'Vaikrita Group' with sediments and magmatism of various ages including Precambrian to Phanerozoic, have been finally remobilized during Oligocene–Miocene period.

Reviewing the geochronological data (Sinha and Bist, 1986) the granitic bodies confined in Vaikrita zone should be classified in three major categories:

1. 15.2 ± 1.5 Ma, 20.7 ± 0.1 Ma and 27.8 ± 0.6 Ma (Lower Miocene group)
2. 37 ± 0.5 Ma; 37.6 ± 1.1 Ma; 39.2 ± 1 Ma; 48.6 ± 3.4 Ma (Eocene group)
3. 87.3 ± 1 Ma; 90.6 Ma (Upper Cretaceous group)

These three groups of granitic bodies show good cross-cutting interrelationship with each other and the country rocks. Upper Oligocene age of the Ladakh pink porphyritic granite is being

Table 7.4. Geochronology of the Jutogh rocks

S. No.	Sp. No.	Rock type	Dating of	Location	Age (Ma)
1.	17–64	Biotite chlorite schist	Whole-rock	Simla	23 (Lower Miocene)
2.	35–19	Garnetiferous schist	Biotite	Chaupal	30
3.	92–52	Crystalline schist	Muscovite	Chaur + Didar Naura	37
4.	80–64	Chlorite graphite schist	Whole-rock	Simla	42
5.	129–44	Garnetiferous schist	Biotite	Kulu	85
6.	13–64	Biotite Boileaugunge quartzite	Whole-rock	Simla	86
7.	10–64	Muscovite Boileaugunge quartzite	Whole-rock	Simla	120
8.	16–64	Biotite tourmaline Boileaugunge quartzite	Whole-rock	Simla	123 (Lower Cretaceous)

Table 7.5. Geochronology of the crystalline rocks from the Simla Himalayan zone

S. No.	Sp. No.	Rock type	Dating of	Location	Age (Ma)	Remarks
1.	70	Biotite gneiss	Biotite	Sarahan, North of Rampur, Sutlej valley	13	Middle Miocene: I
2.	15	Migmatized quartz biotite crystalline schist	Biotite	Manali, Beas valley	37 ± 3	II
3.	116–41	Biotite schist	Biotite	Manali, Beas valley	49 ± 3	
4.	72–40	Migmatized biotite gneiss	Biotite	North Sutlej valley	50 ± 3	
5.	20–13	Gneiss	Whole-rock	Chor mountain, Simla	148 ± 5	III
6.	33	Garnet mica schist	Whole-rock	Chor mountain, Simla	160 ± 14	
7.	153/56	Granite	Muscovite	Manali synform	179 ± 6	
8.		Garnet gneiss	Whole-rock	Chor mountain, Simla	250	IV
9.	48–A	Granite	Whole-rock	Chor mountain, Simla	295 ± 13	
10.	80/30	Granitized schist	Biotite	Pandoh, Chail nappe	338 ± 0	(Lower Carboniferous)

also considered here, but actually it falls in the territory beyond the Tethyan zone. However, this event is also dated with the events in the Central Crystalline.

The 37 to 48 Ma acid magmatic events are marked by the granitic bodies with kyanite and sillimanite.

The Upper Cretaceous magmatic events is a prominent event since in Badrinath area there is a profuse penetration of these rocks along the deep-seated root-zones with calc-silicate rocks, as well as these granitic bodies from the top of Nilkanth and other Central Crystalline peaks. However, the sample of granitoid collected from the Choptachatti area in the ridge between Gopeshwar and Okhimath provides rather paradoxical age picture. K/Ar figure is 90.6 Ma but Rb/Sr figure is 1500 Ma (personal communication, Prof. V. B. Bhanot).

The other two dates 26 ± 5 Ma and 740 ± 24 Ma obtained from Gori and Darma valleys on gneissic bodies with finer to larger porphyroblasts points to much higher age of Middle Permian to Upper Riphean.

Considering the published Rb/Sr date of crystalline rocks from Munsyari area in Gori Ganga valley an age of 1890 ± 155 Ma and that of Dhakuri area 2326 ± 43 Ma has been obtained (Bhanot *et al*., 1980). These figures along with the 1058 Ma (Powell *et al*., 1979) from Rohtang area further northwest confirms the Precambrian elements present in the Central Crystalline Vaikrita Complex.

Table 7.7 presents the K/Ar geochronology of mainly amphibolitic rock complex confined either along the tectonic contact or closer to it (see Figures 102 and 103). The age values could be broadly grouped in three categories: 1–25.0; 42.7 Ma; 2–91.3, 147.3 Ma, 180 Ma; and 3–371.7 (Mid-Devonian). Besides the first two figures closer to Himalayan orogenic events, the 91.3, 143.3 and 180.0 Ma are Mesozoic events dating Upper Cretaceous to Lower Jurassic dates. These Mesozoic dates coming up from different zones of Himalaya require a serious attention for explanation. But it is clear from the concentration of higher dates of Mesozoic age along the 'So-sa–Jhamrigaon Reverse fault' (Figures 103 and 104) that it was active during Jurassic–Cretaceous period too.

Table 7.8 presents a comprehensive picture of ophiolite geochronology from the thrust mass of

Table 7.6. Data on K–Ar isotopic dates of granitoids from the Central Crystalline zone of Higher Central Himalaya

S. No.	Sp. No.	Name of the rock Locality of the sample	K(%)	^{40}Ar (g/g . 10^{-6})	^{40}Ar (% rad.)	^{40}Ar (cm^3 g/g . 10^{-6})	^{40}K (g/g . 10^{-9})	^{40}Ar/^{40}K (10^{-3})	Age (Ma)	Mean value (Ma)	Remarks
1.	364–19/76Exp	Tourmaline granite, Mapang Goriganga valley	3.18 3.18 3.18	3.88 3.88 3.88	31.0 12.0 35.6	1.81 1.62 2.12	3.24 2.90 3.78	0.83 0.75 0.97	14.8 13.4 17.4	15.2 ± 1.5	Mid Miocene (whole-rock)
2.	5–2/75Exp	Garnetiferous granite, Urthing Darma valley	3.55 3.55	4.33 4.33	44.4 41.3	2.28 2.78	5.03 4.96	1.16 1.15	20.8 20.6	20.7 ± 0.1	Lower Miocene (whole-rock)
3.	LKG/51/77	Pink porphyritic granite, Hemiya, Ladakh	3.44	4.20	49.5	3.58	6.38	1.52	27.2	27.8 ± 0.6	Upper Oligocene (whole-rock) Further north of Tethyan zone Upper Eocene
4.	14–3/75Exp	Kyanite sillimanite granite, Sela, Darma valley	3.49 3.49	4.26 4.26	60.9 63.9	5.01 4.86	8.94 8.67	2.10 2.04	27.4 36.3	37 ± 0.4	Upper Eocene (whole-rock)
5.	157–20/75Exp	Garnetiferous Tourmaline granite, Budhi, Kali valley	3.00 3.00	3.66 3.66	47.3 49.3	4.20 4.48	7.49 8.00	2.05 2.18	36.5 38.7	37.6 ± 1.1	Lower Oligocene (whole-rock)
6.	158–20/75Exp	Granite, Budhi, Kali valley	3.00 3.00	3.66 3.66	53.1 56.5	4.40 4.64	7.86 8.26	2.75 2.26	38.3 40.2	39.2 ± 1	Upper Eocene (whole-rock)
7.	23–4/75Exp	Kyanite, sillimanite, garnetiferous granite, Baling, Darma valley	5.77 5.77 5.77	7.04 7.04 7.04	56.0 79.2 64.6	11.75 9.86 10.62	20.97 17.61 18.95	2.98 2.50 2.69	53.6 44.4 47.7	48.6 ± 3.4	Lower Eocene (whole-rock)
8.	2AKS/74Exp	Aplitic granite, Malari	4.28 4.28	5.22 5.22	81.8 71.0	12.1 12.3	21.6 22.0	4.14 4.21	73 74	73.5	Upper Cretaceous (whole-rock)
9.	1–1/76EXP	Biotite–tourmaline granite, Badrinath	3.81 3.81	4.65 4.65	71.5 77.7	12.79 13.12	22.83 23.43	4.91 5.04	86.2 88.5	87.3 ± 1.1	Upper Cretaceous (whole-rock)
10.	11–5/76Exp	Augen gneiss, Choptachatti, Garhwal	4.27 4.27	5.21 5.21	64.3 81.0	15.05 15.05	26.86 26.86	5.16 5.16	90.6 90.6	90.6	Upper Cretaceous (whole-rock) Rb/Sr 1500 Ma (personal communication Prof. V. B. Bhanot)
11.	26–16/76Exp	Muscovite–biotite gneiss, Suringghat, Gori valley	3.48 3.48	4.24 4.24	84.3 92.5	36.40 37.80	64.98 67.50	15.30 15.91	256.6 256.1	261	Middle Permian (whole-rock)
12.	80–13/75Exp	Porphyritic gneiss, Khet Darma valley	3.19 3.19 3.19 3.19	3.89 3.89 3.89 3.89	98.9 98.6 98.9 99.8	118.65 108.54 108.09 104.30	211.79 193.75 192.94 186.17	54.44 49.81 49.58 47.84	788 733 730 708	740 ± 24	Upper Riphean (whole-rock)

1. 364–19/76Exp Tourmaline/microcline granite. Texture: holocrystalline, hypidiomorphic, perthitic. Mineral constituents: quartz, feldspar, plagioclase, orthoclase and plenty of microcline sometimes kaolinized. Tourmaline and apatite as accessory.
2. 5–2/75Exp Garnetiferous granite. Quartz, plagioclase, muscovite, biotite, garnet, subordinate potash feldspar (microcline), crudely foliated.
3. LKG 51/77 Pink porphyritic granite from Hemiya, Ladakh Himalaya (Refer Sharma *et al.*, 1978, for details).
4. 14–3/75Exp Kyanite–sillimanite bearing granite, other minerals quartz and plagioclase.
5. 157–20/75Exp Garnetiferous tourmaline granite. High percentage of large microcline tabular plates. Microcline also in the ground mass. Subordinate constituent observed. Magmatism in the zone of deep faults.
6. 158–20/75Exp Coarse grained. Nonfoliated texture: porphyritic/porphyroblastics. Phenocryst of potash feldspar (microcline and orthoclase). Effect of shearing visible. Magmatism occurred in the zone of deep fault.
7. 23–4/75Exp Kyanite–sillimanite garnetiferous granite. Mineral constituents: quartz, plagioclase, K-feldspar, muscovite–biotite, garnet, sillimanite, kyanite.
8. 2AKS/74Exp Aplitic tourmaline granite. Plagioclase, quartz, K-feldspar, tourmaline. Deformation twinning in plagioclase common; many of them bent and fractured. Quartz in thin section highly sheared and stretched.
9. 1–1/76Exp Biotite–tourmaline granite of Badrinath. Medium grained granite with plagioclase, K-feldspar, quartz, biotite and muscovite.
10. 11–5/76Exp Augen gneiss. Large porphyroblasts of K-feldspar (orthoclase)—pethitic, altered matrix mainly consisting of quartz, plagioclase, biotite and muscovite. Quartz—euhedral colourless grains with low relief; Feldspar—euhedral laths with medium plagioclase relief. Polysynthetic twinning. Orthoclase—colourless, subhedral grains. Biotite, muscovite and apatite in accessory. Texture—holocrystalline hypidiomorphic.
11. 26–16/76Exp Muscovite–biotite gneiss. Quartz—20%, undulating extinction. Plagioclase—(15%), albite twinning. Orthoclase—colourless, R.I. moderate. Feldspar with inclusions of mica and magnetite, and in places sericitization is visible. Pleochroism pale brown to deep brown. Biotite—(15%). Muscovite—(20–25%). Accessory—apatite, zircon. Texture—holocrystalline hypidiomorphic. Structure—gneissic and myrmekitic.
12. 80–13/75Exp Porphyritic gneiss. Foliated-gneissic texture. Porphyroblasts of K-feldspar (orthoclase) surrounded by myrmekite rim. Porphyroblasts of larger size (5 cm × 2 cm). Slight cataclastic effect visible (for chemical, silicate and spectral analysis—refer to Sp. no. 25–2/78Exp).

Table 7.7. Geochronology of the amphibolite complex of Higher Himalayan Central Crystalline zone

S. No.	Sp. No.	Name of the rock and sample locality	K (%)	^{40}Ar (g/g . 10^{-6})	^{40}Ar (% rad.)	^{40}Ar (cm^3 g/g . 10^{-6})	^{40}Ar (g/g . 10^{-9})	$^{40}Ar/^{40}K$ (10^{-3})	Age (Ma)	Mean value (Ma)	Remarks
1.	166–21/75Exp	Amphibolite Samari, Southern thrust contact of Sirdang zone, Kali valley	0.40 0.40	0.488 0.488	— —	0.336 0.430	0.60 0.77	1.23 1.58	22.0 28.0	25.0	Oligocene–Miocene Boundary
2.	5–3/73Exp	Amphibolite with sphere and relict pyroxene, Helang, Southern contact of MCT, Alaknanda valley.	0.50	0.61	15.0	0.82	1.46	2.46	42.7	42.7	Middle Eocene
3.	31–7/78Exp	Amphibolite, Jhamrigaon, Southern contact Sirdang zone, Darma valley	0.60 0.60 0.60	0.73 0.73 0.73	11.6 14.2 14.2	2.07 2.17 2.17	3.69 3.88 3.88	5.05 5.29 5.29	88.5 92.6 92.6	91.3	Upper Cretaceous
4.	172–21/75Exp	Amphibolite, Sirkha, Central part Sirdang zone, Kali valley	1.09 1.09	1.33 1.33	19.0 13.98	6.22 6.51	11.1 11.62	8.35 8.74	144.0 150.5	147.3	Upper Jurassic
5.	21–6/78Exp	Amphibolite, Sosa, below Southern contact zone of Sirdang	0.99 0.99 0.99	1.21 1.21 1.21	29.0 45.3 32.0	6.94 7.60 6.81	12.38 13.57 12.15	10.25 11.23 10.06	175.3 191.2 172.2	180.0	Lower Jurassic
6.	79–13/75Exp	Amphibolite, Jhandadhura, Southern contact Gam Malla zone	0.82 0.82	1.00 1.00	53.0 60.9	12.83 12.93	22.90 23.80	22.89 23.08	370.3 373.0	371.7	Middle Devonian
7.	58–10/78Exp	Amphibolite, Belchidhura	0.26								
8.	48–8/78Exp	Amphibolite, Nyu	0.27	The rocks contain insignificant K.							
9.	49–8/78Exp	Amphibolite, Nyu	0.38	Age not determined							

All analyses on whole-rock.

Table 7.8. K–Ar isotopic data on absolute age determination of ophiolite rock complex from Higher Himalayan Tethyan zone

S. No.	Sp. No.	Name of the rock Locality of the sample	K (%)	^{40}K (g/g . 10^{-6})	^{40}Ar (% rad.)	^{40}Ar (cm^3 g/g . 10^{-6})	^{40}Ar (g/g . 10^{-9})	$^{40}Ar/^{40}K$ (10^{-3})	Age (Ma)	Mean age (Ma)	Remarks
1.	99–17/73Exp	Spilite (amygdaloidal, associated with ophiolitic melange, Balcha Dhura Pass, Malla Johar, Central Tethys Himalaya	1.21 1.21	1.40 0.48	21.6 27.38	4.79 5.36	8.57 9.59	5.81 6.48	102 113	107.5	Boundary between Lower and Upper Cretaceous (whole-rock)
2.	118–17A/73Exp	Spilite (amygdaloidal, associated with ophiolitic melange, Balcha Dhura Pass, Malla Johar, Central Tethys Himalaya	1.335 1.335	1.65 1.65	32.2 31.9	3.63 4.24	6.50 7.49	3.99 4.60	69 81	75 ±	Campanian—Maestrichtian (whole-rock)
3.	119–17A/73Exp	Spilite (amygdaloidal, associated with ophiolitic melange, Balcha Dhura Pass, Malla Johar, Central Tethys Himalaya	1.435 1.435	1.75 1.75	20.4 39.2	4.17 3.97	7.46 7.11	4.26 4.06	74 72	73 ±	Campanian—Maestrichtian (whole-rock)
4.	96–17/7Exp	Spilite (amygdaloidal, associated with ophiolitic melange, Balcha Dhura Pass, Malla Johar, Central Tethys Himalaya	0.47					—	—	—	Unworkable, a lot of gases
5.	139–21/73Exp	Spilite (amygdaloidal, associated with ophiolitic melange, Balcha Dhura Pass, Malla Johar, Central Tethys Himalaya	0.118					—	—	—	Insignificant K, age not determined
6.	CH 7/76	Dras volcanics associated with ophiolitic melange, Chiktan Nala section, Ladakh Himalaya	0.905 0.905	1.10 1.10	27.6 52.3	2.76 2.69	4.93 4.79	4.46 4.34	78.5 76.5	77 ±	Upper Cretaceous (Campanian) (whole-rock)
7.	LKG 204/77	Shyok volcanics (basal part), Khardung, Ladakh Himalaya	2.92 2.92 2.92	3.56 3.56 3.56	23.6 54.1 52.3	4.59 4.06 4.14	8.19 7.25 7.38	2.30 2.03 2.07	40.9 36.1 36.9	38 ±	Upper Eocene (whole-rock)

Malla Johar below 'exotic blocks'. Five specimens of basic volcanites collected from Balcha Dhura Pass (6040 m) were investigated at first instance followed by two more specimens from the Ladakh Suture zone in geochronological laboratory of Armenian Academy of Sciences, U.S.S.R.

The geological information on the Higher Northern zone can be obtained from the works of Griesbach (1891, 1893); Diener (1898); von Kraft (1904); Heim and Gansser (1939a); Gansser (1964); and Shah and Sinha (1974c). Out of the ophiolite specimens three values have come up from the Malla Johar area: 73, 75 and Ma ranging from boundary between Lower and Upper Cretaceous to Campanian–Maestrichtian. Those dates from Ladakh area 77 ± 1 (Upper Cretaceous) and 38 ± 2 (Upper Eocene) age. An overview of figures could be summarized by an Upper Cretaceous age of ophiolitic complex with another younger phase of 38 Ma in Shyok area.

Microscopically the rocks from Malla Johar area are spilitic in nature, having fluidic texture indicating flowage and amygdales of calcite and virulitic zeolites. Gansser (1964) mentions them perhaps as amygdaloidal porphyrites. The albite oligoclase microlites show alteration and rather smaller crystal of clinopyroxene (augite) are also present. The opaques are in sufficient quantity scattered throughout the mass. Sometimes the amygdales increase in volume up to 45–50 per cent and brownish in colour. They are incompact due to a higher percentage of amygdales whereas other specimens are compact in nature, having greenish grey tint and a fine-grained texture.

Three major phases of ophiolitic emplacement or eruption are obvious with 107.5 Ma at the boundary between Lower and Upper Cretaceous followed by 73, 75 and 77 Ma phases, and finally the 38 ± 2 Ma phase in the Shyok Valley.

Some of these phases are closer to the acidic and basic magmatism in the crystalline zone of Central Himalaya (Tables 7.1, 7.2 and 7.6).

Chapter 8

Petrochemistry

This chapter incorporates the summarized data on the petrochemical and petrographical study of different rock types from the different structural facies zones of the Central Himalaya. Special emphasis has been laid on the crystalline metamorphic and magmatic rock types of the Vaikrita group from the Higher Central Himalayan zone as well as their overthrust klippen masses in the southern territory. In the northern Tethyan zone ophiolite specimens collected from the Kiogarh thrust mass have been also geochemically, petrographically, and geochronologically studied. However, the major part of this northern zone is out of Indian territory, and hence out of the author's research area. Specimens were collected by the author all along the different traverses as shown in the geological map (Figures 12 and 13), with oriented locations as shown in the traverse figures.

The specimen collection was made in the field, keeping in mind the structural profile for elemental, spectral and silicate analyses. Radiometric dating of suitable samples was also envisaged. Geochemical behaviour of elements under different parameters of tectonic regime and process was also taken into consideration while collecting the specimens.

In the text there has been summarized the results of some two hundred thin sections studied under microscope; sixty-five geochemical silicate analyses amongst them thirty-five analyses were repeated for control check, eighty-two spectral analyses amongst them forty analyses were repeated for control check in different laboratories. Analyses were done in the following geochemical laboratories: Wadia Institute of Himalayan Geology (WIHG), Dehra Dun, India; Institute of Geology of Ore Deposits, Petrography, Mineralogy and Geochemistry (IGEM), Academy of Science of USSR, Moscow; Plasma Laboratory of Geological Institute (GIN), Academy of Sciences of USSR, Moscow; and Austrian Geological Survey, Vienna.

Besides the above-mentioned unpublished data, fifty published geochemical analyses pertaining to the southern lesser zone of Himalaya, from flyschoidal formation, tectonic thrust sheets of Chail and Jutogh (Sinha, 1977a, 1978a; Ashgirei *et al*., 1977) have also been taken into consideration. A review of geochemical analyses and their corresponding petrographic study led the author to classify in broader perspective the magmatic and metamorphic rocks of the crystalline zone of the Central Himalaya. This leads to important conclusions pertaining to the distribution of certain elements showing deep-seated tectonics along the Main Axial zone (MAZ), and perspective for future search of mineral deposits, besides the already discovered barite–polymetallic mineralization.

All the analytical data has been tabulated in Tables 8.1, 8.2, and 8.3. Table 8.1 presents the silicate analyses done by plasma spectrophotometry. Many of the specimens were repeated in Wadia Institute of Himalayan Geology (WIHG) laboratory. Determinations were done with AAS and UV-spectrophotometer. There was observed not much divergence in the results. Hence, WIHG data are not being produced separately. But all the semiqualitative spectral analyses done at IGEM laboratory, Moscow (Table 8.2) were rechecked and some of them were repeated in the source laboratory itself. Moreover, WIHG lab performed some control check analyses of those analysed at IGEM, on the Perkin–Elmer Atomic Absorption installation. Hence, the qualitative data in ppm has been reproduced in Table 8.3. It is to be noted that all the semiquantitative data in Table 8.2 has been converted to ppm range to keep up the modern practice in geochemistry.

The classification of Levinson–Lessing is followed for interpretation with picrite basalt 46 per cent to Liparite 76 per cent, as common type series. The most important inferences can be summarized as follows:

Calc-silicate rocks from different areas but situated on the Main Axial Zone (MAZ), *viz*. sp. 79–16/78 Exp, Pindar valley), sp. 115–16/75 Exp (Kali valley) and other similar type from Alaknanda Gori and Darma valleys, give high concentration of Pb-100 ppm, Zn-272 to 295 ppm, Sr–1433 to 1740 ppm. The high concentration of the above-mentioned elements are indication of the deep-seated hydrothermal activity of juvenile liquid from as deep as boundary between crust and mantle i.e. lower part of Moho discontinuity (65–70 km, H. Gupta, 1973). So the author's contention of the existence of a deep-seated divergence root-zone in Central Higher Himalaya is supported by these geochemical data too. Moreover, the deep basic character in picrite–granulitic type of composition is also confirmed by less SiO_2 41.49 per cent (78–16/78 Exp, Pinder MAZ).

The younger granites intruding the older metamorphics of Vaikrita group have been classified into three distinct categories on the basis of their K–Ar dating. Also according to their geochemical characteristics they are categorized in the same way. It is most interesting to note that the coarse grained pegmatitic granite (35–19/76; 158–20/75 Exp; 2AKS/74 Exp; 84–17/78 Exp) with tourmaline show very high boron (from 1000 up to 6000 ppm) indicating that there should be the last stage of granitic differentiation intrusion having the high potassium and sodium variation with the development of pegmatitic field. And the study in this direction would lead to the discovery of rare earth minerals association with them in the Higher Himalayan tectonic zone. Although, the discovery of barite–polymetallic mineralization is established to be associated with them but particularly the rare earth discovery would be the most coveted achievement.

The other type of granite of Badrinath area (1–1/76 Exp) with garnet and Na_2O = 0.49323 per cent, K_2O = 0.14736 per cent and boron level to 7–9 ppm differs from the above-mentioned type. K–Ar data closer to 87.3 ± 1.1 Ma. Not perspective for rare-earth prospecting.

Third type of granite occur with kyanite–sillimanite and garnet (14–3/75 Exp) with date closer to 37 ± Ma. The presence of kyanite and sillimanite points to post-genetic increased pressure phenomenon after the formation of main granitic fluid. Since this type of granitic bodies are confined in the MAZ area, hence it could be speculated that increased pressure phenomenon possibly might have happened in course of impregnation of a granitic body in the deep-zone of root-zone divergence itself (Figures 109 and 110).

Another type of interesting magmatic rock is the big porphyroblastic (5 cm × 3 cm) gneiss and smaller porphyroblastic gneisses. The first is that a Khet and Pangu type and second one is Joshimath–Askot–Rungling type. Geochemically again the big porphyritic gneiss from Khet give very high boron (4000–6000 ppm). It is rather anomalic to this type of rock and may lead to speculation that their location along the alignment of tectonic zone of outer Sirdang–Gam Malla might be a pointer to another deep-seated pneumatolitic phenomenon there the root-zone of lower thrust sheet of Chail (?) should be searched. Most interesting is the presence of blue quartz associated with the gneissic rocks below this horizon in Kali valley near Tawaghat (Figure 90). Similar blue quartz are also associated with Chail tectonic unit in Sutlej valley. The occurrence of amphibolite with profuse garnet as found in Gori valley (32–18/76 Exp, Bugdier) and other in Badrinath area is closer to Iceland type basalt in their primary composition. After metamorphism under high pressure spessartite is formed: but pyrope is not available which is low presence type. This is also indicative of MAZ deep tectonic phenomenon.

Other type of amphibolite (52–8/76 Exp) occurring near the MCT zone in Darma valley present a picture of area of stress type pressure leading to relative concentration of the femic (Fe, Mn) component due to liberation of light mobile salic (Na, K) component from primary rocks. Hence, the amphibolitic sills emplaced along thrust zone should also carry kyanite–sillimanite as found in the country acidic gneissic complex under the stress condition. This observation is yet to be made.

Other typical petrochemical type rocks are mica schists, quartzite and quartz sericite schists

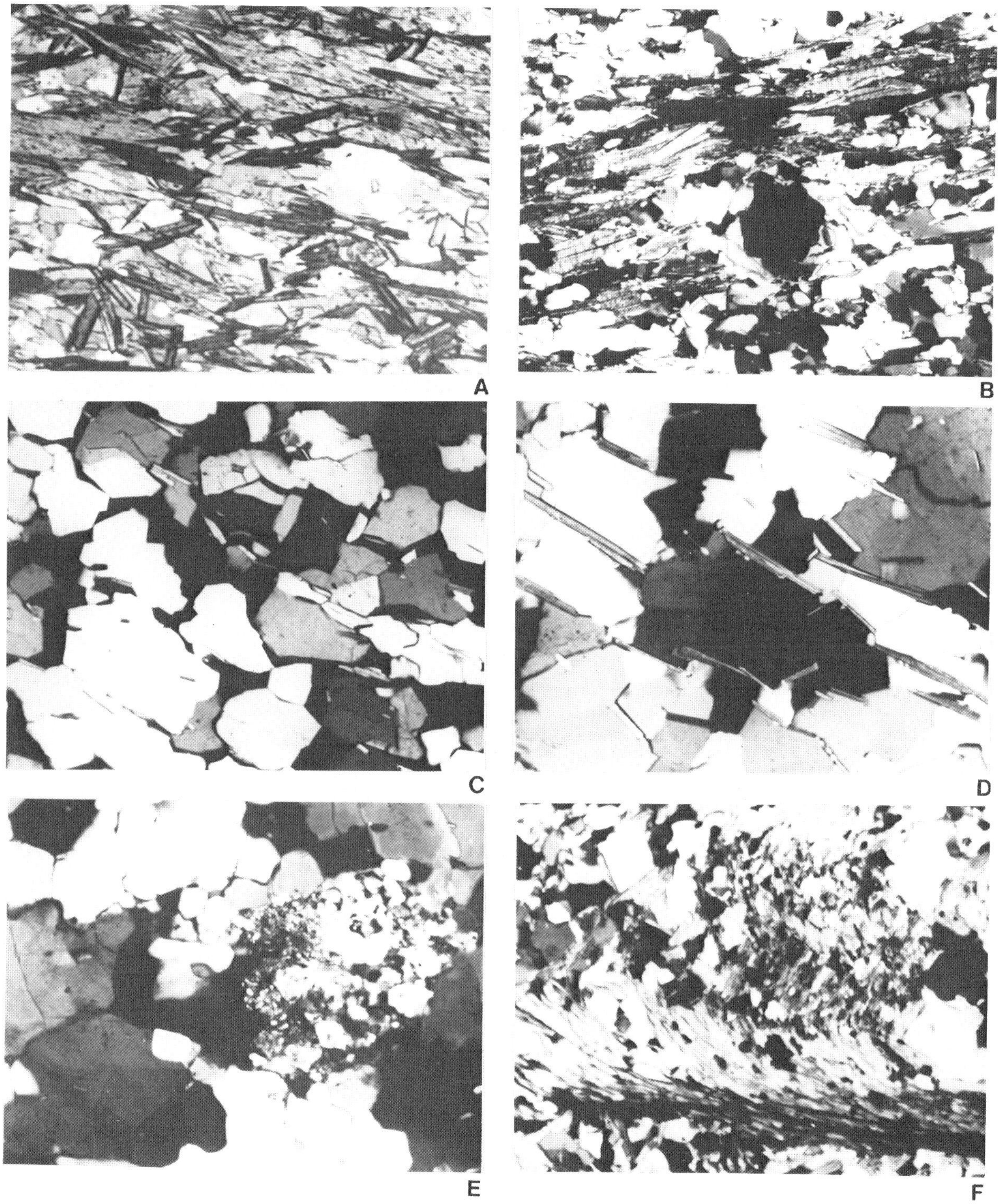

Figure 109. Photomicrographs of crystalline rocks from Higher Central Himalaya. Magnification ×63, Nicols crossed. A. Chloritoid–chlorite schists of Chail tectonic unit. Chloritoids are highly pleochroic in bluish tinge. Chlorites occur in blobs of green colour. Other associated minerals are muscovite, epidote, actinolite and sphene. Locality: Chamoli, Alaknanda valley. B. Quartz sericite schist, in the autochthonous unit below the Main Central Thrust. Locality: Chamoli, Alaknanda valley. C. Quartz–feldspathic gneiss in the Sirdang–Gam Malla tectonized zone. Locality: Khet, Darma valley. D. Schistose quartzite from the Sirdang–Gam Malla Sedimentary zone. Locality: Sirdang, Kali valley. E. Porphyritic gneiss, K-Ar date 740 ± 24 Ma. Locality: Khet, Darma valley. F. Chlorite–biotite schist, showing refolded schistosity. Locality: Jueti, Gori Ganga valley

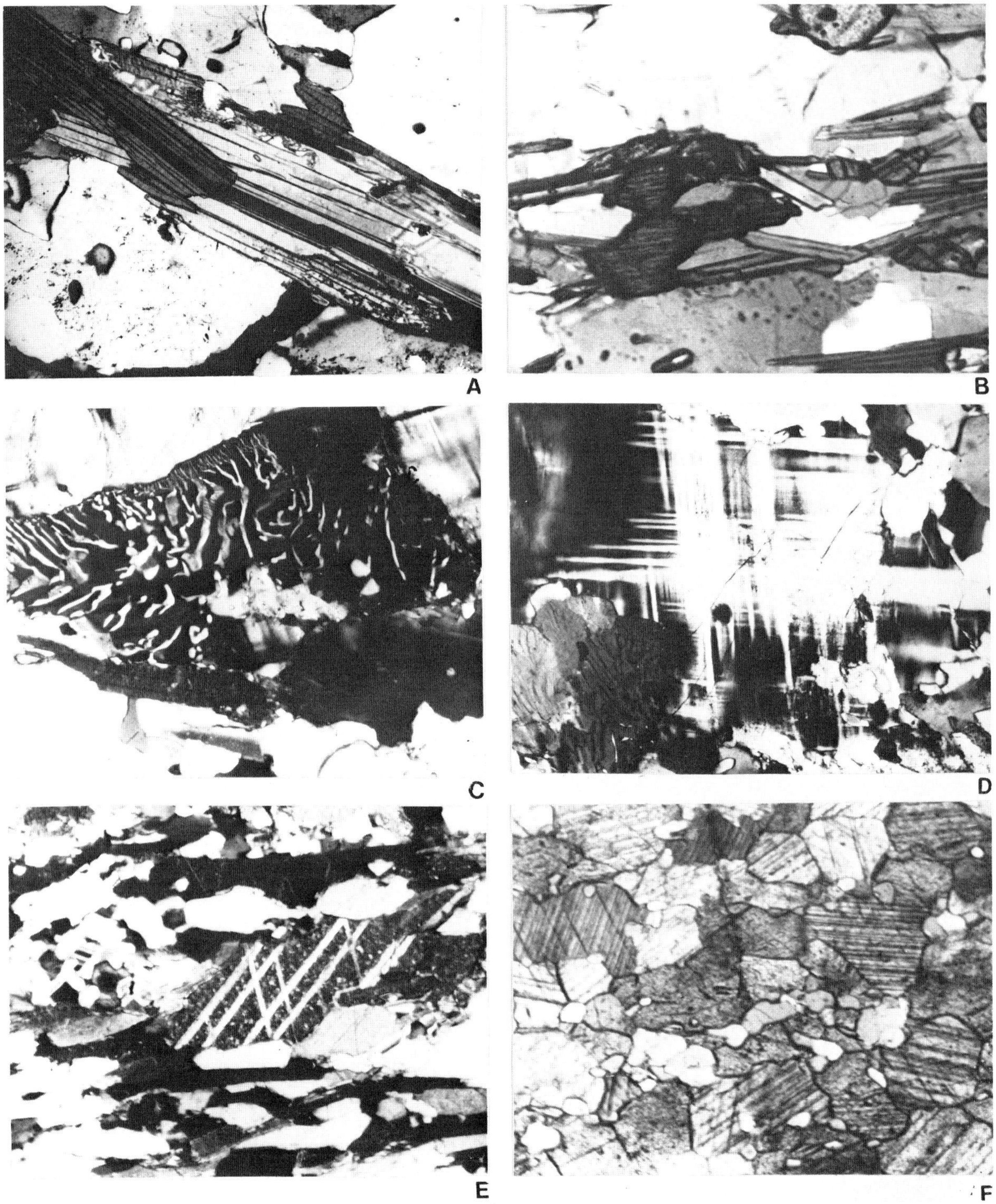

Figure 110. Photomicrographs of some magmatic and metamorphic rocks of Central Crystalline Vaikrita group of Higher Himalayan zone. Magnification ×63. Nicols crossed for A to E; polarized light for F. A. Porphyritic gneiss. K/Ar date: 261 ± 5 Ma. Locality: Suringghat Gori valley. B. Garnetiferous mica schist in the Vaikrita group. Locality: Malpa, Kali valley. C. Granitic gneiss showing the myrmekitic intergrowth of feldspar. Locality: Jipti, Kali valley. D. Granitic gneiss (= C) shows the profuse growth of microcline cross-hatching laths. Locality: Jipti, Kali valley. E. Calc-silicate rocks are the important component of Vaikrita with diopside crystals. It is of deep-seated origin and geochemically represent basic picritic composition. Locality: Budhi, Kali valley. F. Marblized limestone in the tectonized sedimentary zone of Sirdang. Locality: Jueti, Kali valley

and chlorite sericite schists from the Higher Central Himalaya.

The author sincerely acknowledges the consultation with Dr B. P. Zolotariov (GIN, Academy of Sciences of USSR, Moscow) for help in interpretation of the geochemical data. It is obviously admitted that a large volume available and yet unworked data for final interpretation is being presented in this chapter. It is hoped that a far fetched result would be derived by further studies, especially related with genetic aspects and crustal composition.

Analysis by the spectral laboratory of Institute of Geology of Ore Deposits, Petrography, Mineralogy and Geochemistry (IGEM), Academy of Sciences of USSR, Moscow. Analyst—Mrs V. V. Dobronichenko, by the method of spectrophotometry.

Explanation of specimens (Except those which has been covered in the list of silicate analyses):

Explanation of specimen analysed to Table 8.1:

1. 2AKS/74Exp:	Tourmaline pegmatitic granite. Loc. Malari, Dhauli valley.
2. 36–19/75Exp	Calc–silicate rock with calcite diopside and granulitic texture. Loc. Mapang, Gori Valley.
3. 110–16/75Exp:	Garnetiferous mica, granite. Loc. Lamari, Kali valley.
4. 115–16/75Exp:	Marblized calc-silicate rock. Loc. Budhi, Kali valley.
5. 157–20/75Exp:	Garnetiferous tourmaline granite, large microcline phenocrysts and also in ground mass. Subordinate minerals; tourmaline + sphene (?), garnet, microperthite, apatite. Loc. Budhi, Kali valley.
6. 160–20/75Exp:	Quartz–mica schist. Loc. Malipa, Kali valley.
7. 162–21/75Exp:	Quartz–mica schist. Loc. Shinkhola, Kali valley.
8. 165–21/75Exp:	Porphyritic gneiss. Loc. Samari Rungling, MCT Zone.
9. 169–21/75Exp:	Chlorite–sericite schist. Loc. Tithal Pass, Kali valley.
10. 1–1/76Exp:	Medium grained tourmaline granite with plagioclase, K-feldspar, muscovite, biotite. Loc. Badrinath MAZ zone, Alaknanda valley.
11. 2–1/76Exp:	Calc-silicate granulitic rock. Loc. Badrinath, Alaknanda valley.
12. 5–1/76Exp:	Granitogneiss. Loc. Joshimath, Alaknanda valley.
13. 9–4/76Exp:	Gneiss. Loc. Chamoli, Alaknanda valley.
14. 10–4/76Exp:	Quartz–sericite schist. Loc. Pipalkothi, Alaknanda valley.
15. 11–5/76Exp:	Granitoid or biotite gneiss. Holocrystalline hypidiomorphic texture, gneissic texture. Quartz, plagioclase, orthoclase, biotite, apatite, chlorite in mineral constituents. Loc. Mandal, Alaknanda valley.
16. 12–5/76Exp:	Granitoid. Loc. Mandal.
17. 15–7/76Exp:	Feldspathic augen gneiss, augen of altered K-feldspar sometimes also of plagioclase. Loc. Askot Klippe, Didihat.
18. 17–9/76Exp:	Graphitized amphibolitic rock. Loc. Thal, Ramganga valley.
19. 25–16/76Exp:	Amphibolite. Loc. Suring-ghat, Gori Ganga valley.
20. 27–16/76Exp:	Quartz biotite schist. Loc. Suring-ghat, Gori Ganga valley.
21. 32–18/76Exp:	Garnetiferous amphibolite with hornblende, plagioclase, calcite and sphene. Loc. Bugdiar, Gori Ganga valley (transported blocks from right bank).
22. 33–18/76Exp:	Porphyritic gneiss with egg-shaped porphyroblasts. Loc. Bugdiar (transported blocks from right bank of Gori Ganga valley).
23. 35–19/76Exp:	Tourmaline granite. Loc. Chhirkan, Gori Ganga valley.
24. 87–33/76Exp:	Streaky gneiss. Loc. Kimghat, Gori Ganga valley.
25. 22–6/78Exp:	Big porphyritic gneiss (Fig. 91), porphyroblasts 10 cm × 6 cm of plagioclase. Loc. Pangu, Kali valley.
26. 25–7/78Exp:	Big porphyritic gneiss like (25). Loc. Khet, Darma valley.
27. 27–7/78Exp:	Tectonized quartz–sillimanite gneiss from southern fault zone of Gammalla sedimentary zone. Loc. Jhamrigaon, Darma valley.
28. 28–7/78Exp:	Mica schist. Loc. Jhamrigaon above tectonite.
29. 30–7/78Exp:	Amphibolite emplaced in inverted southern wing of sedimentary zone syncline. Loc. Jhamrigaon, Darma valley.
30. 52–8/78Exp:	Amphibolite. Loc. Nyu, Darma valley, northern tectonic contact zone of sedimentary window.
31. 84–17/78Exp:	Tourmaline granite. Loc. Phurkia, Pindar valley, Main Axial Zone.
32. 73–14/78Exp:	Porphyritic gneiss. Loc. Dhakuri pass, Pindar valley.
33. 77–15/78Exp:	Porphyritic gneiss. Loc. Jignigaon, Pindar valley.
34. 78–16/78Exp:	Calc-silicate with picrite–granulitic composition. Loc. Phurkia, Pindar valley, Main Axial Tectonic Zone.
35. 79–16/78Exp:	Calc-silicate with pyrite specks. Loc. Phurkia Pindar valley, MAZ area.

36. 89–17/78Exp: Porphyritic gneiss. Loc. Khati, Pindar valley.
37. 92–17/78Exp: Sugary quartzite. Loc. Loharkhet, Saryu valley.

Explanation of specimens which have not been covered in the list of silicate analyses but included in Table 8.2:

Sp. no.	*Tentative name and location*
1. 8–8/76–Exp:	Biotite-granitoid. Loc. Gwaldham, Gwaldham–Baijnath Klippe zone.
2. 11–5/78–Exp:	Carbonaceous mica schist. Loc. Sirkha, Sirdang sedimentary zone.
3. 14–6/76–Exp:	Biotite schist. Loc. Askot Klippe zone.
4. 164–21/75Exp:	Biotite schist, schistocity developed by the well-aligned flakes of mica (biotite), other minerals quartz, plagioclase, K-feldspar and epidote as accessory. Loc. Rungling, Kali valley.
5. 59–10/78–Exp:	Porphyritic gneiss. Loc. Sobala, Darma Valley.
6. 117–17/73–Exp:	Diabase, amygdaloidal in ophiolitic suite. Loc. Kiogarh–1, Malla Johar.
7. 90–15/75–Exp:	Bluish limestone. Loc. Sirdang sedimentary window zone, Kali valley.
8. 25–6/75Exp:	Garnetiferous mica schist (granitized), rock is sheared to considerable extent, being in the zone of thrust. Quartz with wavy extinction, biotite and garnet have been partly chloritized, feldspars sericitized and mica is *bele* and wavy.
9. 13–3/75Exp:	Two mica granite or poorly foliated granitoid. Holocrystalline texture with intertonguing boundaries of the grains and a poorly developed foliation shown by biotite–muscovite laths. Clouding of feldspar and quartz is quite common. Quartz, K-feldspar perthite–plagioclase, biotite and muscovite have common deformed growth sphene; zircon, sericite and chlorite are very rare. Loc. Sela, Darma valley.
10. 15–5/73Exp:	Garnetiferous amphibolite; mineral constituents hornblende, plagioclase, quartz, biotite; foliated. Loc. Hanumanchatti (Badrinath), Alaknanda valley.
11. 163–21/75–Exp:	Porphyritic gneiss. Loc. Shinkhola camp, Kali valley.
12. 118–17A/75Exp:	Amygdaloidal spilitic rock. Loc. Kiogarh, Balcha Dhura Pass, Malla Johar (for details see *Table 6.8 of chapter 6*).
13. 160–20/75Exp:	Quartz mica schist. Loc. Malpa, Kali valley.
14. 20–4/75–Exp:	Pegmatitic granite, deformed and slightly kaolinized. Loc. Baling, Tectonic zone in Darma valley.
15. 158–20/75Exp:	Tourmaline granite intruding the gneiss-schists–calc-silicate complex as apophyses. Quartz seen squeezing along the fracture, tourmaline, coarse grained, texture porphyroblastic, phenocryst of potash-feldspars (microcline and orthoclase), shearing and subsequent recrystallization. Loc. Budhi, Tectonic zone of Kali valley. K/Ar data 39.2 ± 6 Ma.
16. 5–2/75Exp:	Carboniferous granite; quartz, plagioclase, muscovite/biotite, garnet, subordinate potash feldspar (microcline), crudely foliated. Loc. Urthing, Darma valley. K/Ar data 20.7 ± 01 Ma.
17. 165–21/75Exp:	Porphyritic gneiss. Loc. Rungling top, Kali valley.
18. 159–20/75–Exp:	Porphyritic feldspathic gneiss, Microcline perthite, quartz, plagioclase, muscovite, microcline replacing the ground mass. Loc. Budhi, Kali valley.
19. 162–21/75Exp:	Quartz mica schist. Loc. Shinkhola camp, northern right bank of tributary to Kali River.
20. 36–19/75–Exp:	Calc-silicate rock. Loc. Chhirkan, Gori Ganga valley.

Table 8.1. Silicate analysis of specimens from Central Higher Himalaya

Field specimen No.		2AKS/ 74Exp*	36–19/ 75Exp†	110–16/ 75Exp†	115–16/ 75Exp*	157–20/ 75Exp*	160–20/ 75Exp†	162–21/ 75Exp*†	165–21/ 75Exp†
1.	SiO_2	76.732	73.642	75.348	29.458	68.815	69.961	69.420	71.244
2.	TiO_2	0.04814	0.09622	0.15550	0.42004	0.13342	0.84733	0.51034	0.16149
3.	Al_2O_3	13.139	15.391	13.132	6.793	15.159	15.667	14.158	14.743
4.	Fe_2O_3	1.670	1.113	1.504	2.742	1.301	6.374	3.849	1.782
5.	MnO	0.15937	0.47423	0.39340	2.435	0.05386	2.039	0.04229	0.01926
6.	MgO	0.15937	0.47423	0.39340	2.435	0.05386	2.039	1.146	0.38899
7.	CaO	0.40	1.123	0.14268	33.779	0.21724	2.289	1.615	0.36899
8.	P_2O_5	0.11032	0.10057	0.08346	0.05629	0.02626	0.19229	0.22217	0.39203
9.	Na_2O	4.928	4.865	1.925	0.70006	2.765	2.853	3.327	0.11567
10.	K_2O	1.604	3.486	5.449	0.34956	8.754	3.225	4.285	3.412
11.	BaO	0.00160	0.02100	0.01691	0.01874	0.02619	0.05924	0.07995	0.03695
12.	SrO	0.00145	0.01822	0.00	0.13069	0.00647	0.02338	0.00918	0.00
13.	Cr_2O_3	0.00	0.00	0.00	0.1384	0.00	0.00	0.00	0.00
14.	NiO	0.00960	0.00328	0.00	0.00676	0.00065	0.00698	0.00425	0.00240

Table 8.1 (*continued*)

Field specimen No.		169–21/ 75Exp†	1–1/ 76Exp*†	2–1/ 76Exp*†	5–1/ 76Exp*†	9–4/ 76Exp*†	10–4/ 76Exp*†	11–5/ 76Exp*†	12–5/ 76Exp*†
1.	SiO_2	70.444	74.606	73.033	72.897	64.87	77.059	71.89	71.17
2.	TiO_2	0.54295	0.06992	1.138	0.17183	0.76	0.45573	0.25	0.32
3.	Al_2O_3	14.379	14.248	10.476	14.329	14.62	11.887	13.25	13.52
4.	Fe_2O_3	4.145	0.66749	6.301	1.442	5.78	2.455	2.73	3.27
5.	MnO	0.04411	0.00	0.05064	0.00	0.00	0.02909	0.05	0.03
6.	MgO	1.258	0.06568	2.878	0.29712	1.37	0.78763	0.14	0.48
7.	CaO	2.128	0.00	0.65	0.02291	2.32	0.40892	0.84	0.71
8.	P_2O_5	0.11512	0.08983	0.10947	0.11005	0.08	0.12079	0.05	0.03
9.	Na_2O	2.560	0.49323	2.013	3.098	3.062	0.13548	3.590	3.311
10.	K_2O	3.194	0.14736	2.965	3.821	4.415	4.372	5.029	4.811
11.	BaO	0.08105	0.04189	5.517	0.01571	0.08	0.04233	0.014	0.03
12.	SrO	0.00614	0.00	0.00599	0.00215	0.02145	0.00	0.00435	0.00805
13.	Cr_2O_3	0.00	0.00	0.00	0.00	0.00	0.00	0.02	0.00
14.	NiO	0.00	0.01646	0.03528	0.00235	0.00	0.00647	0.02	0.26

Field specimen No.		15–7/ 76Exp*†	17–9/ 76Exp*†	25–16/ 76Exp*†	27–16/ 76Exp*†	32–18/ 76Exp*†	33–18/ 76Exp*†	35–19/ 76Exp*†	87–33/ 76Exp*†
1.	SiO_2	79.630	47.524	49.992	67.494	46.229	72.078	72.261	69.692
2.	TiO_2	0.08230	0.23006	1.621	0.52114	0.67625	0.11523	0.11760	0.51114
3.	Al_2O_3	14.018	5.719	13.300	15.852	13.606	15.891	15.194	13.087
4.	Fe_2O_3	1.328	1.944	14.740	4.012	21.632	0.91661	2.747	1.002
5.	MnO	0.05739	0.02894	0.19629	0.04749	0.38627	0.01554	0.00837	0.03962
6.	MgO	0.43829	9.823	6.482	1.026	3.150	0.48412	0.28891	0.60219
7.	CaO	0.26495	13.719	9.507	2.343	14.205	1.274	0.00(0.56)	0.31484
8.	P_2O_5	0.23536	0.06837	0.17843	0.13266	0.30014	0.09238	0.11700	0.11567
9.	Na_2O	4.193	0.23243	2.235	3.064	0.65178	4.910	3.558	3.034
10.	K_2O	2.567	1.773	0.53328	4.613	0.24358	3.705	5.522	4.527
11.	BaO	0.01034	0.01763	0.00574	0.00708	0.02813	0.02504	0.00894	
12.	SrO	0.00114	0.00560	0.01345	0.02042	0.00157	0.02590	0.00409	0.00177
13.	Cr_2O_3	0.00	0.00	0.00	0.00	0.00	0.00	0.00	0.00
14.	NiO	0.12305	0.00696	0.00796	0.00009	0.01036	0.00	0.00	0.01235

Field specimen No.		22–6/ 78Exp*†	25–7/ 78Exp*†	627–7/ 78Exp*†	28–7/ 78Exp*†	30–7/ 78Exp*†	52–8/ 78Exp*	84–17/ 78Exp*	73–14/ 78Exp*†
1.	SiO_2	69.450	67.529	62.078	74.703	52.748	51.84	75.562	69.938
2.	TiO_2	0.21537	0.28604	0.76421	0.04079	1.318	1.56	0.03683	0.20142
3.	Al_2O_3	15.138	15.815	16.127	14.701	14.070	14.07	14.224	14.877
4.	Fe_2O_3	1.864	2.140	8.733	0.49844	14.021	11.93	0.64152	1.531
5.	MnO	0.01445	0.00	0.10309	0.00	0.09527	0.16	0.00	0.01925
6.	MgO	0.54397	0.38980	3.672	0.20351	4.534	6.07	0.07505	0.15748
7.	CaO	0.64941	0.01730	0.76567	0.00	9.733	9.62	0.63	0.46312
8.	P_2O_5	0.12524	0.06306	0.10892	0.02958	0.023294	0.16	0.01879	0.12926
9.	Na_2O	2.066	2.271	0.85343	3.742	2.320	2.735	5.550	2.294
10.	K_2O	8.229	10.462	0.00	5.279	0.68533	0.42195	2.111	7.503
11.	BaO	0.04655	0.10015	0.03537	0.00936	0.01452	0.005	0.00161	0.06606
12.	SrO	0.00751	0.00948	0.05150	0.00882	0.02731	0.02715	0.00080	0.00848
13.	Cr_2O_3	0.00	0.00	0.00	0.00	0.00	0.00	0.00	0.00
14.	NiO	0.01521	0.00138	0.00506	0.00	0.01682	0.02	0.00940	0.05933

Table 8.1 (*continued*)

	Field specimen No.	77–15/ 78Exp*†	78–16/ 78Exp*	79–16/ 78Exp*†	89–17/ 78Exp*†	92–17/ 78Exp*†
1.	SiO_2	74.618	41.49(+0.6)	42.560	69.9	92.883
2.	TiO_2	0.1033	1.55	0.30177	0.62	0.14874
3.	Al_2O_3	13.488	15.84	10.681	13.26	1.982
4.	Fe_2O_3	1.601	12.36	1.980	2.18	1.757
5.	MnO	0.02166	0.04	0.05105	0.005	0.00
6.	MgO	0.73343	13.83	1.592	2.24	0.03983
7.	CaO	0.42858	0.00	22.762	2.96	0.60
8.	P_2O_5	0.12039	0.12	0.04729	0.67	0.00
9.	Na_2O	2.116	5.083	1.557	3.239	0.18148
10.	K_2O	5.484	0.15559	3.732	5.357	0.67073
11.	BaO	0.00024	0.002	0.02559	0.018	0.00128
12.	SrO	0.00	0.01028	0.13733	0.1332	0.00
13.	Cr_2O_3	0.00	0.00	0.00	0.00	0.00
14.	NiO	0.00317	0.00	0.1047	0.002	0.01164

* Spectral analysis also done (see Table 8.2.).
† Silicate analysis repeated at Wadia Institute of Himalayan Geology Laboratory. Determination with AAS and UV-visible spectrophotometer.
Analyses done by the Spectral-Laboratory of the Institute of Geology and Ore Deposits, Petrography, Mineralogy and Geochemistry (IGEM), Academy of Sciences of USSR, Moscow. Analyst – Mme V. V. Dobronichenko. Method: Spectrophotometry.

Table 8.2. Semi-quantitative spectral analyses of rock specimens from Central Higher Himalaya

		Field Sp. no.								
Sl. no.	Elements in ppm	27–7/ 78Exp†	8–8/ 76Exp	11–5/ 78Exp	15–7/ 76Exp*	10–4/ 76Exp*	110–16/ 75Exp*	32–18/ 76Exp	17–9/ 76Exp*	5–1/76 76Exp*
1.	Be	4–6	7–9	7–9	7–9	3	1–3	7–9	1–3	7–9
2.	As							100		
3.	Te									
4.	Sc	10–30	10	10–30	10–30	–	10–30	10	10–30	
5.	P	1000	400–600	300	400–600	700–900	400–600	–	300	400–600
6.	Sb	–	–	–						
7.	Ta	–	–	–						
8.	Tl	–	–	–						
9.	Mn	400–600	200–300	70–90	100	100	100–300	1000–3000	100–300	100
10.	Pb	10–30	70–90	100	40–60	30	40–60	–	70–90	100
11.	Sn	1–3	7–9	4–6	10–30	–	4–6	–	–	10–30
12.	Nb	10	30	30	30	10–30	10–30	10–30	?	30
13.	W	–	–	–	–	–	–	–	–	
14.	Ga	10	70–90	30	100	10	10	4–6	10	10–30
15.	Ge									
16.	In									
17.	Bi									
18.	Mo	2–3	–	2–3	–	–	–	1–3	7–9	–
19.	V	100	7–9	10	7–9	10–30	10	40–60	40–60	10–30
20.	Li	–	40–60	–	40–60	–	30	–	–	40–60
21.	Cu	40–60	10	7–9	10	10	3	10	30	10–30
22.	Cd									
23.	Ag									
24.	Na	7000–9000	40000–60000	10000–30000	70000–90000	1000–3000	10000–30000	7000–9000	4000–6000	10000
25.	Zn	100–300	40–60	–	–	70–90	40–60	40–60	–	100
26.	Ti	1000–3000	1000	700–900	70–90	1000–3000	700–900	1000–3000	1000–3000	10000
27.	Co	10–30							10–30	
28.	Ni	40–60	–	10	–	10	–	10	10–30	7–9
29.	Zr	100	100	400–600	70–90	100–300	100	100	300	100–300
30.	Mg	10000	700–900	400–600	4000–6000	7000–9000	1000–3000	40000–60000	70000–90000	4000–6000
31.	Si	70000–90000	70000–90000	70000–90000	70000–90000	70000–90000	70000–90000	70000–90000	70000–90000	70000–90000
32.	Al	70000–90000	70000–90000	70000–90000	70000–90000	70000–90000	70000–90000	70000–90000	70000–90000	70000–90000
33.	Fe	70000–90000	10000–30000	10000	10000	10000	7000–9000	7000–9000	10000	7000–9000
34.	Cr	100	10	10	10	10–30	10	10–30	40–60	10
35.	Ca	4000–6000	7000–9000	10000	4000–6000	1000–3000	4000–6000	40000–60000	70000–90000	7000–9000
36.	Sr	100	100–300	10	100–300	100	100–300	100–300	100–300	100–300
37.	Ba	300	400–600	100–300	100–300	400–600	100–300	100	100–300	300
38.	La	–	40–60	100	–	10–30	10–30	–	40–60	–
39.	Ce	–	–	–	–	–	–	–	–	–
40.	Y	10–30	100–300	100–300	10–30	30	40–60	–	30	10–30
41.	Yb	–	1–3	1–3	–	–	–	–	–	–
42.	Hg	–	–	–	–	–	–	–	–	–
43.	B	10–30	10	7–9	10–30	100–300	10–30	4–6	40–60	100

Table 8.2 (*continued*)

Sl. no.	Elements in ppm	78–16/ 78Exp*	87–33/ 76Exp*	14–6/ 76Exp	11–5/ 76Exp*	89–17/ 78Exp*	164–21/ 75Exp*	115–16/ 75Exp*	10–3/ 75 AKS/Exp*	2–1/ 76Exp*
1.	Be	1–3	4–6	4–6	7–9	4–6	7–9	1–3	4–6	1–3
2.	As	–	–	–	–	–				
3.	Te	–	–							
4.	Sc	10–30	10	10–30	–	10–30	10–30	10–30	–	10–30
5.	P	400–600	400–600	400–600	400–600	700–900	700–900	700–900	400–600	1000
6.	Sb	–	–	–	–	–	–	–	–	–
7.	Ta	–	–	–	–	–	–	–	–	–
8.	Tl	–	–	–	–	–	–	–	–	–
9.	Mn	40–60	300	100–300	100	300	100–300	1000	100	100–300
10.	Pb	10–30	70–90	10–30	100–300	100	30	–	40–60	10
11.	Sn	7–9	4–6	–	–	–	7–9	–	30	–
12.	Nb	40–60	30	10–30	40–60	10–30	10–30	40–60	30	10–30
13.	W	–	–	–	–	–	–	–	–	–
14.	Ga	10–30	30	10–30	30	40–60	10–30	4–6	40–60	10–30
15.	Ge	–	–	–	–	–	–	–	–	–
16.	In	–	–	–	–	–	–	–	–	–
17.	Bi	–	–	–	–	–	–	–	–	–
18.	Mo	–	–	–	3	–	1–3	–	1–3	–
19.	V	40–60	40–60	40–60	10	30	40–60	40–60	4–6	100
20.	Li	–	30	–	–	–	–	–	40–60	–
21.	Cu	1–3	1–3	1–3	1–3	7–9	40–60	10	10	30
22.	Cd	–	–	–	–	–	–	–	–	–
23.	Ag	–	–	–	–	–	–	–	–	–
24.	Na	70000–90000	10000–30000	40000–60000	10000–30000	40000–60000	40000–60000	1000	40000–60000	10000–30000
25.	Zn	–	40–60	40–60	40–60	40–60	100	100–300	70–90	10–30
26.	Ti	4000–6000	1000–3000	1000–3000	1000	3000	4000–6000	1000	700–900	7000–9000
27.	Co	–	–	10–30	–	–	10	–	–	10–30
28.	Ni	10–30	10	40–60	–	10	10–30	10–30	–	30
29.	Zr	700–900	100–300	100–300	400–600	300	300	100–300	100	700–900
30.	Mg	10000–30000	7000–9000	10000–30000	400–600	3000	10000	10000–30000	3000	10000
31.	Si	70000–90000	70000–90000	70000–90000	70000–90000	70000–90000	70000–90000	70000–90000	70000–90000	70000–90000
32.	Al	70000–90000	70000–90000	70000–90000	70000–90000	70000–90000	70000–90000	70000–90000	70000–90000	70000–90000
33.	Fe	7000–9000	10000	70000–90000	10000	10000–30000	40000–60000	70000–90000	10000	20000–30000
34.	Cr	100	10–30	100	–	10–30	30	–	–	100
35.	Ca	40000–60000	10000	10000–30000	7000–9000	10000–30000	10000–30000	70000–90000	4000–6000	7000–9000
36.	Sr	300	100–300	100–300	100–300	100–300	100–300	300	100–300	100–300
37.	Ba	10–30	400–600	40–60	300	700–900	400–600	70–90	40–60	300
38.	La	100	10	100	100–300	40–60	40–60	100	10–30	100
39.	Ce	–	–	–	–	–	–	–	–	–
40.	Y	–	–	–	1–3	–	–	–	–	–
41.	Yb	–	–	–	1–3	–	–	–	–	–
42.	Hg	–	–	–	–	–	–	–	–	–
43.	B	10–30	10	10	10	10	10–30	4–6	100	7–9

Sl. no.	Elements in ppm	30–7/ 78Exp*	22–6/ 78Exp*	52–8/ 78Exp*	77–15/ 78Exp*	25–7/ 78Exp*	1–1/ 76Exp*	73–14/ 78Exp*	59–10/ 78Exp*	117–17/ 73Exp
1.	Be	3	10	1–3	3	4–6	3	4–6	10–30	7–9
2.	As	–	–	–	–	–	–	–	–	–
3.	Te	–	–	–	–	–	–	–	–	–
4.	Sc	40–60	10	10	–	10–30	10	10	10–30	10–30
5.	P	1000–3000	700–900	1000–3000	400–600	400–600	400–600	400–600	1000–3000	1000–3000
6.	Sb	–	–	–	–	–	–	–	–	–
7.	Ta	–	–	–	–	–	–	–	–	–
8.	Tl	–	–	–	–	–	–	–	–	–
9.	Mn	1000–3000	100–300	1000	300	100–300	100–300	100–300	300	700–900
10.	Pb	10	40–60	10–30	70–90	100	40–60	30	70–90	10–30
11.	Sn	–	–	–	4–6	7–9	7–9	10	–	–
12.	Nb	10–30	10–30	10–30	10–30	10–30	10–30	10–30	10–30	10
13.	W	–	–	–	–	–	–	–	–	–
14.	Ga	10–30	10–30	30	30	30	10–30	40–60	40–60	30
15.	Ge	–	–	–	–	–	–	–	–	–
16.	In	–	–	–	–	–	–	–	–	–
17.	Bi	–	–	–	–	–	–	–	–	–
18.	Mo	1–3	1–3	1–3	–	1–3	–	1–3	–	1–3
19.	V	300	10–30	300	10–30	10–30	10–30	10–30	30	100–300
20.	Li	–	–	–	40–60	–	40–60	–	–	100
21.	Cu	100–300	1–7	10–30	30	40–60	4–6	4–6	4–6	10
22.	Cd	–	–	–	–	–	–	–	–	–
23.	Ag	–	–	–	–	–	–	–	–	–
24.	Na	10000–30000	10000–30000	30000	40000–60000	40000–60000	70000–90000	70000–90000	70000–90000	10000
25.	Zn	100–300	40–60	100	100–300	100–300	40–60	70–90	100	300
26.	Ti	10000	1000	10000	1000	1000–3000	1000	1000–3000	3000	7000–9000
27.	Co	70–90	–	40–60	–	–	–	–	–	40–60
28.	Ni	70–90	–	40–60	10	10	10	10	10–30	100
29.	Zr	100–300	100–300	100	100	100–300	–	100	100–300	100
30.	Mg	40000–60000	10000	40000–60000	1000–3000	4000–6000	1000–3000	1000–3000	1000–3000	70000–90000
31.	Si	70000–90000	70000–90000	70000–90000	70000–90000	70000–90000	70000–90000	70000–90000	70000–90000	70000–90000
32.	Al	70000–90000	70000–90000	70000–90000	70000–90000	70000–90000	70000–90000	70000–90000	70000–90000	70000–90000
33.	Fe	70000–90000	10000–30000	70000–90000	10000	10000–30000	7000–9000	10000–30000	70000–90000	70000–90000
34.	Cr	30	10–30	70–90	10	30	10	10–30	10–30	100–300
35.	Ca	40000–60000	4000–6000	40000–60000	7000–9000	10000	10000	10000–30000	30000	10000–30000
36.	Sr	300	100	100–300	100–300	100–300	100–300	100	100–300	100
37.	Ba	300	700–900	70–90	300	400–600	400–600	400–600	100	100–300
38.	La	40–60	–	–	–	10–30	10–30	40–60	100	–
39.	Ce	–	–	–	–	–	–	–	400–600	–
40.	Y	40–60	10–30	10–30	10	10–30	10	30	100	–
41.	Yb	–	–	–	–	–	–	–	–	–
42.	Hg	–	–	–	–	–	–	–	–	–
43.	B	7–9	7–9	7–9	7–9	4000–6000	7–9	7–9	10	700–900

Table 8.2 (*continued*)

Sl. no.	Elements in ppm	84–17/ 75Exp*	79–16/ 78Exp*	27–7/ 78Exp*	90–15/ 75Exp	25–6/ 73Exp	162–21/ 75Exp*	12–5/ 76Exp*	13–3/ 75Exp	15–5/ 73Exp
1.	Be	7–9	1–3	4–6	3	3	1–3	10	1–3	3
2.	As	–	–	–	–	–	–	–100	–	–
3.	Te	–	–	–	–	–	–	–	–	–
4.	Sc	10	10–30	10	10	10–30	10–30	10–30	10	30
5.	P	300	400–600	–	400–600	1000	1000	1000	300	1000
6.	Sb	–	–	–	–	–	–	–	–	–
7.	Ta	–	–	–	–	–	–	–	–	–
8.	Tl	–	–	–	10	–	–	–	–	–
9.	Mn	100–300	100–300	100	100	400–600	300	400–600	70–90	1000–3000
10.	Pb	100	100–300	40–60	100	30	100	100	100	10
11.	Sn	10–30	–	7–9	10–30	–	7–9	10	4–6	–
12.	Nb	70–90	–	10–30	30	10–30	10–30	100	30	10
13.	W	–	–	–	–	–	–	–	–	–
14.	Ga	30	–	–	–	–	30	40–60	10–30	10
15.	Ge	–	7–9	10–30	30	10–30	–	–	–	–
16.	In	–	–	–	–	–	–	–	–	–
17.	Bi	–	–	–	–	–	–	–	–	–
18.	Mo	–	–	–	–	1–3	–	1–3	–	–
19.	V	10	40–60	10	10–30	70–90	40–60	10	10–30	100–300
20.	Li	–	–	–	–	–	–	–	–	–
21.	Cu	4–6	10–30	4–6	10–30	70–90	10	7–9	10–30	10–30
22.	Cd	–	–	–	–	–	–	–	–	–
23.	Ag	–	–	–	–	–	–	–	–	–
24.	Na	70000–90000	4000–6000	10000–30000	10000–30000	10000–30000	70000–90000	70000–90000	30000	10000–30000
25.	Zn	40–60	–	40–60	100–300	100	100	100–300	40–60	100
26.	Ti	300	1000–3000	700–900	1000–3000	4000–6000	4000–6000	3000	1000	10000–30000
27.	Co	–	–	–	–	30	10	–	–	40–60
28.	Ni	–	30	–	10	40–60	10–30	10	–	70–90
29.	Zr	40–60	100–300	–	100	100–300	300	400–600	40–60	100
30.	Mg	400–600	40000–60000	1000	1000	10000–30000	10000	700–900	400–600	40000–60000
31.	Si	70000–90000	70000–90000	70000–90000	70000–90000	70000–90000	70000–90000	70000–90000	70000–90000	70000–90000
32.	Al	70000–90000	70000–90000	70000–90000	70000–90000	70000–90000	70000–90000	70000–90000	70000–90000	70000–90000
33.	Fe	4000–6000	70000–90000	7000–9000	10000	70000–90000	30000	40000–60000	4000–6000	70000–90000
34.	Cr	–	10	–	10	100	30	–	10	40–60
35.	Ca	4000–6000	70000–90000	3000	4000–6000	10000	10000–30000	10000–30000	3000	40000–60000
36.	Sr	100	300	100–300	100	100–300	100–300	100–300	70–90	100–300
37.	Ba	40–60	100	100–300	100	700–900	700–900	100	100–300	70–90
38.	La	40–60	–	10–30	–	–	70–90	100–300	–	–
39.	Ce	–	–	–	–	–	–	1000	–	–
40.	Y	30	10–30	10–30	10–30	30	40–60	100	10–30	10–30
41.	Yb	–	–	–	–	–	–	1–3	–	–
42.	Hg	–	–	–	–	–	–	–	–	–
43.	B	1000	10	10	1000	7–9	7–9	7–9	7–9	7–9

Sl. no.	Elements in ppm	2AKS/ 74Exp*	163–21/ 75Exp	118–17A/ 73Exp	9–4/ 76Exp*	7–16/ 76Exp*	25–16/ 76Exp*	35–19/ 76Exp*	160–20/ 75Exp	20–4/ 75Exp
1.	Be	1	4–6	1–3	3	4–6	1	10	1–3	10–30
2.	As	–	–	–	–	–	–	–	–	–
3.	Te	–	–	–	–	–	–	–	–	–
4.	Sc	10	10–30	10–30	10–30	10–30	30	–	10	10
5.	P	300	400–600	1000–3000	1000	1000	1000–3000	–	400–600	400–600
6.	Sb	–	–	–	–	–	–	–	–	–
7.	Ta	–	–	–	–	–	–	–	–	–
8.	Tl	–	–	–	–	–	–	–	–	–
9.	Mn	70–90	100	1000–3000	400–600	400–600	100	70–90	400–600	100–300
10.	Pb	40–60	70–90	10–30	40–60	40–60	10–30	70–90	40–60	100
11.	Sn	4–6	7–9	–	7–9	–	–	10–30	–	30
12.	Nb	10–30	10–30	40–60	10	10–30	10–30	10	10–30	10–30
13.	W	–	–	–	–	–	–	–	–	–
14.	Ga	10–30	10–30	10–30	30	10–30	10–30	30	10–30	40–60
15.	Ge	–	–	–	–	–	–	–	–	–
16.	In	–	–	–	–	–	–	–	–	–
17.	Bi	–	–	–	–	–	–	–	–	40–60
18.	Mo	–	–	–	–	1–3	1–3	–	–	1–3
19.	V	4–6	10–30	70–90	40–60	30	100–300	10	10–30	10–30
20.	Li	–	–	–	–	–	–	–	–	70–90
21.	Cu	4–6	10–30	10	7–9	10	30	3	10	7–9
22.	Cd	–	–	–	–	–	–	–	–	–
23.	Ag	–	–	–	–	–	–	–	–	–
24.	Na	70000–90000	40000–60000	40000–60000 70000–90000	10000–30000	40000–60000	10000–30000	10000–30000 40000–60000	10000–30000	40000–60000
25.	Zn	100	100–300	40–60	–	100	100	40–60	40–60	400–600
26.	Ti	40000–60000	10000	10000–30000	4000–6000	3000	4000–6000	700–900	1000	700–900
27.	Co	–	–	10–30	10–30	10	40–60	–	–	–
28.	Ni	–	–	10–30	10–30	10	40–60	–	–	–
29.	Zr	–	100	300	100–300	100–300	100	400–600	100–300	40–60
30.	Mg	70–90	700–900	10000	4000–6000	7000–9000	40000–60000	1000	3000	1000–3000
31.	Si	70000–90000	70000–90000	70000–90000	70000–90000	70000–90000	70000–90000	70000–90000	70000–90000	70000–90000
32.	Al	70000–90000	70000–90000	70000–90000	70000–90000	70000–90000	70000–90000	70000–90000	70000–90000	70000–90000
33.	Fe	4000–6000	10000	70000–90000	10000–30000	40000–60000	70000–90000	40000–60000	7000–9000	7000–9000
34.	Cr	–	10	–	10–30	10–30	10–30	10	10	10
35.	Ca	40–60	7000–9000	40000–60000	10000–60000	10000–60000	40000–60000	4000–6000	10000–30000	10000
36.	Sr	–	100	300	100–300	100–300	100–300	100	100	100
37.	Ba	–	100–300	100–300	400–600	400–600	100	100–300	300	100–300
38.	La	10–30	10–30	40–60	10–30	70–90	–	–	30	10–30
39.	Ce	–	–	–	–	–	–	–	–	–
40.	Y	–	10	10–30	40–60	10–30	10	10–30	40–60	10
41.	Yb	–	–	–	–	–	–	–	–	–
42.	Hg	–	–	–	–	–	–	–	–	–
43.	B	4000–6000	10–30	10	7–9	–	7–9	700–900	10–30	10–30

Table 8.2 (*continued*)

Sl. no.	Elements in ppm	158–20/ 75Exp‡	5–2/ 75Exp‡	165–21/ 75Exp	159–20/ 75Exp	15–7/ 76Exp*‡	2–21/ 75Exp*‡	87–33/ 76Exp*‡	92–17/ 78Exp*‡	36–19/ 76Exp
1.	Be	7–9	4–6	1–3	4–6	3	1–3	4–6	–	4–6
2.	As	–	–	–	–	–	–	–	–	–
3.	Te	–	–	–	–	–	–	–	–	–
4.	Sc	–	–	10	–	–	40–60	–	10	10–30
5.	P	400–600	300	400–600	400–600	300	–	300	700–900	400–600
6.	Sb	–	–	–	–	–	–	–	–	–
7.	Ta	–	–	–	–	–	–	–	–	–
8.	Tl	–	–	–	–	–	–	–	–	–
9.	Mn	100	100–300	100–300	100	100	1000–3000	70–90	10–30	300
10.	Pb	70–90	30	40–60	100	40–60	10–30	40–60	10	40–60
11.	Sn	–	–	40–60	10	10–30	–	–	–	10
12.	Nb	30	10–30	30	30	30	10–30	10–30	30	10–30
13.	W	–	–	–	–	–	–	–	–	–
14.	Ga	10–30	30	40–60	40–60	40–60	10–30	30	1–3	10–30
15.	Ge	–	–	–	–	–	–	–	–	–
16.	In	–	–	–	–	–	–	–	–	–
17.	Bi	–	–	–	40–60	–	–	–	–	–
18.	Mo	–	–	1–3	1–3	1–3	3	–	1–3	–
19.	V	7–9	10–30	10–30	10	7–9	300	10–30	4–6	30
20.	Li	–	40–60	40–60	70–90	10–30	–	10–30	–	–
21.	Cu	3	1–3	7–9	1–3	4–6	100–300	7–9	10	40–60
22.	Cd	–	–	–	–	–	–	–	–	–
23.	Ag	–	–	–	–	–	–	–	–	–
24.	Na	40000–60000	40000–60000 70000–90000	10000–30000	40000–60000 70000–90000	40000–60000	10000–30000	40000–60000 70000–90000	700–900 –	10000–30000 40000–60000
25.	Zn	70–90	40–60	70–90	40–60	40–60	–	–	40–60	–
26.	Ti	400–600	700–900	1000	400–600	400–600	7000–9000	1000–3000	400–600	4000–6000
27.	Co	–	–	–	–	–	70–90	–	–	10
28.	Ni	–	–	–	–	–	100	–	–	30
29.	Zr	40–60	70–90	70–90	40–60	40–60	400–600	100	100	100–300
30.	Mg	1000–3000	1000–3000	3000	1000–3000	1000–3000	40000–60000	1000–3000	400–600	1000–3000
31.	Si	70000–90000	70000–90000	70000–90000	70000–90000	70000–90000	70000–90000	70000–90000	70000–90000	70000–90000
32.	Al	70000–90000	70000–90000	70000–90000	70000–90000	70000–90000	70000–90000	70000–90000	70000–90000	70000–90000
33.	Fe	7000–9000	7000–9000	7000–9000	4000–6000	4000–6000	70000–90000	7000–9000	7000–9000	10000–30000
34.	Cr	10	10	10	10	10	100–300	10	10	70–90
35.	Ca	10000	10000	4000–6000	10000–30000	4000–6000	40000–60000	10000–30000	400–600	70000–90000
36.	Sr	100	100	70–90	10–30	100	100–300	100–300	30	300
37.	Ba	100–300	300	100	100–300	100–300	400–700	400–700	30	300
38.	La	40–60	–	30	10–30	10–30	?	40–60	10–30	40–60
39.	Ce	–	–	–	–	–	–	–	–	–
40.	Y	–	10	10–30	10–30	10–30	40–60	10–30	10	40–60
41.	Yb	–	–	–	–	–	–	–	–	–
42.	Hg	–	–	–	–	–	–	–	–	–
43.	B	1000–3000	7–9	400–600	10–30	7–9	7–9	10–30	40–60	10

Sl. no.	Elements in ppm	33–18/ 76Exp*	28–7/ 78Exp*	157–20/ 75Exp*	22–6/ 78Exp*†
1.	Be	4–6	10–30	400–600	1–3
2.	As	–	–	–	–
3.	Te	–	–	–	–
4.	Sc	–	10–30	–	–
5.	P	–	1000	1000–3000	–
6.	Sb	–	–	–	–
7.	Ta	–	–	–	–
8.	Tl	–	–	–	–
9.	Mn	40–60	400–600	10000–30000	40–60
10.	Pb	40–60	40–60	70–90	40–60
11.	Sn	7–9	10	70–90	–
12.	Nb	10–30	30	10–30	10–30
13.	W	–	–	–	–
14.	Ga	10–30	10–30	100	300
15.	Ge	–	–	70–90	–
16.	In	–	–	–	–
17.	Bi	–	–	–	–
18.	Mo	–	–	1–3	–
19.	V	30	70–90	40–60	10–30
20.	Li	–	–	–	–
21.	Cu	10	4–6	40–60	10
22.	Cd	–	–	–	–
23.	Ag	–	–	–	–
24.	Na	10000–30000	10000–30000	40000–60000	10000–30000
25.	Zn	–	70–90	–	40–60
26.	Ti	1000–3000	3000	700–900	700–900
27.	Co	–	10–30	–	–
28.	Ni	–	70–90	–	–
29.	Zr	100	100–300	100	100
30.	Mg	10000–30000	1000–3000	1000–3000	1000–3000
31.	Si	70000–90000	70000–90000	70000–90000	70000–90000
32.	Al	70000–90000	70000–90000	70000–90000	70000–90000
33.	Fe	3000	40000–60000	10000–30000	3000
34.	Cr	10–30	100–300	10–30	10
35.	Ca	10000–30000	10000	40000–60000	4000–6000
36.	Sr	400–600	100–300	100–300	100
37.	Ba	300	300	70–90	700–900
38.	La	10–30	40–60	–	10–30
39.	Ce	–	–	–	–
40.	Y	–	70–90	100	10–30
41.	Yb	–	1–3	1–3	–
42.	Hg	–	–	–	–
43.	B	10	100–300	7–9	4–6

* See Table 8.1

† Silicate analyses done with plasma-spectrophotometer Jobin Yvon (France) at Geological Institute of Academy of Sciences, USSR, Moscow, under the supervision of Dr. D. Y. Choporov.

‡ K/Ar Sp. No. 46–39.2 ± 1 Ma; Sp. No. 47—20.7 ± 01 Ma; Sp. No. 57—37.6 ± 1.1 Ma.

Table 8.3. Elemental quantitative geochemical analyses at WIHG Laboratory (Perkin–Elmer atomic absorption) (in ppm)

Sl. no.	Sample no.	Mn	Li	Cd	Pb	Co	Cu	Cr	Ti	Sr	Zn
1.	16–5/78*	108.0	5.4	2.85	50	60	15	214.0	7140	202	114
2.	52–8/78*	145.0	18.2	2.8	125	60	25	320	4286	168	166
3.	162–21/75*	580	52.9	2.8	100	100	110	386	7140	940	293
4.	89–17/78*	340	36.4	2.8	75	60	28	142.8	7140	236	120.8
5.	1–1/76*	75.0	9.0	2.8	50	20	18	171.4	1428	302	28.0
6.	77–15/78*	95.0	45.4	2.8	40	40	20	198	2856	56	74
7.	9–4/76*	425	14.6	2.8	50	60	17.5	142.8	10000	346	230
8.	79–16/78*	280	10.9	2.8	100	60	22.5	114	4286	1740	252.4
9.	30–7/78*	2140	3.64	2.8	50	140	97.5	314	14286	340	252.4
10.	84–17/78*	110	3.6	1.4	75	140	40	940	15714	426	288
11.	28–17/78*	88	10.9	Nil	75	20	17.5	171.4	Nil	156.8	157.2
12.	73–14/78*	128	28	1.4	75	20	142.8	1428.6	134.4	20	142.8
13.	22–6/78*	92.5	20	Nil	100	40	35	700	1428	123.2	156.4
14.	88–33/76*	393	72	4.2	75	40	30	360	4286	89.6	156.4
15.	35–19/76*	72.5	9.0	2.86	100	Nil	25	286	1428	89.6	174.6
16.	165–21/75	132.5	40	Nil	125	Nil	30	186	Nil	44.8	226
17.	160–20/75	375	43.6	1.42	75	60	17.6	114	7140	224	268
18.	157–20/75*	305	154.6	2.8	74	60	17.6	528	4286	536	141.8
19.	10–4/76*	187.5	Nil	Nil	50	20	25	389	4286	33.6	261.4
20.	27–7/78*	472.5	40	Nil	50	60	30	157.2	7140	56	254
21.	11–5/78*	165	14.6	Nil	125	Nil	22.5	300	1428	56	180
22.	110–15/75*	262.5	14.6	Nil	75	Nil	16.5	100	2856	54.8	162.8
23.	92–17/78*	57.5	Nil	Nil	25	Nil	22.5	756	1428	22.4	180
24.	36–19/76	107.5	47.2	Nil	75	Nil	20	172	Nil	324	232
25.	2–1/76*	412.5	27.2	2.86	25	40	35	514	11428	123.2	249
26.	17–9/76	235	12.8	2.86	200	40	72.5	242.8	4286	145.6	258
27.	25–16/76*	1440	2.6	2.8	50	140	40	260	12860	336	374
28.	27–16/76*	325	27.2	–	50	40	42.5	300	4286	336	160
29.	25–7/78*	142.5	20	–	75	–	25	260	2858	156.8	160
30.	2AKS/74*	230	7.2	–	150	–	15	300	1428.6	–	160
31.	S–1/76	125	67.2	4.2	75	–	55	242.8	1428.6	100.8	268
32.	115–16/75*	1125	3.6	–	100	60	22.5	328.6	4286	1433.6	295.2
33.	10–3/75*	460	23.6	–	200	40	325	357.2	5714	604.8	640
34.	32–18/76*	3030	7.2	4.2	75	80	775	414	5714	100.8	276
35.	169–21/75†	102.5	27.2	–	100	–	147.5	357.2	–	112	272
36.	33–18/76*	115	54.6	–	100	–	20	200	1428.6	425.6	144.6
37.	15–7/76*	252.5	50.9	–	75	–	40	142.8	5714	112	145.4
38.	12–5/76*	207.5	14.5	–	125	–	20	900	2857.2	67.2	172

Analyst—Drs M. S. Rathi and P. P. Khanna (WIHG Geochemical Lab.).
* Specimen with repeated spectral analysis of IGEM LAB.
1. 16–5/78–Exp Micaceous graphitic schist, location Sirkha, Sirdang sedimentary zone, Kali valley.
2. S–1/76 Metabasic, location Rudraprayag.
3. 169–21/75Exp Chlorite–sericite schist (phyllitic), location Tithal Pass, Sirdang sedimentary zone, Kali valley.

Chapter 9

Barite–Polymetallic Mineralization

9.1 Historical Development

The Tethyan zone of the greater part of Central Himalaya was considered to be barren, without any reported occurrences of mineralization since the first geological reports by the end of last century. This information was based on expedition traverses. Road track building activities in early sixties provided well exposed fresh rock cut walls for observation by the author's expedition 1973 (Shah and Sinha, 1974b; Sinha, 1974b, 1978c). It was the promising quality of barite analysed for the first time with 64.2 per cent BaO which promoted the author to *take up detailed prospecting and exploration work in the area in the following* field season of 1974, engaging a survey party from the Survey of India, Dehra Dun. Continuation of a long-term project in the Greater and Tethyan zone of Central Himalaya of Kumaun and Garhwal and successive expeditions in the Darma and Gori Ganga valleys in 1975 and 1976 enabled the author to establish further the continuation of mineralized zones (Figure 111) to the southeastern.

9.2 Geological Setting

Tethyan sediments in the Higher Central Himalaya of Garhwal and Kumaun zone is situated in the north of Central Crystalline zone in tectonic contact with Vaikrita Group of metamorphic and igneous rocks (Figures 13, 111 and 112). The lowermost part of the thick Precambrian–Palaeozoic and Mesozoic sequence is represented by phyllite and carbonaceous slates of Martoli. This thick sediment group (over 5,000 m) of the Martoli formation is unconformably overlain by a continuous sequence of Ralam Conglomerate–Garbyang carbonate-terrigenous–Shiala carbonate-argillaceous–Yong reefal biothermal limestone–'Variegated' arenaceous carbonate–argillaceous formation and Muth quartzite carbonate series representing Precambrian, Cambrian, and Silurian–Devonian periods. After a marked Hercynian gap, the rock series continues with Mesozoics to lower Tertiary (Shah and Sinha, 1974b; Sinha, 1980c; Mehrotra and Sinha, 1981).

9.3 Barite–Polymetallic Mineralization Occurrences

The first observed occurrence of the barite–polymetallic mineralization vein is on the right bank of the Girthi Ganga gorge at Barmatiya, about 70 km as the crow flies east of the Badrinath holy Shrine (Figure 12), and 16 km from Malari (75°59′ : 30°47′). Further southeast, approximately 70 km as the crow flies from the first abovementioned deposit, in the left bank of Gonkha Gad, a tributary of Gori Ganga, about 2 km upstream from village Milam, near Lakhuri Bhel (80°10′ : 30°26′30′), we come across another polymetallic mineralization occurrence with Cu–Fe sulphide ores associated with quartz vein in the phyllitic rocks below the Martoli–Ralam unconformity.

A similar third occurrence is situated a further 50 km as the crow flies, in the gorge of Darma Ganga river along the southeast prolongation at Lasser Yangti (80°32′ : 30°18′) and approximately 6 km downstream from Rama camp site in the left bank of the gorge (Figure 88). The barite vein (5–10 m) with malachite–azurite stain occur cross-cutting the northeast-dipping beds and along the contact zone of Ralam quartzite and

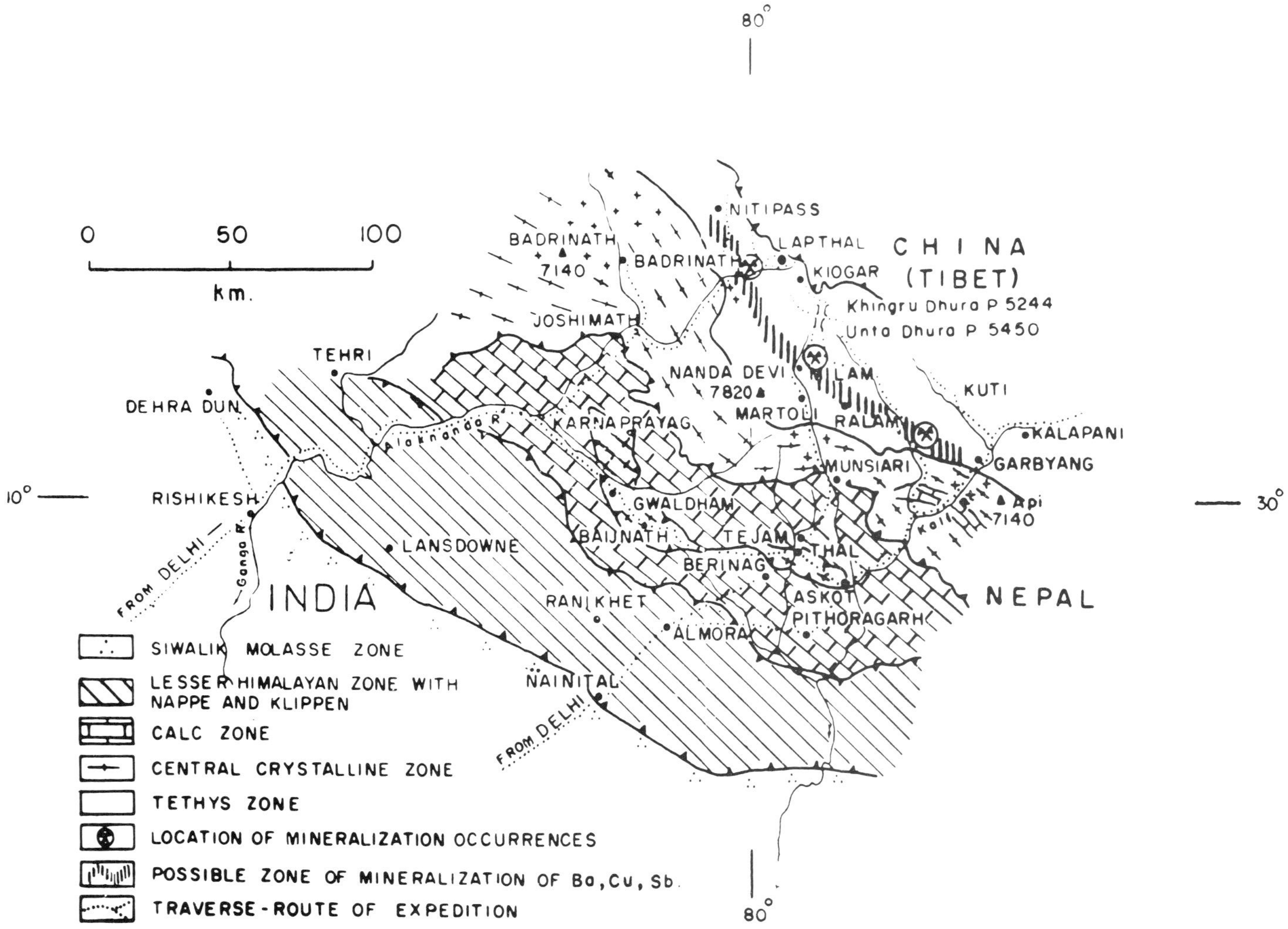

Figure 111. A generalized tectonic zonation of Central Himalaya showing the mineralized belt in Tethyan Zone

dolomitic limestone of Garbyang formation (Figure 13). The country rocks are limonitized dolomitic limestone showing thin partings and dipping at 60° northeast. The bartie veins with varying thickness (15 to 50 cm) are intimately related with quartz vein (SSE–NNW dip-80° SSW) which show copper and iron sulphide mineralization.

9.4 Barmatiya Deposit

The country rock in the mineralized district of Barmatiya on Girthi Valley is mainly dolomitic limestone of the Garbyang formation (Figure 112) which becomes more arenaceous in the ascending lithologic succession (Figure 113). The base of the sequence is quartzite which is followed by dark bluish silty shale and sericite schist. Sometimes it is calcareous and quartzitic. The mineralization starts with the old workings of copper and pyritic ore bodies 8 km from Malari upstreams along the Girthi valley. One can see the abandoned adits in the wall opposite the 8 km bridge over the Girthi while climbing in the right bank. The concentration of barite increases along the right valley of 300 m in the right bank over a spur situated east of the Shiva temple built on road (Figure 115a). The actual widespread exposure is in the eastern slope of spur along old track. This location (Figures 116 and 117) is usually overlooked by travellers since it remains inaccessible due to new track downside (Figure 114) at a height of 30 m from the Girthi water level.

Folding pattern of country rock is highly complicated due to gravity tectonics playing a major role and back folding with toothpaste style and gliding down of structure. The fold type commonly controlling the mineralization in the mineralized district is with axial plane of 40° inclination and azimuth 280°N plunging 25°N (Figure 43).

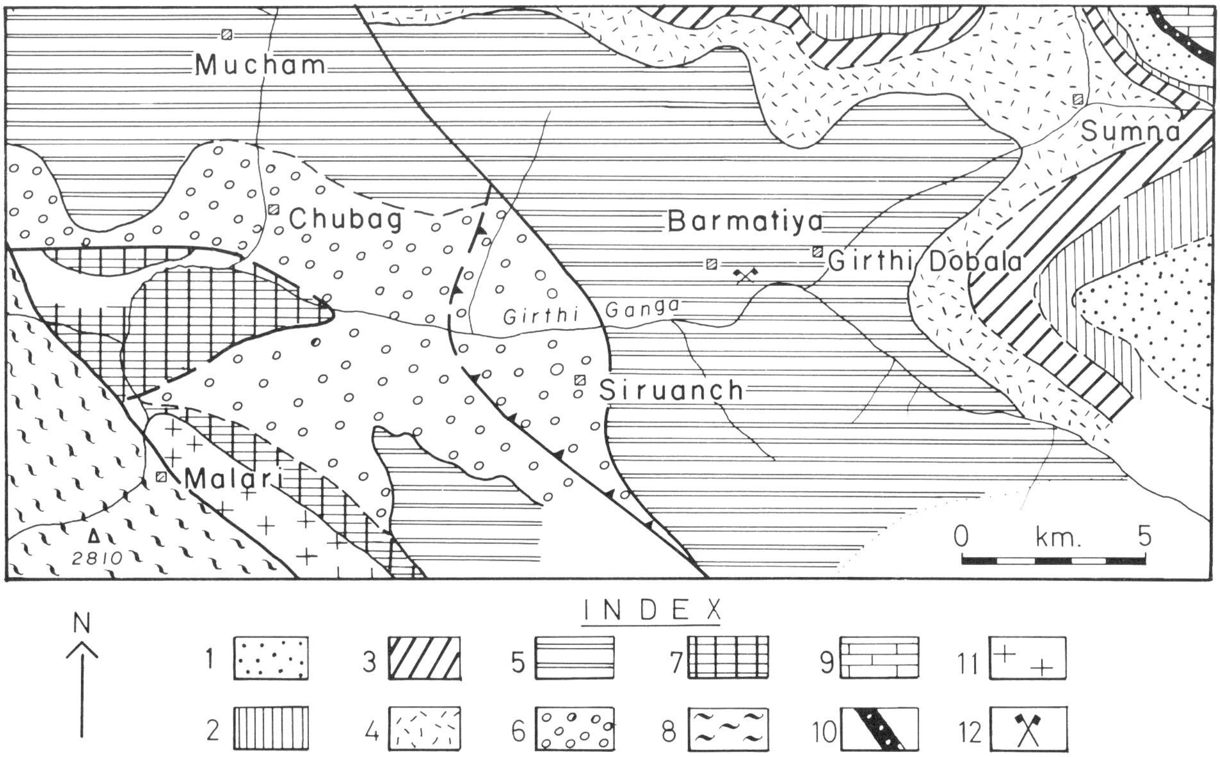

Figure 112. Geological map of the Barmatiya area in Tethyan Zone of Higher Garhwal Himalaya. 1. Muth quartzite (Devonian); 2. Variegated formation (Silurian); 3. Yong limestone (Ordovician-Silurian); 4. Shiala formation (Ordovician); 5. Garbyang formation (Cambrian ?); 6. Ralam formation (Precambrian/Cambrian ?); 7. Martoli Group (Precambrian ?); 8. Vaikrita Group; 9. Kuti shale (Triassic); 10. Kuling shale (Permian); 11. Tertiary granite; 12. Mineralization locality

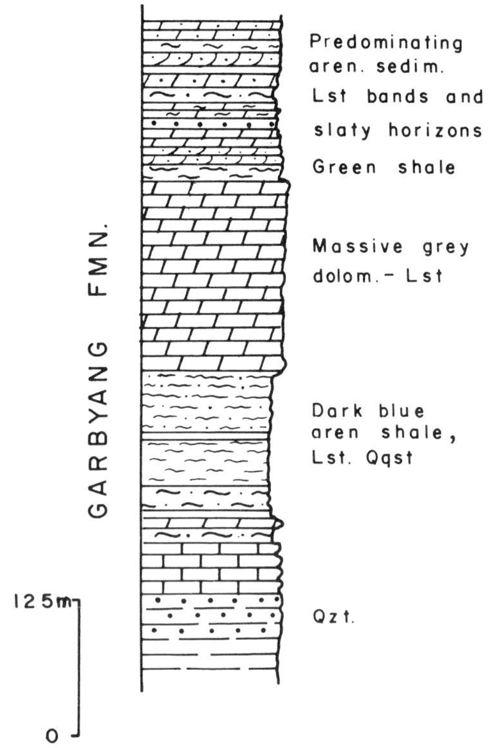

Figure 113. Stratigraphic column of Garbyang in mineralized area

The structural control of mineralization is self explanatory in the detail map drawn on 1 : 1000 scale (Figure 115a), and section (Figures 116 and 117). Each limb of the synclinal structure is further complicated by minor folds and flexures with inversion. The major joint pattern controlling the barite and associated polymetallic mineralization are 1.45°N, 2.50° azimuth 340°N, 3.90° azimuth 50°N. Such barite pattern is successively intersected by schistosity of 60°N azimuth with dip inclination of 20° (Figure 118). The cross-cutting veins (Figures 121 and 122) are to illustrate the nature of their disposition with the aim to explore the possibility of economic deposit exploratory work for reserve finding as well as the academic study of barite at suitable sites keeping the norm of investigation (Krasulin, 1967). Fifteen trenches (Figure 115b) were dug different length and depth (see next column).

Fig. 115a
Fig. 115b
Fig. 122

The major concordant barite body which was exposed in T_1, T_2 and T_6 could not be countered in the trenches in upper part of slope northward. The structural control along the apex of anticline in the eastern flank of the area could be observed whereas the 3.5 m thick vein is confined in the plunging syncline along the synclinal structure.

	Length (m)	Average width (m)
T_1	38	1
T_2	24	0.94
T_3	12	0.83
T_4	11	1.50
T_5	4.50	1
T_6	3.50	
T_7	2.50	
T_8	3.0	
T_9	10 m	
T_{10}	4 m	
T_{11}	7 m	
T_{12}	3.80	1.30
T_{13}	2.30	
T_{14}	2.30	
T_{15}	2.10	
T_{16}	2.50	

Other cross-cutting veins are not of major significance except the vein shown in Figures 115a and 123. Systematic furrow and spot samples were collected from the trenches for studies under thin and polished sections. Geochemical analysis of BaO and other silicates as well as spectral analysis of different major and trace elements has been carried out of the well documented specimens.

9.5 Study of Barite under Thin and Polished Sections

9.5.1 Thin section study

In the thin section three main types of paragenetic mineral assemblages have been established: (I) Monomineralic barite aggregate; (II) Quartz (Calcite)–barite and quartz calcite association; (III) Calcite monomineralic aggregate.

Barite aggregate: amongst the barite aggregate the following is observed:

(1) Medium grained aggregate with conspicuous banded schistose structure. The elongated crystal of barite are oriented in one direction, sometimes in between them occurs thin sericitic flakes which are partly elongated.

The possible processes of formation of oriented barite aggregate are: (a) Metasomatism with inherited schistose structure; (b) Recrystallization during the process of tectonic deformation which is supported by the development of carbonate.

(2) Aggregate of large tabular barite grains with more or less sharply expressed deformational

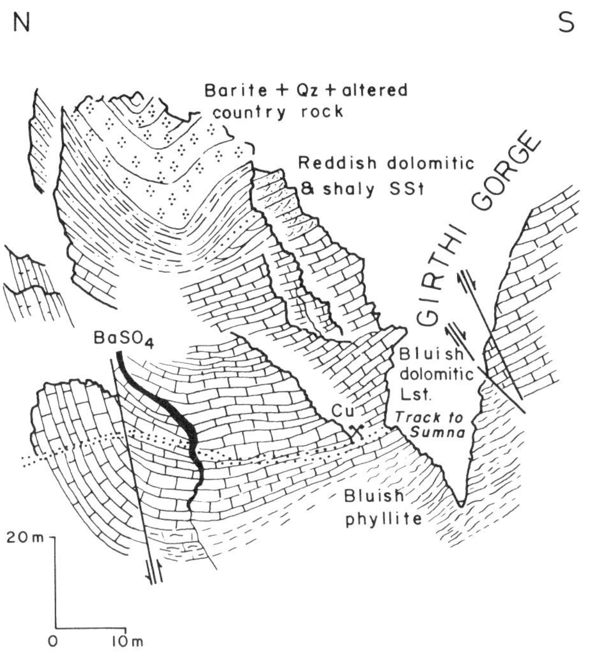

Figure 114. Girthi gorge showing the barite mineralization and position of the old copper working

reworking. Strongly deformed aggregate are represented by the intense growth of twinned grains (with the twinning of polysynthetic type), elongated twinned strips, aggregates of cataclasized and recrystallized grains, with veinlets of fine grains barite amidst quartz grain of less deformed types.

(3) Aggregate of barite without any sign of appreciable deformation.

Dissimilar in the quantitative relationship among different mineral components, quartz and calcite constitute 30 per cent of the mineral mass. The relationship amongst the above minerals shows a much later calcite and quartz grain formation with respect to large grain of barite, paragenetically associated with the small grain barite of second generation. Expected quantity of BaO in such specimens are 40 to 60 per cent.

In their own characters the quartz barite aggregate are closer to the observed sulphide epithermal mineral deposits; the others are similar to hypothermal mineral deposits. Quartz as a rule is well crystallized and forms medium grain aggre-

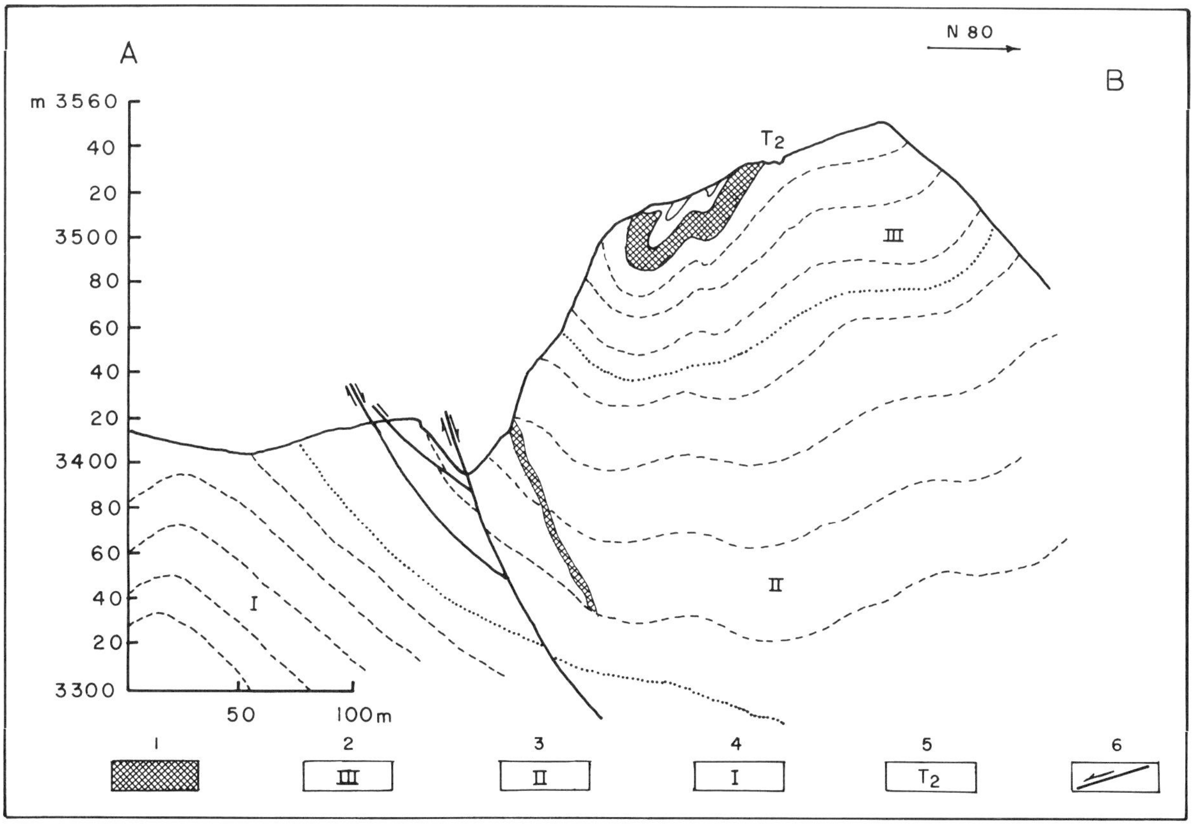

Figure 116. Geological cross-section along A–B of Figure 115. Key: 1. Barite vein; 2. Arenaceous limestone; 3. Dolomitic limestone, sandstone with cross and graded bedding structure, shale and quartzite; 4. Pyritiferous quartzite and slate; 5. Trench; 6. Fault

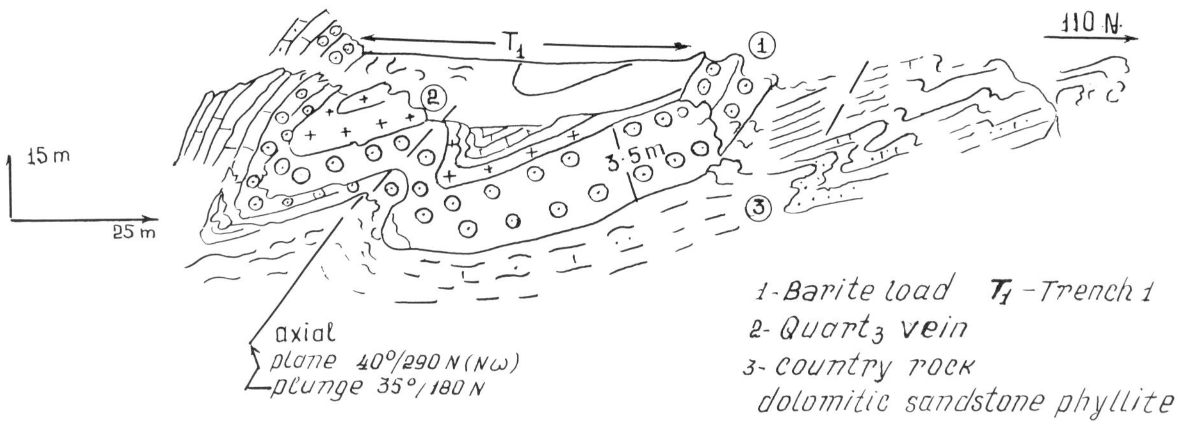

Figure 117. Concordant barite vein (3.5 m)

gates with idiomorphic features in relation to barite. Calcite is not observed in systematic segregation with smaller veinlets. In some of the slides barites are not met with. Here quartz-calcite coalescence with well recrystallized quartz and calcite grains are observed. In some slides epidote minerals are observed to grow over the calcite.

Calcite aggregates are composed of grains of crystalline growth. Intensely deformation in aggregate with copious gliding twinning and superimposition of quartz, sericite-radiating fibrous forms, similar to the dispersion of vitrite, are also found. Some slides represent obviously the rock type of quartzitic sandstone. Considering the above data, it is inferred that in this zone of hydrothermal mineralization there is also the occurrence of sulphide mineralization (pyrite, copper sulphide completely oxidized).

To judge the prospect of mineralization of the area it is advisable to study the petrogeochemical characteristics of country rocks, their alteration and geochemical properties.

9.5.2 Polished-section study

The ten selected specimens were studied in polished section under the microscope (Figure 124). The barite are mainly of two types: white

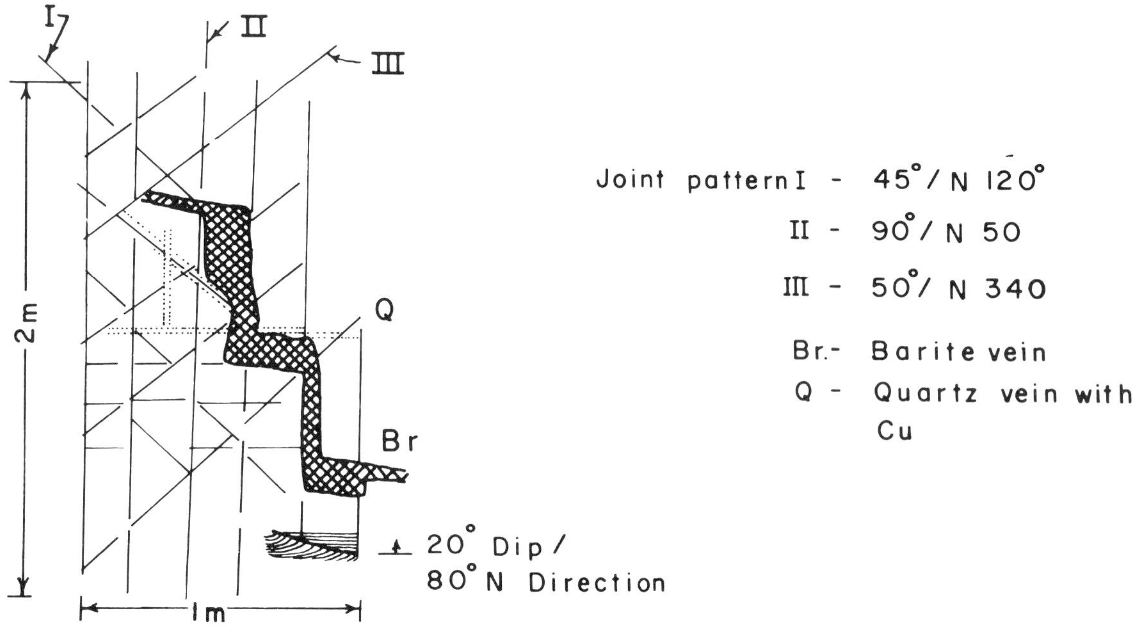

Figure 118. Structurally-controlled mineralization along partings

Figure 119. Exposed barite lode in the surface. Ice-axe and hammer should be used as scale

Figure 120. Barite invading the country rock

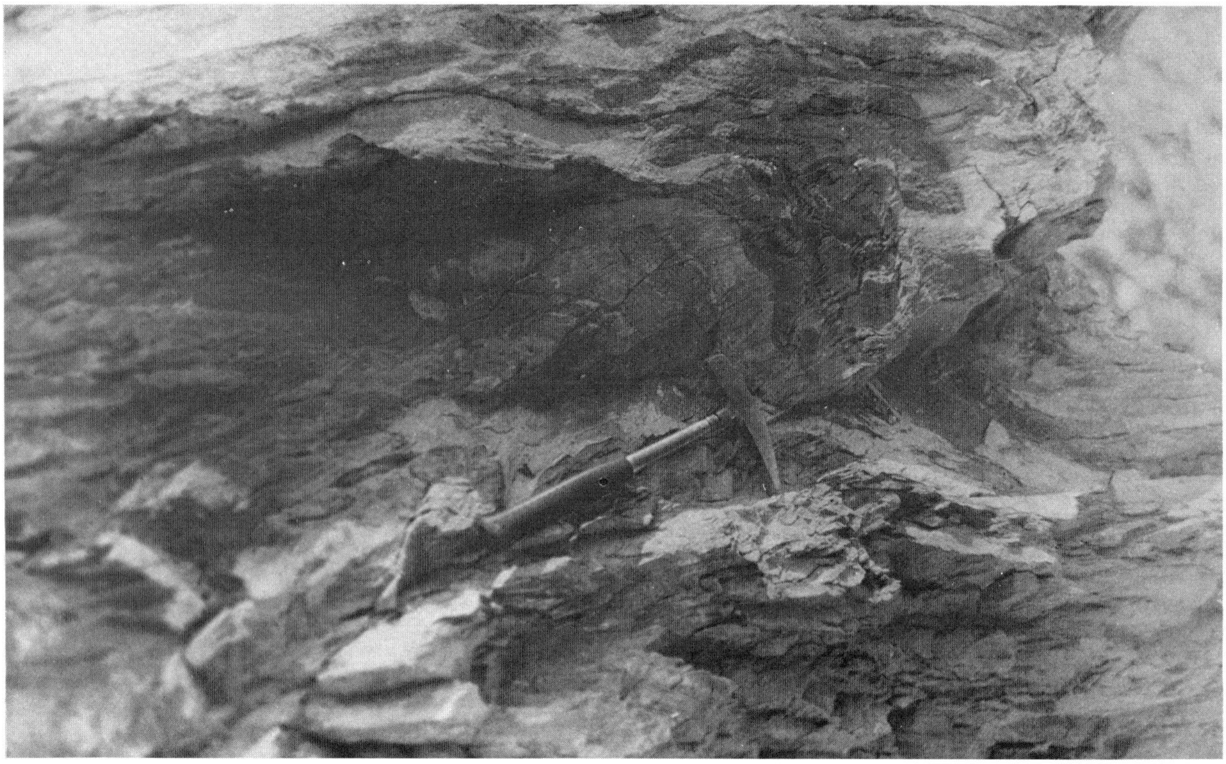

Figure 121. Barite hydrothermal solutions are controlled by structures in country rock

and massive, and radiating bladed type with associated impurities. As described earlier a large number of adit holes of old copper workings are dotted on either side of the Girthi gorge along the 8 km length of track on the way to Sumna via Barmatiya. The mineralization has resulted in the hydrothermal assemblage of chalcopyrite ($CuFeS_2$), chalcocite (Cu_2S), covellite (CuS), bornite ($CuSF_2S_4$), cuprite (Cu_2O), malachite ($CuCO_3$, $Cu(OH)_2$, azurite (2 $CuCO_3$, $Cu(OH)_2$, pyrite (FeS_2), pyrrhotite (Fe_nS_{n+1}), galena (PbS), sphalerite (ZnS), stibnite (Sb_2S_3), calcite ($CaCO_3$), and quartz (SiO_2). *The systematic petrographic investigations and ore-microscopy thus reveal a barite-calcite and barite-polymetallic mineralization.*

The host rocks are intruded by sulphides in the fissures and granular interspaces (Figure 124). Sulphides are replacing the gangue minerals and thus the mineralization appears to be hydrothermal in origin (Figure 124). The mineralization associated with hydrothermal quartz appears to be structurally controlled by epigenetic phase, which is ultimately deformed and broken into pieces. The presence of idiomorphic hydrothermal quartz also supports its hydrothermal origin (Figure 124) which is later on by supergene sulphide enrichment. The other base metal sulphides have not been observed frequently in the polished sections. Therefore, the source of fluid cannot be reliably inferred.

The fine to coarse grained highly fractured rocks are intruded with sulphides in the fissures and intergranular spaces. The dominating ore is chalcopyrite with fine granular dissemination of pyrite and pyrrhotite. The other copper sulphides, chalcocite, cuprite, bornite etc. are of secondary origin derived after chalcopyrite and occurs along the cracks and fissures in chalcopyrite. In few sections these fractured sulphides remarkably reveal cleavable and twinned chalcocite and secondary malachite. Pyrite and pyrrhotite both occurs as fine inclusions in chalcopyrite. Chalcopyrite also occurs as inclusions in chalcopyrite which appears to be relict. Chalcocite is transverse by several thin veinlets of malachite. At a few places chalcopyrite reveals fine inclusion of galena (Figure 124).

Some thin sections reveals dominance of chalcocite, may be of supergene enrichment, but the presence of idiomorphic hexagonal coarse quartz grain imbedded in massive sulphides and restricted along the weak planes indicate a hydrothermal source.

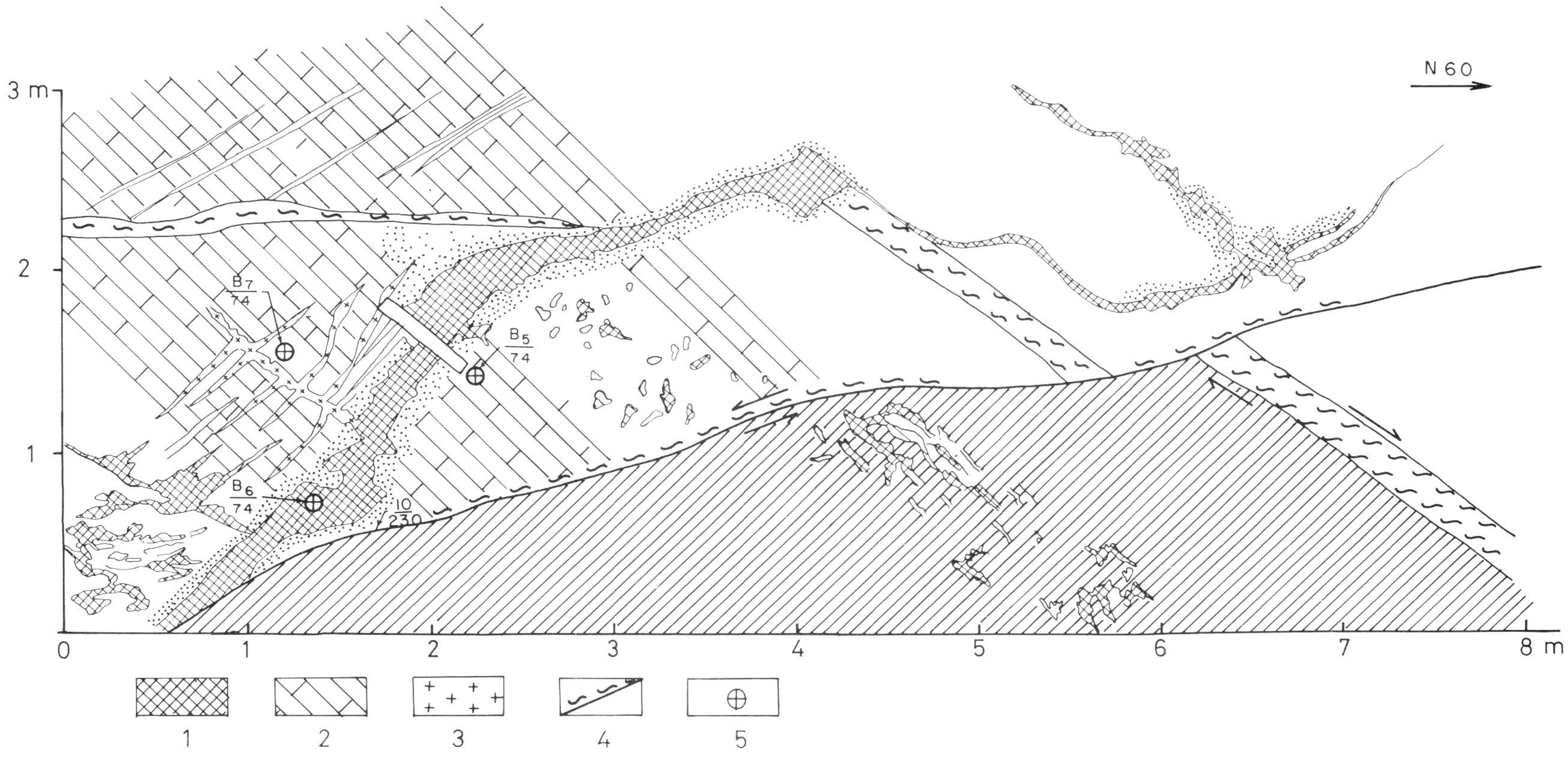

Figure 123. Documentation of barite vein to show interrelation with country rock and two phases of mineralization. 1. Barite vein; 2. Dolomitic limestone with phyllite; 3. Quartz vein with Cu–Fe mineralization; 4. Fault gouge; 5. Sample location

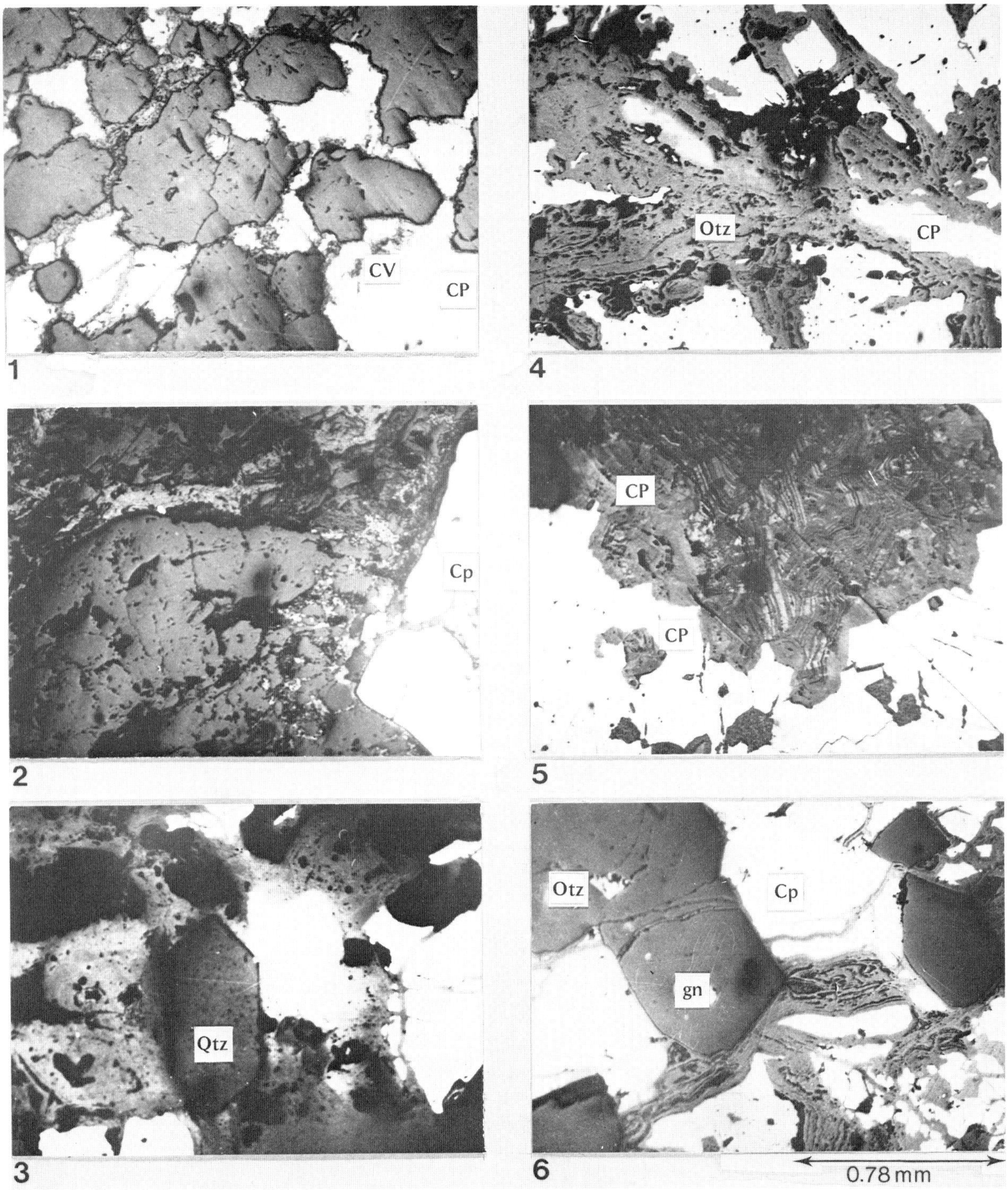

Figure 124. Photominerograph of polished sections (× 42); Abbreviation: Qtz–Quartz, Cp–Chalcopyrite, CV–Covellite, gn–Galena; 1. Chalcopyrite being replaced by zoned covellite; 2. Chalcopyrite being replaced by galena and both have inclusions of hydrothermal quartz; 3. Hexagonal idiomorphic quartz, chalcopyrite, and chalcocite supporting hydrothermal origin; 4. Chalcopyrite filled in the grains interspaces of quartz; 5. Chalcocite cross-cut with zones veinlets of limonite along fractured/sheared zone filled with chalcopyrite; 6. Galena and chalcopyrite in association with idiomorphic hydrothermal quartz. One quartz crystal showing inclusion of galena X

The majority of polished sections reveal the replacement of chalcopyrite by chalcocite, covellite, and malachite in fractured calcareous rocks. In addition the replacement of a few grains of bornite by covellite is also clearly evident. It is further observed that coarse-grained chalcopyrite is being replaced by chalcocite and limonite and the replacement starts from the peripheral boundaries, whereas pyrite is also being replaced by limonite. These sulphides are highly fractured at places and cross-cut with veinlet of limonite.

The above finding comprises chalcopyrite, chalcocite, malachite, pyrrhotite, pyrite, covellite, bornite, cuprite, azurite and galena as ore minerals. The massive sulphides are widespreadly replacing the gangue minerals. It could be safely inferred that mineralization is structurally controlled, and has suffered deformation. In addition the presence of idiomorphic hydrothermal quartz supports view of hydrothermal origin of deposits.

9.6 Geochemistry of Barite and Polymetallic Mineralization

The authoritative work on the geochemistry of barium is the monograph of W. von Engelhardt (1936). Barium is an element that is highly concentrated in the earth's crust.

The most common barium ore occurring in nature is barite. It is deposited by fluids with a high oxidation potential, where sulphur is present as sulphate. Suitable conditions of this kind occur close to the earth's surface or in subsurface areas where sulphate solutions mix with reduced Ba-containing waters. In solution, Ba migrates to the region of sulphate stability and this often bound to narrow zone close to the earth's surface (Puchelt, 1972). The detail work on the mode of barite formation and possibilities of $BaSO_4$ transport in solutions has been carried out by Puchelt (1972), Ba contains in all cases, certain amounts of isomorphous strontium. It has been argued that probably magmatic hydrothermal fluids originally do not contain any significant amount of Ba. But acquire through the process of leaching of wall rock. Barite in this deposit is associated with copper–lead–zinc mineralization which is hydrothermally introduced along deep-seated fault, fractures, or thrust planes. Since barite is at the end of the paragenetic sequence of mineralization and commonly physically separate from the main mass of sulphide mineralization. Barium has the highest energy of absorption among the alkali earth metals: Ba–Sr, Ca–Mg. Absorption of barium by manganese hydroxides is remarkable, containing up to 10 per cent Ba. It has been observed that the characteristics of Ba migration in the oxidation zones of sulphide deposits where barite occur, are different. Ba does not migrate in the sulphate environment. The largest concentration of barium leading to economic deposits, are in association with the sulphate barrier. It has been documented in the sketches of trenches that these are pockets of fine-grained loose barite sands contacting with the massive barite and country rock or barite with polymetallic mineralization. Perhaps this occurrence is described as 'friable pocket' (Parel'man, 1977, p. 214).

Field examination of occurrence of barite and collection of specimens for the purpose of geochemical evolution study showed an interesting association of Ba with many other metals especially Cu, Pb, Zn, Sb, As, Sc, Nb, Ga, Mo, V, Ag, Ti, Zr, B and Ca, Mg, Sr, Cr, Si, etc. (Table 9.3). Documented specimen of barite vein (Figure 122) situated near the wall rock shows Pb (40 to 60) Cu (400–600 ppm). Ti (1000 to 3000 ppm), Sr (10,000–30,000 ppm) and B–30 ppm (Table 9.3) specimen Br/74 (Fig. 122) representing the altered country rock shows 30 ppm, Sb and Ni 4–6 ppm whereas the pure barite specimen B2/74 of the same vein does not show the presence of above-mentioned trace elements presents in wall rock of vein.

Examining another vein (Figure 123) and the silicate and spectral analyses of the collected samples it is apparent that the Ba is associated with appreciable strontium and copper besides other common elements (Table 9.3, 9.1). The quality of barite is extremely good with BaO–61.25 per cent and SO_3–33.51 per cent. The analysis of specimen collected from the trench I (T1) (Figure 115b, Table 9.1, 9.2) have shown the concentration of $BaSO_4$ 95.53 per cent (T_1–3/74) to 68–79 per cent (T_1–18/74). The Cu-content varies in from 3410 ppm to 57 ppm. The later vein quartz (T_1–71/74) with 2.51 per cent Ba is rock in Cu 3410 ppm. T_1–19/74 of country rock with 22.62 MgO and 21.42 CaO confirms its dolomitic nature of the altered country rock (Table 9.1) with 0.04 per cent Ba and 60 ppm Cu. Trench 2 (T_2) analysed specimen shows the BaO–50.97 per cent with Cu 81 ppm. The Ba in altered country rock (T_2–2/74 and T_2–7/74) is 10,000 ppm to 90,000 ppm. Specimen T_3–2/74 is the fine grained friable barite sand called 'barite *sypuchka* (in

Table 9.1. Silicate and major elemental analyses of barite–polymetallic deposit specimens from the Girthi and Darma valleys

		Field Sp. no.							
		T1–13/74	T1–15/74	T1–18/74	T1–19/74	T2–6/74	T3–5/74	T4–1/74	B–6/74
1.	SiO_2, %	15.70	89.65	14.16	7.23	5.44	62.67	66.08	0.97
2.	TiO_2, %	0.01	0.01	0.01	0.03	0.03	0.21	0.05	0.02
3.	Al_2O_3, %	0.50	0.01	2.20	6.40	0.10	18.70	8.10	0.10
4.	Fe_2O_3 t, %	0.05	1.00	0.10	8.60	0.05	1.50	1.75	0.15
5.	MgO, %	0.08	0.73	0.15	22.62	0.14	1.62	1.58	0.06
6.	CaO, %	0.35	0.56	0.30	21.42	0.12	0.77	2.57	0.25
7.	BaO, %	47.92	2.14	45.05	0.04	50.97	6.47	8.16	61.25
8.	SrO, %	2.02	0.21	1.94	0.01	9.53	0.49	0.44	2.48
9.	SO_3, %	33.62	1.12	23.74	0.21	32.53	4.17	4.86	33.51
10.	$GV_{1000°C}$, % Ignition loss	0.27	1.58	0.26	32.73	0.17	3.05	4.31	0.64
11.	Cu, ppm	58	1900	106	60	81	460	2010	90
12.	Mn, ppm	3	18	4	15	3	15	16	9
13.	$BaSO_4$, 7	73.62		68.79		83.50			94.76

Analyst—Dr Peter Klein, Austrian Geological Survey, Vienna by courtesy Dr G. Fuchs. Analyses performed on Atomic Absorption (Perkin–Elmer).
Except the specimens 152–20/75, 46–9/75 from the Darma valley (Fig. 88), the rest have been collected from the Barmatiya area of the Girthi Ganga valley. 'T' indicates trench samples; the rest are grab samples.

Russian) under spectral analysis show as 100–300 ppm. Cu 1000 ppm and Ba up to 6000 ppm and Si up to 90,000 ppm T_3–4/74 contain up to 90,000 ppm Ba with 10 ppm Pb and Boron (B). Specimen T_4–1/74 from trench 4 representing a quartzite–barite is high with Si (90,000 ppm) and Ba up to 6000 ppm and Cu up to 30,000 ppm. Thus vein quartz is invariably mixed with barite posing a question of its separation while mining to maintain the grade. The other analysis contained in Table 9.2 and described in Tables 9.1 and 9.3 can be used for more academic importance to study the geochemistry of Ba and other major and minor elements.

In Table 9.4 five representative specimens of barite were chosen for specific gravity determination and it has been found that average specific gravity of 4.3 is fairly high for good commercial variety of barite to be used in various industries.

9.7 Preliminary Estimated Reserve and Future Scope

Studying carefully the chemical analysis and field relationship it is plausible to infer that these are low temperature (50°–200°C) hydrothermal deposits. As a rule they are so far away from the magmatic chamber that no intrusives are found within the ore-bearing areas. The granitic intrusion (73 Ma—Sinha and Bagdasarian, 1977) of Himalayan orogeny are as far as 5 km or more from the mineralization occurrence.

Much of the world's barite production has been used as an ingredient of circulation muds in rotary drilling of oil and gas wells. The consumption in the oil industry goes up to 75 per cent of total output. Rest of the world's annual supply is utilized into the manufacture of glass as a flux and for imparting brilliance and clarity to the product: into the paint and rubber industries as filler and

Table 9.2. Chemical analyses of barite specimen

	Sp. No.		
Radicals	B—2/74	50—6/74	36—9/74
SO_3	34.17	34.07	32.83
BaO	63.62	61.96	56.67
SrO	1.48	2.34	4.00
CaO	0.24	0.18	0.20
MgO	0.02	0.01	0.02
Fe_2O_3	0.03	0.13	0.07
CO_2	0.00	0.38	0.16
Total	99.72	99.60	99.95

Analysis—Mms Arapova and Tikhomirova, Central Chemical Laboratory, Institute of Geology of Ore Deposits, Academy of Sciences of USSR, Moscow.
Sample location details:
1. B–2/74Exp – Barite: From barite vein exposed on the track between Barmatiya to Sumna (Figure 119).
2. 50–6/74Exp – Barite: 10 m further north from Trench I, from exposed vein between T_1 & T_2.
3. 36–9/75Exp – Barite: Sample from mineralized zone between Go and Rama in Darma valley (Figure 88).

Table 9.3. Spectral analyses of barite specimen in association with polymetals and country rocks

Sl. no.	Field Sp no. / Elements in ppm \ Lab. no.	T_3–2/74 1/77	B5/74 2/77	T_2–7/74 4/77	69–11/74 5/77	F–3/74 6^a/77	36–9/75 6/77	36–9/75* 7/77	43–3/74 8/77
1.	Be	1	–	10–30	–	–	–	–	–
2.	As	100–300	–	100–300	–	–	100	100	–
3.	Te	–	–	–	–	–	–	–	–
4.	Sc	–	–	–	–	10	–	–	–
5.	P	–	–	400–600	–	–	–	–	–
6.	Sb	–	–	–	–	–	–	–	–
7.	Ta	–	–	–	–	–	–	–	–
8.	Tl	–	–	–	–	–	–	–	–
9.	Mn	100	4–6	70–90	40–60	3000	40–60	7–9	300
10.	Pb	–	–	–	10–30	10–30	30	30	–
11.	Sn	–	–	7–9	–	–	–	–	–
12.	Nb	–	–	40–60	–	–	–	–	–
13.	W	–	–	–	–	–	–	–	–
14.	Ga	3	–	10–30	–	–	–	–	–
15.	Ge	–	–	–	–	–	–	–	–
16.	In	–	–	–	–	–	–	–	–
17.	Bi	–	–	–	–	–	–	–	–
18.	Mo	1–3	–	–	–	–	–	–	–
19.	V	10–30	–	100	–	–	–	–	–
20.	Li	–	–	–	–	–	–	–	–
21.	Cu	1000	10–30	1000–3000	10–30	100–300	100–300	100	100–300
22.	Cd	–	–	–	–	–	–	–	–
23.	Ag	–	–	–	–	–	–	–	–
24.	Na	1000	–	4000–6000	1000–3000	1000–3000	–	–	–
25.	Zn	–	–	70–90	–	–	–	–	–
26.	Ti	100–300	10–30	4000–6000	10–30	300	10–30	10–30	10–30
27.	Co	–	–	10–30	–	–	–	–	–
28.	Ni	10–30	–	40–60	–	10–30	–	–	–
29.	Zr	–	–	70–90	–	100–300	–	–	–
30.	Mg	4000–6000	100–300	10000	1000–3000	70000–90000	1000–3000	40–60	400–600
31.	Si	70000–90000	400–600	70000–90000	1000–3000	70000–90000	70000–90000	10000–60000	1000–3000
32.	Al	10000	700–900	70000–90000	1000–3000	4000	1000	1000	1000–3000
33.	Fe	10000	100–300	10000–30000	700–900	40000–60000	1000	400–600	1000–3000
34.	Cr	30	40–60	1000–3000	400–600	100	100	10–30	10
35.	Ca	7000–9000	4000–6000	4000–6000	70000–90000	70000–90000	4000–6000	1000	3000
36.	Sr	100	10000–60000	1000–3000	3000	3000	40000–60000	40000–90000	40000–60000
37.	Ba	4000–6000	70000–90000	10000–90000	10000–50000	10000–60000	70000–90000	70000–90000	70000–90000
38.	La	–	–	–	–	–	–	–	–
39.	Ce	–	–	–	–	–	–	–	–
40.	Y	–	–	–	–	–	–	–	–
41.	Yb	–	–	–	–	–	–	–	–
42.	Hg	–	–	–	–	–	–	–	–
43.	B	40	–	400–600	40–60	10	–	–	–

Table 9.3 (*continued*)

Sl. no.	Field Sp. no. / Elements in ppm / Lab. no.	T_2–2/74‡ 9/77	T_{15}–3/74 10/77	T_4–1/74‡ 11/77	T_{14}–7/74 12/77	B_2/74*† 13/77	30–3/74 14/77	T_3–4/74‡ 15/77	50–6/74*†‡ 16/77
1.	Be	1–3	1–3	1–3	–	–	–	–	–
2.	As	–	–	–	–	–	–	–	–
3.	Te	–	–	–	–	–	–	–	–
4.	Sc	–	–	–	–	–	–	–	–
5.	P	–	300	400–600	–	–	–	–	–
6.	Sb	–	–	–	–	–	–	10	–
7.	Ta	–	–	–	–	–	–	–	–
8.	Tl	–	–	–	–	–	–	–	–
9.	Mn	100	100	100–300	10–30	4–6	10	3000	40–60
10.	Pb	–	10–30	40–60	10–30	–	30	10	10–30
11.	Sn	–	–	–	–	–	–	–	–
12.	Nb	10	10	10	–	–	–	–	–
13.	W	–	–	–	–	–	–	–	–
14.	Ga	4–6	7–9	4–6	–	–	–	4–6	–
15.	Ge	–	–	–	–	–	–	–	–
16.	In	–	–	–	–	–	–	–	–
17.	Bi	–	–	–	–	–	–	–	–
18.	Mo	–	–	–	–	–	–	4–6	–
19.	V	10–30	10–30	10–30	–	–	–	4–6	–
20.	Li	–	–	–	–	–	–	–	–
21.	Cu	1000–3000	10000	10000–30000	40–60	7–9	40–60	100–300	100–300
22.	Cd	–	–	–	–	–	–	–	–
23.	Ag	–	–	4–6	–	–	–	–	–
24.	Na	1000–3000	1000–10	100–3000	–	–	–	–	–
25.	Zn	–	–	–	–	–	–	–	–
26.	Ti	1000	1000	400–600	–	–	–	40–60	40–60
27.	Co	–	10	–	–	–	–	–	–
28.	Ni	10–30	10	10–30	–	–	–	10–30	–
29.	Zr	1000	70–90	40–60	–	–	–	–	–
30.	Mg	3000	3000	4000–6000	400–600	70–90	100	3000	300
31.	Si	70000–90000	70000–90000	70000–90000	1000	1000	1000–30000	4000–6000	4000–6000
32.	Al	10000–30000	10000–30000	10000–30000	100	100	700–900	700–900	700–900
33.	Fe	1000	10000	10000–60000	1000	400–600	700–900	70000–90000	1000
34.	Cr	100–300	100–300	100	10–30	10	10–30	10–30	10–30
35.	Ca	3000	70000–90000	10000	400–600	700–900	1000–3000	30000	
36.	Sr	1000–3000	7000–0000	4000–6000	40000–60000	70000–60000	40000–60000	10000–60000	10000–60000
37.	Ba	10000	10000	4000–6000	90000	90000	90000	90000	90000
38.	La	–	–	–	–	–	–	–	–
39.	Ce	–	–	–	–	–	–	–	–
40.	Y	–	–	–	–	–	–	–	–
41.	Yb	–	–	–	–	–	–	–	–
42.	Hg	–	–	–	–	–	–	–	–
43.	B	–	40–90	40–60	–	–	–	10	

Table 9.3 (*continued*)

Sl. no.	Field Sp. no. / Elements in ppm / Lab. no.	39–9/75‡ 17/77	F_2/74‡ 18/77	65–26/76 19/77	B_4/74‡ 20/77	59–26/76 21/77	B1–74‡ 22/77	F1(T–2)/74 23/77	$T_1$16/74 24/77
1.	Be	4–6	–	–	–	–	–	–	–
2.	As	–	–	–	–	70–90	–	–	–
4.	Sc	–	–	–	–	–	–	–	–
5.	P	300	–	–	–	700–900	–	–	–
6.	Sb	–	40–60	100–300	30	100	–	–	–
7.	Ta	–	–	–	–	–	–	–	–
8.	Tl	40–60	10–30	100–300	–	–	–	–	–
9.	Mn	–	10–30	10–30	40–60	100	100–300	10	4–6
10.	Pb	–	–	–	–	–	40–60	–	–
11.	Sn	10	–	–	–	7–9	–	7–9	–
12.	Nb	–	–	–	–	–	–	–	–
13.	W	10	–	–	–	–	–	–	–
14.	Ga	–	–	–	–	3	–	–	–
15.	Ge	–	–	–	–	–	–	–	–
16.	In	–	–	–	–	–	–	–	–
17.	Bi	–	–	–	–	–	–	–	–
18.	Mo	40–60	–	–	4–6	–	–	–	–
19.	V	–	–	–	–	1–3	–	–	–
20.	Li	700–900	100–300	400–600	–	4–6	–	–	–
21.	Cu	–	–	–	100–300	300	400–600	100–300	10–30
22.	Cd	–	–	–	–	–	–	–	–
23.	Ag	1000–3000	1000–3000	1000–3000	–	–	–	–	–
24.	Na	–	–	–	–	1000	–	–	–
25.	Zn	1000	10–30	10–30	–	–	–	–	–
26.	Ti	10–30	–	–	400–600	70–90	1000–3000	10–30	1000–3000
27.	Co	10–30	–	–	–	–	10–30	–	–
28.	Ni	40–60	–	–	–	30	30	–	–
29.	Zr	–	–	–	40–60	30	100–300	–	–
30.	Mg	3000	700–900	4000–6000	3000	1000	4000–6000	100	40–60
31.	Si	70000–90000	70000–90000	1000–3000	10000–60000	70000–90000	10000–60000	70000–90000	100–300
32.	Al	10000–30000	3000	1000–3000	4000–6000	3000	7000–9000	1000–3000	100–300
33.	Fe	7000–9000	1000	1000	1000–3000	10000	4000–6000	1000–3000	400–600
34.	Cr	100	100	–	10–30	100–300	40–60	40–60	10–30
35.	Ca	1000	1000–3000	70000–90000	4000–6000	1000–3000	7000–9000	1000	700–900
36.	Sr	100	10000–60000	400–600	10000–30000	70–90	10000–30000	40000–60000	10000–60000
37.	Ba	3000	90000	1000–3000	90000	3000	90000	90000	90000
38.	La	–	–	–	–	–	–	–	–
39.	Ce	–	–	–	–	–	–	–	–
40.	Y	–	–	–	–	–	–	–	–
41.	Yb	–	–	–	–	–	–	–	–
42.	Hg	–	–	–	–	–	–	–	–
43.	B	40–60	10–30	–	–	10	30	–	–

Table 9.3 (*continued*)

Sl. no.	Field Sp. no. / Elements in ppm	F(T–1)/74	62–7/74	T_4–3/74‡	152–20/75
	Lab. no.	25/77	26/77	27/77	28/77
1.	Be	–	–	–	–
2.	As	–	4000–6000	–	–
3.	Te	–	–	–	–
4.	Sc	–	10–30	–	–
5.	P	–	–	–	–
6.	Sb	–	40,000–60,000	100–300	–
7.	Ta	–	–	–	–
8.	Tl	–	–	–	–
9.	Mn	30	40–60	100–300	40–60
10.	Pb	–	10	30	–
11.	Sn	–	–	–	–
12.	Nb	–	–	–	–
13.	W	–	–	–	–
14.	Ga	–	–	–	–
15.	Ge	–	–	–	–
16.	In	–	–	–	–
17.	Bi	–	70–90	–	–
18.	Mo	–	40–60	–	–
19.	V	–	–	–	–
20.	Li	–	–	–	–
21.	Cu	40–60	40000–90000	1000	10
22.	Cd	–	100–300	–	–
23.	Ag	–	70–90	–	–
24.	Na	–	–	–	–
25.	Zn	–	10000	–	–
26.	Ti	10–30	10–30	100	10–30
27.	Co	–	30	–	–
28.	Ni	–	–	–	–
29.	Zr	–	–	–	–
30.	Mg	70–90	4000–6000	4000–6000	10000–60000
31.	Si	40000–60000	70000–90000	10000	100–300
32.	Al	700–900	3000	1000–3000	700–900
33.	Fe	1000–3000	70000–90000	3000	400–600
34.	Cr	30	10–30	10–30	10–30
35.	Ca	1000	4000–6000	40000–60000	70000–90000
36.	Sr	40000–60000	100	10000–60000	700–900
37.	Ba	90000	1000	90000	3000
38.	La	–	–	–	–
39.	Ce	–	–	–	–
40.	Y	–	–	–	–
41.	Yb	–	–	–	–
42.	Hg	–	400–600	–	–
43.	B	–	10	–	–

* Silicate analysis. † Specific gravity. ‡ Repeated analysis.

Analyses by Mme A. B. Galubtsina, Special Analysis Laboratory of Institute of Geology of Ore Deposits, Academy of Sciences of USSR, Moscow. Courtesy—Dr Yu. G. Safonov.

Besides the normal field collection numbers, 'T' indicates trench samples, 'F' the furrow specimens, and 'B' for the isolated vein grab samples, as indicated in (Figs. 109, 112, 114, 119 and 120).

pigment, and the chemical industry for the preparation of various compounds. In India these industries consumed over 30 per cent of our total output. An analysis of the trend in the production of barite in the country and available reserve indicates that the known reserve may not last very long. With the passage of time, keeping present rate of consumption, the known reserve of grade + 90 per cent $BaSO_4$ would deplete by 1989 (Barite—A Market Survey 1973: 0.40).

Table 9.4. Specific gravity of barites

S. no.	Sp. no.	Sp. gr.	Location
1.	50–6/74Exp	4.25	See Table 9.2
2.	B–2/74Exp	4.41	See Table 9.2
3.	T1–4/74Exp	4.32	See Fig. 8.16
4.	B–6/74Exp	4.16	See Fig. 8.14
5.	48–6/74Exp	4.35	See Fig. 8.4

Determination courtesy Dr Kuldip Chandra, KDM Institute of Petroleum Exploration (ONGC), Dehra Dun.

The calculated reserve on a very tentative basis assuming the visible outcrop and the mode of occurrence indicates it to a economic deposit of larger commercial value. A preliminary reserve calculation is in the order of a couple of million tonnes.

Furthermore, the presence of other polymetallic sulphide and their recent chemical analysis by atomic absorption techniques and spectrographic method point to a richer sulphide mineralization. Detailed prospecting work would be taken up to explore the so far unknown mineral potential of this part of Himalaya. It is advisable to explore and prospect the entire suggested belt for more barite and polymetallic deposits.

Chapter 10

Concluding Discussion

Since the early years of the history of geological study of Himalaya it has been felt that the Central Himalayan zone, incorporating mainly the territories of Himachal Pradesh, Garhwal and Kumaun in Uttar Pradesh, holds the key position in understanding the complex riddles of the Himalayan geology. Fortunately I have been associated with the geological study of this region for nearly sixteen years, ever since the beginning of my career as a geologist.

The early research works on the Himalaya inspired me immensely, especially I would like to mention the classics created by C. L. Griesbach (1891) and half a century later by Arnold Heim and Augusto Gansser (1939a).

Before I summarize my research inferences with great reverence to J. B. Auden, I would also like to quote his sentences, as remembered by Gansser in his classic work (p. 233, 1964): 'It is realized that there is often danger in early generalization from incomplete observations such as are made on traverses. The Himalaya is so vast, however, that years will elapse before they are mapped geologically in detail, and it seems desirable to place such observations on record, so as to build a scaffolding for the later formulation of geological structure which will eventually emerge from detailed work. Such results may also have a bearing on the understanding of the geology of the small areas where mapping is at present in hand' (Auden: *Traverses in Himalaya*, 1935).

The sum total of a decade of my research in the selected zone of Higher Central Himalaya is symbolized with the outgrowth of a detail geological map of more than 20,000 km^2 area on 1 : 125,000 scale, besides two other maps and geotraverse profiles prepared in the field. This is the first geological map of the most inaccessible part of Himalaya prepared with the help of systematic traversing and augmenting the essential data with the help of aerial photographs. Thus, it emerged that the Higher Central Himalaya represents a characteristic zone of a complex structural facies zones. The causes of formation of parallel structural facies zones from south to north in the Central Himalaya has been analysed in detail in the text chapters. Main structural facies zone of higher Himalaya is constituted by the rock suites of Tethys Himalaya, and bordered from north by *Great Himalayan Suture* (GHS) or Indus–Tsangpo Suture deep seated structural lineament or the plate boundary of Indian and Asian plates. From the south it is bordered by the *Main Axial Zone* (MAZ), is the so-called transitional tectonic zone dividing the Higher Northern Himalaya from Southern Himalaya. Hence, the Himalayan mountain range is geologically divided into two broad division of south and north along MAZ. It is established that along this border of Higher Central Himalaya there exists a wide synclinorium composed of Phanerozoic sequence of Tethys Himalaya (7000–9000 m) with well established palaeontologic age data control. In the base of section below Cambro-Ordovician sequence there are exposed completely unfossiliferous sequence of the Martoli group (6000 m). These Riphean (Upper Precambrian) little metamorphosed or unmetamorphosed sericite schist, silty phyllite sequence experienced an early Precambrian phase of orogeny giving way to deposition of Ralam conglomerate with a sharp angular unconformity. Perhaps, this gap may coincide with the Cambrian–Precambrian boundary. Over this the Ordovician–Silurian–Devonian sequence has been reworked with the help of megafossil assemblage and discovery of chitinozoa and scolecodonts. Stratigraphic work led to the establishment of a new horizon of 'Yong reefal limestone of Ordovician–Silurian age. Further up a Hercynian epeirogenic gap is recognized with very thin irregular patches of *Fenestella* horizon or the Carboniferous is found to be missing in general. The sequence is followed up to flysch sequence of Malla Johar. In the author's collection the dinoflagellates and coccoliths are discovered giving a Palaeogene age evidence of these strata which were till date recognized as upper Cretaceous. Moreover, in the flysch glauconite is found to replace the radiolarian bodies and also as grains indicating that it is a diagenetic product during the shallowing phase of the depocentre due to the tectonic uplift of the deep marine sediments in this region. This may be possibly the hiatus of Oregonian/Austrian orogenic cycle to be recorded first time in the Tethyan zone of Himalaya and called by the name of Sangcha Malla Pulse. Radiometric dating of ophiolite suite of rocks provide an upper Cretaceous age which are associated with the decollement plane of thrusting.

The entire Tethyan sedimentary pile is found to be tectonically faulted and thrusted to south along the *Tethyan Thrust* (TT) which was earlier regarded as a transitional zone between above mentioned unit and central crystalline metamorphics, TT also assumes a character of lateral movement fault in the northwestern part of the area. In the Central Himalaya part of Himalaya, this thrust becomes a subsidiary tectonic unit of the MAZ due to tremendous squeezing from north. The Main Axial Zone (MAZ) dividing the Southern and Northern Himalaya attaches with it special theoretical and practical significance in the context of recent concept of Himalaya and global tectonics. This deep-seated zone embodies the 'root-zone' of higher crystalline unit, and shows its divergent nature in morphotectonic exposition. Since it divides on either side the Precambrian sediments, hence it is imperative to assume its Precambrian inception. Thus, this transition has played an important role in the history of geological development of Himalaya. After an intensive research it has been established that transition zone of Vaikrita is mainly composed of three rock complexes. In the southern and northern part the exposed rocks are less or unmetamorphosed belonging to Southern and Northern Himalayan zone. And between them is located a complicated system of faults, which confines regionally metamorphosed, predominantly crystalline gneisses and schists, migmatites and magmatic rocks. Higher up there are thick (1000–5000 m) quartzite with distinct sedimentary origin, and over them a complex of calc-silicate rocks associated with schists, gneisses and quartzites. The metamorphism is found to increase upwards as we go to north in normal fashion from chlorite–biotite to kyanite–sillimanite facies. With a tectonic contact in the northern regime the less metamorphosed Precambrian–Phanerozoic sequence is exposed. Often in the Tethyan sedimentary zone we come across second order vertical fault systems. Repeated strata due to thrusting is not uncommon along many parallel thrust surfaces in NW–SE direction. Although there are yet scanty recent geophysical data, but those comparative data obtained from Eastern Alps and Central Himalaya confirms that the transition MAZ zone represents the axial zone of entire Himalayan mountain and assumes an *en echelon* character.

In order to explain the basic structural facies difference between the southern and northern zone, situated on the either side of the MAZ, a large area of southern Lesser Himalaya was studied and mapped. It emerged that between the Main Boundary Fault (MBF) and the Main Central Thrust (MCT) there existed two distinct structural facies zones of Simla–Krol and Shali–Deoban–Tejam from south to north. Both these structural facies zones have their Precambrian basal rock units over the basement which is never exposed in any part of Lesser Himalaya. Study showed that 'Simla–Krol' is a miogeosynclinal regime but the Shali carbonate (Riphean) with immense thickness increasing towards north (5000 m) are associated with basic volcanics. Nevertheless, the characteristic features of Chail tectonic unit with acid metavolcanics and early geosynclinal facies lithological character, and discovery in Tejam, characteristic upper Palaeozoic, is a definite pointer to palaeogeographic regime of *eugeosynclinal* nature in that period. *Hence, to call this part of Himalaya as reactivated Indian platform does not gather any favourable argument*. Therefore, this concept would need a thorough revision, and entire Himalayan region is to be taken as a whole a separate mobile belt. The petrological and geochemical studies helped establish that the calc-silicate rocks confined all along the MAZ zone in the Alaknanda, Pindar, Gori, Darma and Kali valleys exhibit a picrite granulite geochemical nature, showing high concentration of Pb (100 ppm), Sr (1740 ppm), Zn (295 ppm). These are pointers to its deep-seated juvenile hydrothermal origin from lower crustal–upper mantle layers.

These data support the view of deep-seated nature of the MAZ confining with it zone of divergence or root-zone of higher thrust of crystalline thrust sheet. In the extension of more than 200 km a barite–polymetallic mineralized zone has been established and a commercial type economic mineral deposit of barite has been successfully prospected in the northern fringing Tethyan zone. All these hydrothermal fluids have their source from the younger granitic–aplitic bodies intruding along tectonized zone of metamorphic Vaikrita and sedimentary Tethys. Now it has been found that these aplitic granitic bodies with tourmaline has very high percentage of boron (6000 ppm). This leads to infer that there is possibility of pegmatitic differentiation field and the discovery of rare earth mineral deposit in higher Himalaya.

Therefore, it is recommended that prospecting work for magmatic deposits should be carried out in this part of Central Himalaya. The changing concept in tectonic framework of Central Himalaya carries with it tremendous importance from both academic and practical point of view. The concept opens a new vista to visualize Himalaya as a typical Tethyan geosyncline or sediments filled trench having its own characteristic developmental features, and discovery of many more hidden mineral wealth in its womb.

References

Agarwal, P. N. and Kumar, S. (1981). An ostracod assemblage from the Kalapani Limestone, Malla Johar area, Kumaun Himalayas, U.P., in Sinha, A. K. (Ed.), *Contemporary Geoscientific Researches in Himalaya*, **1**, Dehra Dun, India, 217–224.

Allegre, C. J. and Othman, D. B. (1980). Nd–Sr, isotopic relationship in granitoid rocks and continental crust development; a chemical approach to orogenesis. *Nature*, **286**, 335–346.

Allegre, C. J., –others–and Xu, R. (1984). Structure and evolution of the Himalaya–Tibet orogenic belt. *Nature*, **307**, 17–22.

Andrews-Speed, C. P. and Brookfield, M. E. (1981). Middle Palaeozoic to Cenozoic geology and the tectonic evolution of the Northwestern Himalaya, *Tectonophysics*, **82**, 253–275.

Argand, E. (1924). La tectonique de l'Asie, *C.R. 13, Cong. Int.*, 169–371.

Armstrong, R. L. (1966). K–Ar dating of Plutonic and Volcanic rocks in orogenic belts, in Schaeffer, O. A. and Zahrinager, J. (Eds), *Potassium–Argon Dating*, Springer, New York, 117–133.

Armstrong, R. L. and Hansen, K. (1966). Cordilleran interstructure in the eastern Great Basin, *Am. J. Sci.*, **264**, 112–127.

Ashgirei, G. D. (1966). *Structural Geology*, Ijdevo, Moscow State University (in Russian).

Ashgirei, G. D. (1972). The problem of the origin of continental Earth crust, *Int. Geol. Cong., Sect. 3*, Tectonics, Montreal, 3–9.

Ashgirei, G. D. (1975). Thrust tectonics of Himachal Pradesh Himalayas, in Mahel, M. (Ed.), *Tectonic Problems of the Alpine System*, Veda Publishing House, Slovak Acad. of Sci., Bratislava, 241–250.

Ashgirei, G. D. (1977). A new approach for understanding the geological structure of Himalaya and search of mineral deposits related with them, *Him. Geology*, **7**, 1–21.

Ashgirei, G. D., Sinha, A. K., Raaben, M. E. and Dmitrenko, O. B. (1975). New findings on the geology of Lower Himalaya, Himachal Pradesh, India, *Chayanica Geologica*, **1**(2), 143–151.

Ashgirei, G. D., Pande, I. C., Sinha, A. K., and Malik, B. C. (1976). On the history of metamorphism of Western Himalayas, in *Geology and Mineral Deposits of Asian. African and Latin American countries, 1*, Moscow, 19–36 (in Russian).

Ashgirei, G. D., Sinha, A. K., Pande, I. C., and Malik, B. C. (1977a). A contribution to the geology, geochronology and history of regional metamorphism of Himachal Himalaya, *Him. Geology*, **7**, 102–117.

Ashgirei, G. D., Raaben, M. E., Sinha, A. K., and Dmitrenko, O. B. (1977b). New data on the geology of Lesser Indian Himalaya, *Invest. Ac. Sci. USSR, Ser. Geol.*, **5**, 135–198 (in Russian).

Ashgirei, G. D., Umnova, N. I., and Sinha, A. K. (1981). New findings in the geology of Western Himalaya, *Dokladi Ac. Sci. USSR*, **260**(4), 953–955 (in Russian).

Ashgirei, G. D. and Sinha, A. K. (1982). Some basic questions of tectonics, stratigraphy and palaeontology of Himalaya, *Proceedings of International Himalayan Geology Seminar*, New Delhi, 1976. *G.S.I. Misc. Pub.*, **41**, 361–370.

Auden, J. B. (1933). On the age of certain Himalayan granites, *Rec. G.S.I.*, **66**(4), 461–471.

Auden, J. B. (1934). The geology of the Krol belt, *Rec. G.S.I.*, **67**(4), 357–454.

Auden, J. B. (1935). Traverses in the Himalaya, *Rec. G.S.I.*, **69**(2), 123–167.

Auden, J. B. (1937). The structure of the Himalaya in Garhwal, *Rec. G.S.I.*, **71**, 407–433.

Auden, J. B. (1970). Discussion on the paper 'On the Krol Nappe Hypothesis, by A. Ranga Rao', *J. Geol. Soc. Ind.*, **11**(3), 288–302.

Azmi, R. J. (1983). Microfauna and age of the Lower Tal phosphorite of Mussoorie Syncline, Garhwal Lesser Himalaya, India, *Him. Geology*, **11**, 373–409.

Azmi, R. J., Joshi, M. N., and Juyal, K. P. (1981). Discovery of the Cambro-Ordovician conodonts from the Mussoorie Tal phosphorite; its significance in correlation of the Lesser Himalaya, in Sinha, A. K. (Ed.), *Contemporary Geoscientific Researches in Himalaya*, **1**, Dehra Dun, India, 245–250.

Azmi, R. J. and Sinha, A. K. (1981). Early and Middle Triassic Conodonts from the Kalapani Limestone of Malla Johar, Higher Kumaun Himalaya, *Abst. XII WIHG Seminar*, **44**, Dehra Dun.

Azmi, R. J. and Pancholi, V. P. (1983). Early Cambrian (Tommotian) conodonts and other shelly microfauna from the Upper Krol of Mussoorie Syncline, Garhwal Lesser Himalaya with remarks on Precambrian–Cambrian Boundary, *Him. Geology*, **11**, 360–372.

Banerjee, D. M. (1974). Stratigraphy and depositional characteristics of Tethyan sediments in Kuti–Kalapani area, Kumaun, India, *Him. Geology*, **4**, 296–322.

Banerjee, D. M., Tandon, S. K., and Sinha, A. K. (1975). Trace fossils in the Tethyan succession in a part of northern Kumaun Himalaya, *Proceedings, Symp. on Sediments, Sedimentation and Sedimentary Environment*, Univ. of Delhi, 177–185.

Barytes—A Market Survey, Mineral Economics Division. Indian Bureau of Mines, Ministry of Steel and Mines, Nagpur, 1973.

Batten, J. H. (1838). Note of a visit to the Nite pass of the Grand Himalayan Chain, *J. As. Soc. Beng.*, **VII**, 310–316.

Batten, J. H. (1951). *Official Reports on the Province of Kumaun* (includes the classical statistical sketch of Kumaon by G. W. Traill, 1820) (n.p.), Government Publication.

Batten, J. H. and Manson, K. (1842). Journal of a visit (by E. Manson) to the Malum and the Conta Dhoora pass in Juwahir, *J. As. Soc. Beng.*, **XI**, 1157–1181.

Belousov, V. V. (1962). *Fundamental Problems in Geotectonics*, Moscow (in Russian).

Benioff, H. (1958). Circum-Pacific tectonics, *Symposium on the mechanics of faulting*, Int. Un. Geol. and Geoph., Ottawa.

Berthelsen, A. (1951). A geological section through the Himalaya, *Medd. Danak. geol. Foren.*, **12**, 102–104.

Berthelsen, A. (1968). Debatable problem of structure and stratigraphy of Lower and Higher Himalaya in the Simla–Sutlej section, *Izvest. Ac. Sc. USSR, Ser. Geol.*, **3**, Moscow (in Russian).

Besse, J., Courtillot, V., Pozzi, J. P., Westphal, M., and Zhou, Y. X. (1984). Palaeomagnetic estimates of crustal shortening in the Himalayan thrusts and Zangbo suture, *Nature*, **311**, 623–626.

Bhanot, V. B., Kwatra, S. K., Kansal, A. K., and Pandey, B. K. (1978). Rb–Sr whole rock age for Chail Series of NW Himalaya, *J. Geol. Soc. India.*, **19**, 224–225.

Bhanot, V. B., Pandey, B. K., Singh, V. P., and Kansal, A. K. (1980). Rb–Sr ages for some granitic and gneissic rocks of Kumaun and Himachal Himalaya, in Valdiya, K. S. and Bhatia, S. B. (Eds), *Stratigraphy and Correlation of Lesser Himalayan Formations*, Hindustan Publishing Corporation, Delhi, 139–142.

Bhanot, V. B., Singh, V. P., Pandey, B. K., and Singh, R. P. (1981). Rb–Sr isochron age for the gneissic rocks of Askot crystallines, Kumaun Himalaya, U.P., in Sinha, A. K. (Ed.), *Contemporary Geoscientific Researches in Himalaya*, **1**, Dehra Dun, India, 117–119.

Bhargava, O. N. (1972). A reinterpretation of the Krol Belt, *Him. Geology*, **2**, 47–81.

Bhargava, O. N. (1976). Geology of the Krol Belt and associated formations: a reappraisal, *Mem. G.S.I.*, **106**(1), 168–234.

Bhargava, O. N. (1979). Lithostratigraphic classification of the Blaini, Infra Krol, Krol and Tal Formations—a review, *J. Geol. Soc. India.*, **20**, 7–16.

Bhargava, O. N. (1981). Geological setup of the Krol Belts: a review, in Sinha, A. K. (Ed.), *Contemporary Geoscientific Researches in Himalaya*, **1**, Dehra Dun, India, 89–100.

Bhargava, O. N. (1984). Trace fossils from the ? Cambrian Tal Group, Sirmur District, H.P., and proposed redefinition of the Tal, *J. Palaeon. Soc. India*, **29**, 84–87.

Bhatt, D. K. and Kumar, G. (1980). Discovery of conodonts in the Cambrian of Spiti, Tethys Himalaya, *Curr. Sci.*, **49**(a), 357–358.

Bhattacharya, A. R. (1981). Stratigraphy and correlation of the Inner Kumaun Lesser Himalaya: Implications of a mathematical study, in Sinha, A. K. (Ed.), *Contemporary Geoscientific Researches in Himalaya*, **1**, Dehra Dun, India, 137–150.

Bhattacharya, S. C. and Niyogi, D. (1971). Geologic evolution of the Krol Belt in Simla Hills, H.P., *Him. Geology*, **1**, 178–212.

Bingham, D. K. and Klootwijk, C. T. (1980). Palaeomagnetic constraints on greater India's underthrusting of the Tibetan Plateau, *Nature*, **284**, 336–338.

Bird, P. (1978). Initiation of intercontinental subduction in the Himalaya, *J. Geoph. Res.*, **86** (B10), 9320–9322.

Bist, K. S. (1986). *Geology, tectonostratigraphy, structure and metamorphism of the area around Okhimath in Mandakini Valley, Garhwal Himalaya*. Unpublished D.Phil. Thesis, Garhwal University, 101 pp.

Bist, K. S. and Sinha, A. K. (1982). Some observations on the geological and structural setup of Okhimath area in Garhwal Himalaya—a preliminary report, *Him. Geology*, **10**, 467–475.

Bist, K. S. and Sinha, A. K. (In Press). Imbrication of Main Central Thrust and its tectonic implication in the Central Sector of Himalaya, in Sychantavong, S. P. (Ed.), *Crustal Evolution and Orogeny*, Oxford and IBH Publishing Co., New Delhi.

Blandford, H. F. and Salter, J. W. (1865). *Palaeontology of Niti in the northern Himalaya, being descriptions and figures of the Palaeozoic and secondary fossils, collected by the Colonel Richard Strachey*, **8°**, Calcutta.

Bodenhausen, J. W. A. and Egeler, C. G. (1971). On the geology of the upper Kali Gandaki valley, Nepalese Himalayas, *Kouinkl. Nederl. Akad. Van Water., Amsterdam Proc.*, **B74**(5), 526–546.

Bordet, P. (1955). Les elements structuraux de l'Himalaya de l'Arun et de la region de l'Everest (Nepal Oriental), *C.R. Acad. Sci., Paris*, **240**, 102–104.

Bordet, P. (1961). *Researches geologiques dans l'Himalaya du Nepal region du Makalu*, Edit, C.N.R.S. Paris, 275 pp.

Bordet, P., Colchen, M., and Le Fort, P. (1972). Some features of the geology of the Annapurna Range, Nepal Himalaya, *Him. Geology*, **2**, 537–563.

Bouchez, J. L. and Pecher, A. (1981). The Himalayan Main Central Thrust Pile and its quartz-rich tectonites in Central Nepal, *Tectonophysics*, **78**, 23–50.

Bucher, W. H. (1957). *The Deformation of the Earth's Crust*, 2nd Edn., 1933, 518 pp.

Burrard, S. G. and Hayden, H. H. (1907–1908). *Sketch of the Geography and Geology of the Himalaya mountains and Tibet*, Govt. India Print (Revised by Burrard S. G. and A. M. Heron, 1934, 359 pp).

Burtman, V. S. (1975). Structural geology of Variscan Tien Shan, USSR, *Amer. J. Sci.*, **275A**, 157–186.

Chang Chang-Fa and Cheng Hsi-Lan (1973). Some tectonic features of the Mt. Jolmo Lungma area, southern Tibet, China, *Scientia Sinica*, **16**(2), 257–263, Peking.

Chaterji, G. C. and Swaminath, J. (1976). Problems of stratigraphy and structure of the parts of the Simla Himalaya—an assessment, *Mem. G.S.I.*, **106**(1), 1–16.

Chaterjee, G. C. and Swaminath, J. (1977). The stratigraphy and structure of parts of the Simla Himalaya—a synthesis, *Mem. G.S.I.*, **106**(II), 408–488.

Chhabra, N. L., and Sahni, A. K. (1979). A record of Holothurian Sclerites from the Triassic of Malla Johar, Kumaun Himalaya, *Curr. Sci.*, **48** (23), 1038–1039.

Choudhury, R. and Banerjee, K. C. (1966). Phosphorite occurrence in Naini Tal, U.P., *Technology*, **3**(4), 219.

Code of stratigraphic Nomenclature of India (1971). *G.S.I. Misc. Pub.*, **20**, 1–28.

Colchen, M. (1975). Palaeogeographic and structural evolution of the Tibet area of the Nepal Himalaya (Annapurna Region), *Him. Geology*, **5**, 83–103.

Cookson, I. C. and Eisenack, A. (1960). Upper Mesozoic microplankton from Australia and New Guinea, *Palaeontology*, **2**, 243–261.

Cotter, G. De, P. and Coggin Brown, J. (1907). Notes on certain glaciers in Kumaun, *Rec. G.S.I.*, **XXVI**, pt. 2.

Coward, M. P. (1985). Thrust tectonics and the deep structure of the Pakistan Himalaya, *Geology*, **13**, 417–420.

Crawford, A. R. (1974). Indus Suture Line, The Himalayas, Tibet and Gonwanaland, *Geol. Mag.*, **111**(5), 369–383.

Cunningham, A. (1963). *The Ancient Geography of India* (Varanasi Indological Book House), 299–302.

Dalrymple, G. B. and Lamphere, M. A. (1969). *Potassium–Argon Dating*, W. H. Freeman and Co., San Francisco, 258 pp.

Damon, P. E. (1968). Potassium–argon dating of igneous and metamorphic rocks with application to the Basin ranges of Arizona and Sonora, in Hamilton, E. I. and Farquhar, R. M. (Eds), *Radiometric Dating for Geologists*, Interscience Publishers Inc., New York, 1–71.

Dave, V. K. S. and Rawat, B. S. (1968). A preliminary account of the geological traverses in Bhapkund–Malari–Niti areas, Chamoli Dist., U.P., *Res. J. Roorkee Univ.*, **10**(3–4), 97–108.

Davey, R. J., Downie, C., Sarjeant, W. A. S., and Williams, G. L. (1966). Studies on Mesozoic and Cainozoic dinoflagellate cysts, *Bull. Br. Mus. Nat. Hist. (Geol.) Suppl. 3*, 1–248.

Dewey, J. F. and Bird, J. (1970). Mountain belts and new global tectonics, *Geoph. Res.*, **75**, 2625–2647.

Dewey, J. F. and Burke, K. G. A. (1973). Tibetan, Variscan, and Precambrian basement reactivation: products of continental collision, *J. Geol.*, **81**, 683–692.

Diener, C. (1895a). The cephalopoda of Muschelkalk, *Pal. Ind. G.S.I., Ser.*, **15**,2(2), Pl. 1–31, 1–118.

Diener, C. (1895b). Der Geologische Bau der sedimentrazone des Central Himalaya zwischen Milam und dem Niti Pass, *Verh. K. K. Geol. Reichanst.*, **14**, 370–376.

Diener, C. (1895c). Ergebnisse einar geologischan expedition in den Central Himalaya von Johar, Undes und Painkhanda, *Denkach. Adad. Wiss., Wien*, **62**, 533–608.

Diener, C. (1897a). The Permian fossils of the Productus shales of Kumaun and Garhwal, *Pal. Ind. Geol. Surv. Ind. Ser.*, **15**,1(4), Pl. 1–5, 1–54.

Diener, C. (1897b). Cephalopodes of the Lower Trias, *Pal. Ind. Geol. Surv. Ind. Ser.*, **15**,2(1), pl. 1–23, 1–181.

Diener, C. (1898). Notes on the geological structure of the Chitichun region, *Mem. G.S.I.*, **36**(1), 1–27.

Diener, C. (1903). The Permian fossils of the Central Himalaya, *Pal. Ind. G.S.I., Ser.* **15**(1), pl. 1–0, 1–204.

Diener, C. (1906). The fauna of Tropites limestone, Byans, *Pal. Ind. G.S.I., Ser.*, **15**(1), pl. 1–17, 1–201.

Diener, C. (1912). The Trias of the Himalayas, *Mem. G.S.I.*, **36**, 1–175.

Egeler, C. G., Bodenhausen, J. W. A., De Booy, T., and Nijhuis, H. J. (1964). *On the geology of Central West Nepal—a preliminary note*, 22nd Int. Geol. Congr. India, 1964, Pt. 11, 101–122. New Delhi.

Engelhardt, W. von. (1936). *Chemie der Erde*, **10**, 187.

Fonia, K. S. (1977). *Uttrakhand, Garhwal Himalayas*, Revised by Iris Turner, Asian Journals, Dehra Dun, 270 pp.

Frank, W., Hoinkes, G., Miller Ch., Purtecheller, F., Richeter, W., and Thöni, M. (1976). Relations between metamorphism and orogeny in a typical section of the Indian Himalayas. *Techern. Minre. Petr. Mitt.*, **20**(3), 303–332, Wien.

Frank, W., Thöni, M., and Purtscheller, F. (1976). Geology and petrography of Kulu–South Lahaul area, *Coll. Intern. CNRS, No. 268, Himalaya, Sci. de la Terra*, 147–172.

Frank, W., Gansser, A., and Trommsdorff, V. (1977). Geological observations in the Ladakh areas (Himalayas). A preliminary report, *Schweiz petrogr. Mitt.*, **57**, 89–113, Zürich.

Fuchs, G. (1967). Zum Bau des Himalaya, *Denkschr. Osterr. Acad. Wiss., Math. Natw.*, **113**, 1–211, Wien.

Fuchs, G. (1975). Contributions to the geology of the north-western Himalayas, *Abh. geol. B-A*, **32**, 1–59.

Fuchs, G. (1977). Traverse of Zanskar from the Indus to the Valley of Kashmir—a preliminary note, *Jb. geol. B.-A.*, **120**,(2), 219–229.

Fuchs, G. (1979). On the Geology of Western Ladakh, *Jahrb. Geol. B.-A. Band*, **122**, Heft 2, 513–540, Wein.

Fuchs, G. (1981). Outline of the geology of the Himalaya, *Mitt. Osterr. geol. Ges.*, **74/75**, 101–127, Wien.

Fuchs, G. (1982). *Explanations of the geologic–tectonic map of the Himalaya, Geological Survey of Austria*, Vienna, 50 pp.

Fuchs, G. and Sinha, A. K. (1974). On the geology of Naini-Tal (Kumaun Himalaya). *Him. Geology*, **4**, 463–580.

Fuchs, G. and Sinha, A. K. (1978). The tectonics of the Garhwal–Kumaun Lesser Himalaya, *Jb. geol. B-A*, **121**(2), 219–241, Wien.

Gairola, V. K. (1975). On the petrology and structure of the Central Crystallines of the Garhwal Himalaya, U.P., *Him. Geology*, **5**, 455–467.

Gansser, A. (1964). *Geology of the Himalayas*, Inter-

science Publishers—a division of John Wiley & Sons Ltd., London, 289 pp.
Gansser, A. (1966). The Indian Ocean and the Himalayas: a geological interpretation, *Eclogae geol. Helv.*, **67**, 479–507.
Gansser, A. (1974a). The Ophiolitic Melange, a world-wide problem on Tethyan examples, *Eclogae geol. Helv.*, **67**, 479–507.
Gansser, A. (1974b). *Himalaya*, in Spencer, A. M. (Ed.), *Mesozoic–Cenozoic Orogenic Belts—Data for Orogenic Studies. Spec. Publication geol. Soc. Lond.*, **4**, 267–278.
Gansser, A. (1976). The great Suture Zone between Himalaya and Tibet—a preliminary account, *Himalaya Sciences de la terra*, Ed. CNRS, Paris, 181–191.
Gansser, A. (1979). The ophiolitic Suture Zones of the Ladakh and the Kailas region—a comparison, *J. Geol. Soc. Ind.*, **20**(6), 277–281.
Gansser, A. (1980a). The significance of the Himalayan Suture Zone, *Tectonophysics*, **62**, 37–52.
Gansser, A. (1980b). The Peri-Indian Suture Zone, in *Geology of the Alpine Chains born of the Tethys*, 26th Int. Geol. Cong., Paris, 140–148.
Garg, R., Jain, K. P., Singh, I. B., Kumar, S., and Singh, S. K. (1981). Tethyan Cretaceous radiolaria from Malla Johar area, Kumaun Himalaya, U.P., India, *J. Pal. Soc. India*, **25**.
Gerling, E. K. (1961). The contemporary state of the argon method for the determination of age and its application to geology, *Acad. Sci. USSR*, Moscow, 130 pp. (in Russian).
Ghosh, A. K. and Srivastava, S. K. (1962). Microfloristic evidence on the age of Krol Bed and associated formations, *Proc. Nat. Inst. Sci.*, **28A**(5), 710–717.
Girmaz, G. U. and Sarjeant, W. A. S. (1972). Dinoflagellate cysts and acritarchs from the Kimmeridgian (Upper Jurassic) of England, Scotland and France, *Bull. Br. Mus. Nat. Hist.* (*Geol.*), **21**, 171–267.
Griesbach, C. L. (1880). Palaeontological notes on the Lower Trias of the Himalayas, *Rec. G.S.I.*, **13**(2), 94–113.
Griesbach, C. L. (1891). Geology of the Central Himalayas, *Mem. G.S.I.*, **23**, 232 pp.
Griesbach, C. L. (1893) Notes on the Central Himalayas, *Rec. G.S.I.*, **26**(1), 19–25.
Grinlinton, J. L. (1912). Notes on the Poting Glacier, Kumaon Himalayas, *Rec. G.S.I.*, **52**, 102–126.
Grinlinton, J. L. (1914). Notes on some glaciers of the Dhauli and Lesser Valleys, Kumaon Himalaya, *Rec. G.S.I.*, **55**, 280–335.
Gupta, H. K. (1973). Geophysical investigations for the Himalayan region undertaken at the NGRI, *Proceedings Sem. on Geodynamics of the Himalayan region*, 72–96.
Gupta, L. N. and Saxena, M. N. (1978). Corona structures from the Central Crystalline Zone of Badrinath, Central Himalaya, India, *Him. Geology*, **8**(2), 703–716.
Gupta, L. N. (1978a). Reaction skarn and calcareous rocks of the central crystalline complex of Badrinath, Central Himalayas (India), *Bull. Ind. Geol. Assoc.*, **11**(1), 63–72.
Gupta, L. N. (1978b). Petrology of the Central Crystalline rocks of Badrinath area, Central Himalayas, *Him. Geology*, **8**(2), 725–727.
Gupta, V. J. (1977). Stratigraphic position of the Kioto limestone Himalaya, *Sci. de la terre*, Ed. CNRS, Paris, 195–196.
Gupta, V. J. (1981). Tethys biostratigraphy of Spiti, Ladakh and Kashmir, in Sinha, A. K. (Ed.), *Contemporary Geoscientific Researches in Himalaya*, **1**, Dehra Dun, India, 189–200.
Hagen, T. (1960). *Nepal*, Kummarly and Frey, Bern, 119 pp.
Hainrichs, T. K. and Reimer, T. O. (1977). A sedimentary baryte deposit from the early Precambrian, Fig Tree Group, *Econ. Geol.*, **72**, 1426–1441.
Hamet, J. and Allegre, C. J. (1976). Rb–Sr systematics in garnet from Central Nepal (Manasalu) significance of the Oligocene age and high $^{87}Sr/^{86}Sr$ ratio in Himalayan Orogeny, *Geology*, **4**, 470–472.
Hamilton, E. I. (1965). *Applied Geochronology*, Academic Press, London and New York.
Hashimoto, S. (1973). *Geology of the Nepal Himalayas*, Sapporo (Himalayan Committee Hokkaido Univ.), 286 pp.
Hayden, H. H. (1904). The geology of Spiti, *Mem. G.S.I.*, **36**(1), 1–129.
Hayden, H. H. (1908). A sketch of the geography and geology of the Himalayan Mountains and Tibet, *The Geology of the Himalaya*, Pt. 4, Govt. of India Press, Calcutta, 236 pp.
Hayden, H. H. (1913). Notes on the relationship of the Himalaya to the Indo-Gangetic Plain and the Indian Peninsula, *Rec. G.S.I.*, **43**(2), 138–167.
Heim, A. (1937). Structural studies in the Central Himalaya, *Him. J.*, **9**, 38–43.
Heim, A. (1938)., The Himalayan Border compared with the Alps, *Rec. G.S.I.*, **72**(4), 413–421.
Heim, A. (1956). The geological structure of the Himalaya compared with the Alps, *Proc. Nat. Inst. Sci. Ind.*, **22/A**(4), 228–235.
Heim, A. and Gansser, A. (1938). *Thron der Götter*, Erlebnisse der ersten Schweizersichan Himalaya Expedition, Morgarten-verlag Aktiengasellschaft, Zürich und Leipzig, 270 pp. (in German).
Heim, A. and Gansser, A. (1939a). Central Himalaya, Geological observations of the Swiss expedition 1936, *Mem. Soc. Helv. Sci. Nat.*, **73**(1), 1–245.
Heim, A. and Gansser, A. (1939b). *Throne of the Gods*. Macmillan, London, 233 pp.
Herbert, J. D. (1842). Report on the Mineralogical Survey of the Himalayas lying between the rivers Sutlej and Kali, *As. Soc. Beng.*, **XI**, Suppl. pp. I–CL X III, Map XIII, pt. 1, 171–184.
Hughes, T. W. H., and Waagen, W. (1878). Note on a trip over the Milam Pass, Kumaon with a Description of Fossils, *Rec. G.S.I.*, **XI** (2), 182–187.
Ioannides, N. S., Stavrines, G. N., and Downie, C. (1976). Kimmeridgian microplanktons from the Clavells Hard, Dorset, England, *Micropalaeontology*, **22**, 443–478.
Jaeger, E., Bhandari, A. K., and Bhanot, V. B. (1971). Rb–Sr age determinations on biotites and whole rock samples from Mandi and Chor Granites, H. P., India, *Ecl. Geol. Helv.*, **64**, 521.

Jain, A. K. (1971). Stratigraphy and tectonics of Lesser Himalayan Region of Uttarkashi, Garhwal Himalaya, *Him. Geology*, **1**, 25–58.

Jain, A. K. (1972). Structure of Bidhalna–Pharat Windows and Garhwal Thrust Unit, Garhwal, U.P., *Him. Geology*, **2**, 188–205.

Jain, K. P., Garg, R., Kumar, S., Singh, I. B., and Singh, S. K. (1978). Dinoflagellates and radiolarians from the Tethyan sediments, Malla Johar Area, Kumaon Himalaya, A Preliminary Report, *J. Pal. Soc. India.*, **21** and **22**, 116–119.

Jain, K. P., Garg, R., Singh, I. B., and Kumar, S. (1980). Upper Jurassic dinoflagellates and other microfossils from the Spiti Shale, near Lapthal, Malla Johar area, Kumaun Himalaya, U.P., *Abst. XI WIHG Seminar*, 13 pp.

Jeannet, A. (1959). Ammonites permiennes et faunes triasiques de l'Himalaya central (Expedition Suisse Arn. Heim et A. Gansser, 1936), *Mem. G.S.I., N.S.*, **34**(1), 168 pp.

Jixiang, Yin *et al.* (1980). *A Scientific Guidebook to South Xizang (Tibet), Part B. Geology and Geophysics* (Yin Jixiang, Chief Compiler), Academia Sinica.

Kaila, K. L. and Narain, H. (1976). Evolution of the Himalaya based on seismotectonic and deep-seismic soundings. *Int. Sem. Him. Geology*, New Delhi, 30 pp. (preprint).

Khain, V. E. (1979a). *Regional Geotectonics. Outer Alpine Asia and Australia*, Moscow, Nedra, 353 pp. (in Russian).

Khain, V. E. (1979b). Ophiolites and the nappe structure of the North Caucasus Forerange, *Geotectonika*, **4**, 63–80. (in Russian).

Khanna, A. K. (1978). Subathus-stratigraphic status and nomenclature, *Him. Geology*, **8**(1), 209–223.

Khanna, A. K., Sinha, A. K., and Sah, S. C. D. (1985). Yong Limestone of Tethys Himalaya—its stratigraphic status and palynological fossils, *J. Geol. Soc. India.*, **26**(3), 191–198.

Klootwijk, C. T. (1979). A review of palaeomagnetic data from the Indo-Pakistan fragment of Gondwana land, in Farah, A. and Dejong, K. A. (Eds), *Geodynamics of Pakistan*, 41–80. Geological Society of Pakistan, Quetta.

Klootwijk, C. T., Sharma, M. L., Gergan, J. T., Tirkey, B., Shah, S. K. and Agarwal, V. (1979). The extent of greater India, palaeomagnetic data from the Ladakh intrusives at Kargil, North-western Himalayas, *Earth and Planet. Sci. Lett.*, **66**, 47–64.

Kraft, A. von, (1904). Notes on the 'Exotic Blocks' of Malla Johar in the Bhot Mahals of Kumaon, *Mem. G.S.I.*, **32**(3), 127–183.

Krasulin, V. S. (1967). *Guide Book of Techniques of Geologists*, Nedra, Moscow, 488 pp. (in Russian).

Kristan-Tollman, Edith (1984). Trias-Foraminiferen von-Kumaun in Himalaya, *Mitt. Osterr. geol. Ges.*, **77**, 263–329, Vienna.

Krummenacher, D. (1961). Determination d'age isotopique faites sur quelques roches l'Himalayan du Nepal per la methods potassium–argon, *Schweiz, Mineral. Petrogr. Mitt.*, **41**, 273–283.

Krummenacher, D. (1966). Nepal Central: Geochronometrie des series de l'Himalaya, *Schweiz. Mineral. Petrog. Mitt.*, **46**, 43–54.

Krummenacher, D. (1971). Geochronometrie des roches de l'Himalaya, in Bordet, P. *et al.* (Eds). *Researches Geologiques dans l'Himalaya, du Nepal, Region de la Takhola*, CNRS, Paris, 187–202.

Krummenacher, D., Bassett, A. M., Kingery, F. A., and Layne, H. F. (1978). Petrology, metamorphism and K/Ar age determinations in eastern Nepal, in Saklani, P. S. (Ed.), *Tectonic Geology of the Himalaya*. Today and Tomorrow's Printers and Publishers, New Delhi, 151–166.

Kumar, G., Mehdi, S. H., and Prakash, G. (1972). A review of stratigraphy of parts of Uttar Pradesh Tethys Himalaya, *J. Pal. Soc. India.*, **15**, 86–98.

Kumar, G. (1981). Stratigraphy and tectonics of Lesser Himalaya of Kumaun, Uttar Pradesh, in Sinha, A. K. (Ed.), *Contemporary Geoscientific Researches in Himalaya*, **1**, Dehra Dun, India, 61–70.

Kumar, S. Ravindra, and Pande, I. C. (1972). Deformation of rocks of Simla Hills, *Geol. Rundschau*, **61**(2), 430–441.

Kumar, S., Singh, I. B., and Singh, S. K. (1977). Lithostratigraphy, structure, depositional environment, palaeocurrent and trace fossils of the Tethyan sediments of Malla Johar area, Pithoragarh, Chamoli District, U.P., India, *J. Pal. Soc. Ind.*, **20**, 396–435.

Kumar, S., Singh, I. B., and Singh, S. K., (1978). Cone-in-cone structure in the upper flysch series (Upper Cretaceous) Malla Johar area, Pithoragarh dist., U.P., *Curr. Sci.*, **47**(9), 301–302.

Kumar, S. and Tewari, V. C. (1978). Occurrence of *Conophyton garganicus* from the Gangolihat dolomites, Kathpuria Chhina area, dist. Almora, U.P., *J. Geol. Soc. India.* **19**(4), 174–178.

Lakhanpal, R. N., Sah, S. C. D., and Dube, S. N. (1958). Further observations on plant microfossils from a carbonaceous shale (Krols) near Naini Tal, with a discussion of the age of the beds, *The Palaeobot.*, **7**(2), 111–120.

Le Fort, P. (1975). Himalayas: the collided range. Present knowledge of continental arc, *Amer. J. Sci.*, **275–A**, 1–44.

Le Fort, P., Debon, F., and Sonet, J. (1980). The 'Lesser Himalayan' Cordierite Granite belt, typology and age of the pluton of Manserah, *Sp. issue Geol. Bull., Peshwar Univ. Pakistan*, 51–61.

Levinson-Lessing, F. J. (1932). Uber die Raubildung grosser flacher Intrusivalger, *Min. Petr. Mitt., Bd.*, **43**, 271–282.

Li, Tingdong, Xuchang, X, Guangcan, L., Yanlin, C. and Weiqin, Z. (1984). *On the Crustal Evolution and Uplift mechanism of the Qingshai–Tibet plateau*, Chinese Academy of Sciences, Baiwanzhuang, Beijing.

Lombard, A. (1958). Un itineraire geologique dans l'est du Nepal (Massif du Mont Everest), *Denkschr. Schweiz, Waturf. Ges.*, **82**, 1–107.

Lyon-Caèn, H. and Molnar, P. (1983). Constraints on the structure of the Himalaya from an analysis of gravity anomalies and a flexural model of the lithosphere, *J. Geophys. Res.*, **88**, 8171–8191.

MacIntyre, R. M., Cliff, R. A., and Chapman, N. A. (1981). Geochronological evidence for phased volcanic activity in Fife and Caithness necks, Scotland, *Trans. R. S. Eding. E. Sci.*, **72**, 1–7.

Mamgain, V. D. and Sastry, M. V. A. (1975). *Palaeontologica Indica G.S.I.*, XLIII, Pl. 1–10, 1–33.

Medlicott, H. B. (1864). On the geological structure and relations of the southern portion of the Himalayan Range between river Ganges and Ravee, *Mem. G.S.I.*, **3**(2), 1–206.

Mehrotra, N. C. and Sinha, A. K. (1978). Discovery of microplanktons and the evidences of the younger age of Sangcha Malla Formation (upper Flysch) of Malla Johar area in the Tethyan zone of Kumaun Himalaya, *Him. Geology*, **8**(2), 1001–1004.

Mehrotra, N. C. and Sinha, A. K. (1981). Further studies on microplanktons from the Sangcha Malla formation (upper FLysch) zone of Higher Kumaun Himalaya, in Sinha, A. K. (Ed.), *Contemporary Geoscientific Researches in Himalaya*, **1**, 151–160.

Mehta, P. K. (1977). Rb–Sr geochronology of the Kulu–Mandi Belt, its implication for the Himalayan tectogenesis, *Geol. Rundsch*, **66**, 156–175.

Mehta, P. K. (1980). Tectonic significance of the young mineral dates and the rates of cooling and uplift in the Himalaya, *Tectonophysics*, **62**, 205–217.

Mehta, P. K. and Rex, D. C. (1977). K–Ar geochronology of the Kulu–Mandi Belt, NW Himalaya, India, *Neues, Jahrb, Mineral. Monatsh*, **8**, 343–355.

Misra, R. C. and Valdiya, K. S. (1961). The Calc-Zone of Pithoragarh, with special reference to the occurrence of stromatolite, *J. Geol. Soc. India*, **2**, 78–90.

Misra, R. C. and Sharma, R. P. (1973). Geology of the Nandprayag Klippe and Central Crystalline, Kumaon Himalayas, *Bull. Ind. Geol. Assoc.*, **6**(2), 85–96.

Molnar, P. and Tapponier, P. (1975). Cenozoic tectonics of Asia: effects of a continental collision, *Science*, **189**, 419–426.

Molnar, P. and Chen, W. P. (1981). Extensional tectonics in central and eastern Asia: a brief summary, *Phil. Trans. Roy. Soc. London*, **A. 300**, 1454, 403–306.

Molnar, P. (1984). Structure and tectonics of the Himalaya, *Ann. Rev. Earth Planet. Sci.*, **12**, 489–518.

Moorbath, S. (1967). Recent advances in the applications and interpretation of radiometric age data, *Earth. Sc. Rev.*, **3**, 111–133.

Mu An Tze, Wen Shin-Heuan, Wang Yi Kang, Chang Ping-Kao, and Yin Chi-Haieng (1973). Stratigraphy of the Mount Jolmo Lungma Region in Southern Tibet China, *Scientia Sinica*, **16**(1), 96–111, Peking.

Narain, Hari. (1973). Crustal structure of the Indian subcontinent. *Tectonophysics*, **20**, 249–260.

Nautiyal, A. C. (1985). First Record of Middle Algonkian Microorganisms from Naldera Limestone (Basantpur Formation) of Himachal Himalaya with Biostratigraphy, *Geoscience Journal*, **4**(1), 57–72, Lucknow, India.

Nautiyal, S. P. (1953). Almora district, *Rec. G.S.I.*, **79**(1), 375.

Negi, S. S., Sinha, A. K. and Pande, B. K. (1982). A preliminary report on the geology and structure of the Rudraprayag–Tilwara–Mayali area of Garhwal Himalaya, *Him. Geology*, **10**, 211–219.

Negi, S. S., Pande, B. K. and Sinha, A. K. (1983). A study of drainage morphometry with reference to geology in Rudraprayag–Tilwara–Mayali area of Garhwal Himalaya, in Sinha, A. K. (Ed.), *Contemporary Geoscientific Researches in Himalaya*, **2**, 155–158.

Negi, S. S., Pande, B. K. and Sinha, A. K. (1984a). Litho–petrological observations of the rock-types in the Rudraprayag Tilwara area of Garhwal Lesser Himalaya, *Him. Geology*, **12**, 156–159.

Negi, S. S., Pandey, B. K., and Sinha, A. K. (1984b). Slope studies in Rudraprayag–Tilwara–Mayali area of Garhwal Himalaya with special reference to slope stability, *Him. Geology*, **12**, 330–344.

Nier, A. O. (1950). A redetermination of the relative abundance of the isotopes of carbon, nitrogen, oxygen, argon, and potassium, *Phys. Rev.*, **77**, 789.

Nikolaev, V. A. (1959). About the nature and morphological type of deep seated fault (Main Karatau fault taken as example), *Sov. Geologia*, **7** (in Russian).

Odell, N. E. (1943). The so-called 'Axial Granite Core' of the Himalaya: its actual exposure in relation to its sedimentary cover, *Geol. Mag.*, **80**, 148–154.

Oldham, R. D. (1883). Note on the geology of Jaunsar and Lower Himalayas, *Rec. G.S.I.*, **16**, 193–198.

Pande, I. C. and Saxena, M. N. (1968). Birth and development of Himalaya, *Pub. CAS Geol., Panjab Univ., Chandigarh*, **4**, 1–19.

Pande, I. C. and Kumar, S. (1974). Absolute age determination of crystallines of Manali–Jaspa region, NW Himalaya, *Geol. Rundschau*, **63**(2), 539–548.

Pande, T. (1962). *Kumaun Ke Lok Sahitya*, Almora Book Depot, 46 pp. (in Hindi).

Pant, S. D. (1935). *The Social Economy of the Himalaya*, London, George Allen and Unwin Ltd., 59 pp.

Parel'man, A. I. (1977). *Geochemistry of elements in the Supergene Zone* (Translated from Russian, Jerusalem), 210–220.

Pascoe, E. D. (1973–1975). *A Manual of the Geology of India and Burma*, **1**, 434–435, **2**, 819–925, **3**, 1566–1570.

Pati, U. C. and Rao, P. N. (1981). The Main Central Thrust in U.P. Himalaya, in Sinha, A. K. (Ed.), *Contemporary Geoscientific Researches in Himalaya*, **2**, Dehra Dun, 125–129.

Patriat, P. and Achache, J. (1984). India–Eurasia collision chronology has implications for crustal shortening and driving mechanism of plates, *Nature*, **311**, 615–621.

Pécher, A. and Le Fort, P. (1977). Origin and significance of the lesser Himalayan augen gneisses, *Coll. Int. Centre Netl. Rech. Sci.*, **268**, 319–329.

Peive, A. B. (1945). Deep seated fault in the geosynclinal regions, *Ijv. Ac. Sc. USSR Ser. Geol.*, **5** (in Russian).

Peive, A. B. (1947). Asymmetry of deep tectonic structure of Ural–Tien Shan orogen and the origin of its virgation, *Bull. Mos. Ob. Isp. Pri. Std. Geol.*, **22**(5) (in Russian).

Peive, A. B. (1956). Relation of sedimentation, folding magmatism and mineral deposits with the deep seated faults. *Ijv. Ac. Sc. USSR, Ser. Geol.*, **A1.3** (in Russian).

Pickett, J. W., Tell, J. S., Conaghan, P. J., and Powell, C. McA. (1975). Jurassic invertebrates from the Himalayan Central Gneiss. *Alcharinga*, **I**, 71–85.

Pilgrim, G. E. and West, W. D. (1928). The structure and correlation of the Simla rocks, *Mem. G.S.I.*, **53**, 140 pp.

Powar, K. B. (1972). Petrology and structure of the Central crystalline zone NE Kumaun, *Him. Geology*, **2**, 36–46.

Powell, C. McA. and Conaghan, P. J. (1973a). Polyphase deformation in Phanerozoic rocks of the Central Himalayan Gneiss, NW India, *J. Geol.*, **81**(2), 127–143.

Powell, C. McA. and Conaghan, P. J. (1973b). Plate tectonics and the Himalaya, *Earth Planet, Sci. Lett.*, **20**, 1–12.

Powell, C. McA. (1975). Tectonic models of the Tibetan plateau. *Geology*, 727–731.

Powell, C. McA., Crawford, A. R., Armstrong, R. L., Prakash, R. and Wynne-Edwards, H. R. (1979). Reconnaissance Rb–Sr dates for the Himalayan Central Gneiss, NW India, *Ind. J. Earth Sci.*, **6**(2), 149–151.

Prasad, A. K. and Bhatia, M. R. (1975). Fossil tracheids and other micro-fossils from the Blaini tillite horizon, Simla Hills, India, *Bull. Ind. Geol. Assoc.*, **8**, 191–195.

Puchelt, H. (1972). Barium, in Wedespohl, K. H. (Exec. Ed.), *Handbook of Geochemistry*, Chap. 36, Berlin, Springer Verlag.

Qureshy, M. N. (1969). Thickness of a basaltic layer as a possible cause for the uplift of the Himalaya—suggestion based on gravity data, *Tectonophysics*, **7**, 137–157.

Qureshy, M. N., Venkatachalam, S., and Subrahamanyam, C. (1974). Vertical tectonics in the middle Himalayas, an appraisal from recent gravity data, *Geol. Soc. Am. Bull.*, **85**, 921–926.

Raina, V. K. (1963). A note on some glaciological observations in the Garhwal Himalayas, *Ind. Min.*, **17**, 159–63.

Raina, B. N. (1981). Need for re-interpretation of the geology of Simla area, some suggestions, in Sinha, A. K. (Ed.), *Contemporary Geoscientific Researches in Himalaya*, **1**, Dehra Dun, India, 101–106.

Randhawa, M. S. (1970). *The Kumaun Himalayas*, Oxford & IBH Pub. Co., New Delhi, 185 pp.

Ravindra Kumar, S., and Pande, I. C. (1972). Deformation of rocks of the Simla Hills, *Geol. Rundsch*, **61**(2), 430–441.

Ray, S. K. and Naha, K. (1971). Structural and metamorphic history of the 'Simla Klippe': a summary, *Him. Geology*, **1**, 1–24.

Reed, F. R. S. (1912). Ordovician and Silurian fossils of Central Himalaya, *Pal. Ind. GSI.*, *Ser. 15*, **7**(2), 168 pp.

Reid, E. M., and Chandler, M. E. J. (1933). The London Clay flora, *Bull. Br. Mus. Nat. Hist.*, 1–156.

Rickard, M. J. and Belbin, L. (1980). A new continental assembly for Pangaea, *Tectonophysics*, **63**, 1–12.

Rupke, J. (1974). Stratigraphic and structural evolution of Kumaun Lesser Himalaya, *Sedimentary Geol.*, **11**, (2–4), 81–265.

Sah, S. C. D., Venkatachala, B. S., and Lakhanpal, R. N. (1968). Palynological evidence of the age of Krols, *Publ. CAS, Geol. Panjab Univ., Chandigarh*, **5**, 115–120.

Sah, S. C. D., Maithy, P. K., and Bhargava, O. N. (1977). Some significant palynomorphs from B Member of the Jutogh Formation of Simla Hills, *J. Geol. Soc. Ind.*, **18**(3), 139–145.

Sahni, Birbal. (1936). The Himalayan Uplift since the advent of man: its cultural significance, *Curr. Sci.*, **5**(2), 57.

Sahni, A. and Chhabra, N. L. (1974). Middle Triassic conodonts from Kalapani Limestone Malla Johar, Kumaun Himalaya, *Him. Geology*, **4**, 259–295.

Salter, J. W. and Blanford, J. W. H. F. (1865). *Palaeontology of Niti in the Northern Himalaya being Descriptions of Palaeozoic and Secondary Fossils, Collected by Col. Richard Strachey*, R.E., Calcutta, 1–112.

Sanders, J. E. and Friedman, G. M. (1967). Origin and occurrence of limestones, *Develop. in Sedim., A. Carbonate Rocks*, Elsevier, 215 pp.

Sastry, M. V. A., Chandra, A. and Mamgain, V. D. (1970). Note on Pleistocene varved clays and associated microflora from Goting, Niti area of Kumaun Hills, *Ind. Min.*, **24**(2), 131–133.

Sastry, M. V. A. and Mamgain, V. D. (1970). Note on mineral occurrences in Niti area of Kumaun Himalayas, *Ind. Min.*, **24**(3), 282–288.

Sastry, M. V. A. and Mamgain, V. D. (1971). Notes on geological traverses in Niti area, Kumaun Himalayas, *Abst. Sem. Rec. Res. in Him, G.S.I.*, **14**.

Sastry, M. V. A. and Mamgain, V. D. (1975). Notes on geological traverses in Niti area, Kumaun Himalaya, *G.S.I. Mics. Publi.*, **24**, 164–175.

Sastry, M. V. A. and Mamgain, V. D. (1977). Biostratigraphy of the Palaeozoic and Mesozoic sediments of Kumaun Himalayas, *Rec. G.S.I.*, **108**(2), 1–35.

Schaeffer, O. A. and Zahringer, J. (Eds) (1966). *Potassium–Argon Dating*, Springer Verlag, Berlin, 234 pp.

Schuchert, C. (1931). *Outlines of Historical Geology*, 2nd Ed., New York, Wiley, 347 pp.

Schwan, W., Sinha, A. K., and Bagati, T. N. (1985). Structural tectonic observations along a traverse across the Chamba region between Nanikhad and Tisa in the north-western Himalaya, H.P., India. *N. Jb. Geol. Palaont. Mh.* I, Stuttgart, 553–577.

Shackleton, R. M. (1981). Structure of Southern Tibet, Report on a traverse from Lhasa to Kathmandu organised by Academia Sinica, *J. Struc. Geol.*, **3**(1), 97–105.

Shah, S. K. (1977). Indus ophiolite belt and the tectonic setting of the Malla Johar–Kiogad exotics in Himalaya, *Himalaya Sci. de la Terre*, Ed. CNRS, Paris, 361–368.

Shah, S. K. and Sinha, A. K. (1974a). The first record of algal bioherms in the Palaeozoic sequence of the Tethyan Zone of the Kumaun Himalaya, *Curr. Sci.*, **43**(4), 131.

Shah, S. K. and Sinha, A. K. (1974b). A note on the barytes occurrences in Ralam Garbyang sequence of Kumaun Himalaya, *Curr. Sci.*, **43**(13), 410–411.

Shah, S. K. and Sinha, A. K. (1974c). Stratigraphy and tectonics of the 'Tethyan' zone in a part of Western Kumaun Himalaya, *Him. Geology*, **4**, 1–27.

Sharma, K. K., Sinha, A. K., Bagdasarian G. P., and Gukasian R. Ch. (1978). Potassium–argon dating of Dras-volcanics, Shyok volcanics and Ladakh granite, Ladakh, N.W. Himalaya, *Him. Geology*, **8**(1), 288–295.

Sharma, K. K., Saini, H. S., and Nagpaul, K. K. (1978). Fission track annealing ages of apatite from Mandi granite and their application to tectonic problem. *Him. Geology*, **9**(1), 298–312.

Sharma, V. P. (1977). Geology of the Kulu–Rampur Belt, H.P., *Mem. G.S.I.*, **106**(II), 235–407.

Shatski, N. S. (1960). Paragenesis of sedimentary and volcanic rocks and formations, *Ijd. Ac. Sc. USSR, Ser. Geol. N.5*, (in Russian).

Shcherbak, N. P. (1975). *Petrology and Geochronology of Precambrian of Western Part of Ukrainian Shield*, Ijda, Naukova Dumka, Kiev, 261 pp. (in Russian).

Singh, I. B., Kumar, S., Singh, S. K., Jaikrishna, Jain, K. P., Singh, M. P., and Garg, R. (1980). Palaeontology and biostratigraphy of Spiti Shale formation near Lapthal, Malla Johar, Kumaun area, *U.P. Abst. XI, WIHG, Seminar*, 30–31.

Singh, M. P. and Kumar, S. (1978). Microfauna of the Spiti Shales (upper Jurassic) Mala Johar area Kumaun Himalaya, U.P., *Curr. Sci.*, **47**(4), 120–121.

Singh, M. P., Kumar, S., Singh, I. B., and Singh, S. K. (1980). A record of Triassic Foraminifera from the Kuti Shale (Noric) Kumaun Himalaya, U.P., *Geoscience J.*, **1**(1), 89–91, Lucknow.

Singh, R. L. (Ed.) (1971). *India, A Regional Geography*, Silver Jubilee Publication, National Geog. Soc. India, Varanasi, 443–479.

Singh, S. K., Kumar, S. and Singh, I. B. (1980). Isocrinids and the upper age limit of Spiti shale, Malla Johar areas, Dist. Garhwal, *U.P. Abst., XI, WIHG*, Seminar, **31**.

Sinha, A. K. (1972a). Problem of conjunction of tectonic–structure in the Simla–Shali area of Himalaya, *Collection of lectures of VIII Scientific-Technical Conference of Engineering Faculty*, People's Friendship Univ., Moscow, 144–145. (in Russian).

Sinha, A. K. (1972b). Tectonic position, lithology and age of autochthonous Simla and Shali series and problem of tectonic unit, Jaunsar, Blaini, Infra Krol, Krol and Tal in the Himalaya, India, Unpublished Ph.D. thesis, Friendship Univ. Moscow, 103 pp. (in Russian).

Sinha, A. K. (1974a). Clay mineralogy of green shale from Krol B horizon of Solon area in the Himalaya. *Rec. Res. in Geol.*, **2**, Delhi, 64–75.

Sinha, A. K. (1974b). Barite mineralization in the Tethyan zone of northern Kumaun Himalaya, *Abst. W.I.H.G. Sem.*, Srinagar (Garhwal), **36**.

Sinha, A. K. (1975a). The tectono-stratigraphic significance of Blaini–Formation of Simla Hills, *Bull. Ind. Geol. Assoc.*, **8**, 151–161.

Sinha, A. K. (1975b). Calcareous nannofossils from Simla Hills, H.P. (India), with a discussion of their age in the tectonostratigraphic column, *J. Geol. Soc. India*, **16**(1), 69–77.

Sinha, A. K. (1976). Litho-petrological characteristics of autochthonous Simla and Shali formations of Simla Himalaya, *Absts. Ind. Geol. Cong.*, Delhi, 26–27.

Sinha, A. K. (1977a). Geochronology, petrography–petrochemistry and tectonic significance of basic rock suites of N-W Himalaya, with special reference to H.P., India, *Rec. Res. in Geol.*, **3**, Delhi, 478–494.

Sinha, A. K. (1977b). Riphean stromatolites from Western Lower Himalaya, H.P., India, in Flügel, E. (Ed.), *Fossil Algae*, Springer Verlag, Berlin, 86–100.

Sinha, A. K. (Ed.) (1977c). *Himalayan Geology*, Vol. 7 1–482.

Sinha, A. K. (1977d). A discovery of barite and associated polymetallic mineralized zone in the Tethyan zone of Higher Garhwal and Kamaun Himalaya, *Him. Geology*, **7**, 456–463.

Sinha, A. K. (1978a). Para-autochthonous turbiditic–flyschoidal Simla and terrigenous–carbonate Shale formations of Himalaya: their lithopetrography and tectonic setting, *Him. Geology*, **8**, 425–455.

Sinha, A. K. (1978b). Reviews of *Himalaya*, Sciences de le terre, CNRS, Paris, 560 pp. *Him. Geology*, **8**(2), 1007–1008.

Sinha, A. K. (1978c). Non-renewable mineral resources of Himalaya—a preliminary inventory and future strategies for their environmentally sound exploitation, in *Proceed. Nat. Sem. on Resour. Devel. and Env. Him. Reg.*, Dept. of Sci. and Tech., Govt. India, 276–289.

Sinha, A. K. (1979). Analysis of thrust–nappe and deep-seated lineament tectonics in the framework of Himalayan folded belt, *Abst. X. WIHG Sem.* Dehra Dun.

Sinha, A. K. (1980a). Tectono-stratigraphic problem in the Lesser Himalayan zone of Simla region, H.P., in Valdiya, K. S. and Bhatia, S. B. (Eds), *Stratigraphy and Correlation of Lesser Himalayan Formation*, 99–116.

Sinha, A. K. (1980b). Geology and tectonics of the part of Higher Central Himalayan Tethyan Zone, *Spec. Issue Geol. Bull. Peshwar Univ.*, Pakistan (*Proc. Intern. Commit. Geodynamics*, Group 6), **13**, 29–37.

Sinha, A. K. (1980c). Geology and the tectonic design of Higher and Tethyan Zone of Central Himalaya, *Abst. 26th Intern. Geol. Cong., Paris*, **1**, 5 (tectonics), 391 pp.

Sinha, A. K. (1981a). Geological features of Central crystalline complex and adjacent Tethys zone in Central Himalaya, *Abst. 4th Indo-Soviet Symp. on Earth Sci., Ac. of Sci. USSR, IGEM, Moscow, 24–26*.

Sinha, A. K. (Ed.) (1981b). *Contemporary Geoscientific Researches in Himalaya*, **1**, 1–252, Dehra Dun, India.

Sinha, A. K. (1981c). Geology and tectonics of Himalayan regions of Ladakh, Himachal, Garhwal–Kumaun and Arunachal Pradesh: A review, in Gupta, H. K. and Delany, F. M. (Eds), *Zagros–Hindukush–Himalaya Geodynamic Evolution*, **3**, American Geophysical Union, Washington, 122–148.

Sinha, A. K. (1982). Tectonic framework and geologic evolution of Higher Central Himalaya, Unpublished D. Sc. Thesis Moscow, State University, 500 pp.

Sinha, A. K. (Ed.) (1983). *Contemporary Geoscientific Researches in Himalaya*, **2**, Dehra Dun, India, 168.

Sinha, A. K. (1984). Stromatolites from Shali and Deoban calc-zone: their significance in tectono-stratigraphy, *Proceed. V. Geophyto Conf. Lucknow, Spl. Publ.*, **1984**, 66–70.

Sinha, A. K. (1987). Tectonic zonation of Central Himalaya and the crustal evolution of collision and compressional Belts, in Gupta, H. K. (Ed.), *Deep Seated Procession Collision Zones. Tectonophysics*, **134**, 59–74.

Sinha, A. K. (in press). An obducted and thrusted ophiolitic melange complex in association with 'Exotic-blocks' of Higher Central Himalaya. (A. Gansser Comm. Vol., Lahore, Pakistan)

Sinha, A. K. and Jhingran, A. G. (1977). Deep seated lineament structures in Himalaya and Caucasus: their role in the history of geological development and Metallogeny, *Him. Geology*, **7**, 46–64.

Sinha, A. K. and Bagdasarian, G. P. (1977). Potassium–argon dating of some magmatic and metamorphic rocks from Tethyan and Lesser zones of Kumaon and Garhwal Indian Himalaya: and its implication in the Himalayan tectogenesis, *Himalaya Sci. de la terre,* Ed. CNRS, Paris, 387–394.

Sinha, A. K. and Pal, D. (1978). Ecological problems in the high altitude lakes of Naini Tal area: problem and suggestions, *J. Him. Stud. and Reg. Develop.*, **2**, 39–43.

Sinha, A. K. and Srivastava, R. A. K. (1978). On the occurrence of glauconite with radiolarites in the flysch sediments of Malla Johar area in Higher Himalaya and its significance in tectonics and sedimentation. *Him. Geology*, **8**(2), 1042–1048.

Sinha, A. K. and Srivastava, R. A. K. (1979). Mineral deposits of Kumaun Himalaya and their purposeful exploitation, in Pant, S. (Ed.), *Uttariya*, 10–12 (in Hindi).

Sinha, A. K. and Srivastava, R. A. K. (1986). Sangcha Malla orogenic pulse. A new epeirogenic Mid-Cretaceous cycle in the Himalayan tectonism, in Srivastava, R. A. K. (Ed.), *Glauconite: Form and Function*, Today and Tomorrow's Printers and Publishers, 209–219.

Sinha, A. K. and Nanda, A. C. (1979). Biostratigraphic zonation of Tethys zone of Unta-Dhura–Lapthal Region of Malla Johar (Higher Kumaun Himalaya), *Him. Geology*, **9**(1), 273–282.

Sinha, A. K. and Raaben, M. E. (1979). Lower Cambrian stromatolite from the Deoban Limestone of Lesser Himalaya, *Him. Geology*, **9**(1), 317–323.

Sinha, A. K. and Mehrotra, R. C. (1979). On the origin of barite and associated polymetallic mineralization in Tethyan zone of Higher Garhwal and Kumaun Himalaya, *Him. Geology*, **9**(2), 839–844.

Sinha, A. K. and Dmitrenko, O. B. (1983). Coccoliths from flysch of Malla Johar: a first report, *Him. Geology*, **11**, 251–255.

Sinha, A. K., Ashgirei, G. D., and Umnova, N. I. (1983). Carboniferous palynomorphs from Chail formation of Southern Himachal Himalaya, and their tectonic implications, *Him. Geology*, **11**, 415–421.

Sinha, A. K. and Bist, K. S. (1986) K-Ar dating of the Crystalline and Basic Rocks from Garhwal Central Himalaya, *Indian J. Earth Sci.*, **13**(2–3), 209–215.

Sinha-Roy, S. (1982). Himalayan Main Central Thrust and its implication for Himalayan inverted metamorphism, *Tectonophysics*, **84**, 197–224.

Sitholey, R. V., Sah, S. C. D., and Dube, S. N. (1954). Plant microfossils from carbonaceous shale (Krols) near Naini Tal, *J. Sci. and Indus. Res.*, **13B**(6), 450–451.

Spurr, J. E. (1916). The relation of ore deposition to faulting, *Econ. Geol*, **II**.

Srikantia, S. V. (1977). The Sundernagar Group: its geology, correlation and significance as stratigraphically deepest sediment in the Peninsular of Lesser Himalaya, *J. Geol. Soc. Ind.*, **18**(1), 7–22.

Srikantia, S. B. and Bhargava, O. N. (1967). Kakara series: a new palaeo-ocean formation in Simla Hills, *Bull. Geol. Soc. Ind.*, **4**(4), 113–116.

Srikantia, S. V. and Bhargava, O. N. (1974). The Jaunsar Problem in Himalaya—a critical analysis and elucidation, *J. Geol. Soc. India*, **15**(2), 115–136.

Srikantia, S. V. and Sharma, R. P. (1976). Geology of the Shali belt and the adjoining areas, *Mem. G.S.I.*, **106**(1), 31–166.

Srivastava, R. N. and Venkatraman, K. (1975). Palynostratigraphy of the Blaini Formation, *Bull. Ind. Geol. Assoc.*, **8**, 196–199.

Steiger, R. H. and Jägar, E. (1977). Subcommission on geochronology: convention on the use of decay constants in geo- and cosmochronology, *Earth Planet. Sci. Lett.*, **36**, 359–362.

Stöcklin, J. (1980). Geology of Nepal and its regional frame, *J. Geol. Soc., London*, **137**(1), 1–34.

Strachey, R. (1857). On the geology of parts of the Himalaya mountains and Tibet, *Quart. J. Geol. Soc. London*, **1**, 293–310.

Suess, E. (1885). *Das Antlitz der Erde*, Tampsky, Prag and Freytag, Leipzig.

Tandon, S. K. and Bhatia, S. B. (1978). Ichnocoenosis of the Tethyan zone of Kumaun Himalaya, with special reference to Precambrian–Cambrian boundary, *Jhingran Comm. Vol.*, Delhi, 378–398.

Tapponier, P., others, and Hongrong, Q. (1981). The Tibetan side of India–Eurasia collision, *Nature*, **294**, 405–410.

Tapponier, P., Peltzer, G., and Armijo, R. (1982). Propagating extrusion tectonics in Asia: new insight from simple experiments with plasticine, *Geology*, **10**, 611–616.

Tewari, A. P. (1981). Exotic thrust mass of the Padar area in Kashmir, *Tectonophysics*, **73**, 285–294.

Tewari, B. S. (1969). Nannofossils from the Krol, *Bull. Ind. Geol. Assoc.*, **2**, 122–123.

Thakur, V. C. (1976). Divergent isogrades of metamorphism in some parts of higher Himalaya zone, *Sci. de la terre*, CNRS Paris, 433–441.

Thakur, V. C. (1980). Tectonics of the Central Crystallines of Western Himalaya, *Tectonophysics*, **62**, 141–154.

Thakur, V. C. (1981). An overview of thrusts and nappes of western Himalaya, in McClay, K. and Price, M. J. (Eds), *Thrust and Nappe Tectonics*, The Geological Society of London, 381–392.

Thakur, V. C. (1983). Palaeotectonic evolution of Indus–Tsangpo Suture Zone in Ladakh and Southern Tibet, in Thakur, V. C. and Sharma,K. K. (Eds), *Geology of Indus Suture Zone*, 195–204, WIHG, Dehra Dun.

Tripathi, C., Jangpangi, B. S., Bhatt, D. K., Kumar, G., and Raina, B. K. (1984). Early Cambrian brachiopods from 'Upper Tal', Mussoorie Synclines, Dehra Dun District, U.P., India, *Geophytology*, **14**(2), 221–227, Lucknow.

Trivedi, J. R., Gopalan, K., and Valdiya, K. S. (1984). Rb–Sr Ages of granitic rocks within the Lesser Himalayan Nappes, Kumaun, India, *J. Geol. Soc. India*, **25**(10), 641–654.

Uhlig, V. (1903). The fauna of the Spiti shales, *Pal. Ind. G.S.I.*, **4**.

Valdiya, K. S. (1969). Stromatolites of the Lesser Himalayan carbonate formations and the Vindhyans, *J. Geol. Soc. India*, **10**, 1–25.

Valdiya, K. S. (1970). Simla Slates: the Precambrian flysch of the Lesser Himalaya: its turbidites, sedimentary structures and palaeocurrents. *Geol. Soc. Bull.*, **81**, 451–468.

Valdiya, K. S. (1973). The Blaini conglomerates of Himachal Pradesh and Garhwal Himalaya, *Rec. Res. Geology*, **1**, Delhi, 66–81.

Valdiya, K. S. (1975). Lithology and age of the Tal formation in Garhwal and its implication on stratigraphic scheme of Krol Belt in Kumaun Himalaya, *J. Geol. Soc. India*, **16**, 119–134.

Valdiya, K. S. (1976). Himalayan traverse faults and folds and their parallelism with subsurface structures of north Indian plains, *Tectonophysics*, **32**, 353–386.

Valdiya, K. S. (1979). An outline of the structural set up of the Kumaun Himalaya, *J. Geol. Soc. India*, **20**, 145–157.

Valdiya, K. S. (1980a). The two intra-crustal boundary thrusts of the Himalaya, *Tectonophysics*, **66**, 323–348.

Valdiya, K. S. (1980b). *Geology of the Kumaun Himalaya*, WIHG Publication, Dehra Dun, India, 291 pp.

Valdiya, K. S. (1983). Lesser Himalayan geology: crucial problems and controversies, *Curr. Science*, **52**(18), 839–857.

Valdiya, K. S., Gupta, V. J., Powar, K. B., and Banerjee, D. M. (1972). Palaeozoic biostratigraphy of the Tethys Himalaya in N-W Himalaya, *Nature*, **235**, 152–154.

Valdiya, K. S. and Gupta, V. J. (1972). A contribution to the geology of northern Kumaun, with special reference to the Hercynian gap in Tethys Himalaya, *Him. Geology*, **2**, 1–33.

Valdiya, K. S. and Pant, C. C. (1981). On the existence of the Berinag Thrust, Kumaun Lesser Himalaya, in Sinha, A. K. (Ed.), *Contemporary Geoscientific Researches in Himalaya*, Dehra Dun, **1**, 1–11.

Varadarajan, S. (1974). Prehnite-pumpellyite metagraywacke facies of metamorphism of the metabasites of Bhimtal–Bhowali area, Naini Tal, *Him. Geology*, **4**, 581–599.

Varadarajan, S. (1978). K–Ar ages of the Amritpur granite, dist. Naini Tal, Kumaun Himalaya, and its stratigraphic position, *J. Geol. Soc. India*, **19**, 380–382.

Virdi, N. S., Thakur, V. C., and Azmi, R. J. (1978). Discovery and significance of Permian microfossils in the Tso Morari crystallines of Ladakh, J. K., India, *Him. Geology*, **8**(2), 993–1000.

Voskesenkii, I. A., Kravchenko, K. N., Movschovich, E. B., and Sokolov, B. A. (1971). *Sketch Geology of Pakistan*, Izd. Nedra, Moscow, 1–168 (in Russian).

West, W. D. (1939). The structure of the Shali 'Window' near Simla. *Rec. G.S.I.*, **74**(1), 133–163.

William, H. (1950). Volcanoes of the Panicutin region, Mexico, *U.S. GS, Bull.*, **965–B**.

Windley, B. F. (1983). Metamorphism and tectonics of the Himalaya. *J. Geol. Soc., London*, **140**(6), 849–865.

Yanshin, A. L. (1965). *Tectonic Framework of Eurasia*, Geotektonika, N 5.

Yanshin, A. L. (Ed.) (1966). *Tectonic map of Eurasia* (1 : 5000 000) Ministry of Geology, Moscow (in Russian).

Zavaritski, A. N. (1946). Some facts to be taken into consideration while making tectonic reconstruction, *Ijd. Ac. Sc. USSR, Ser. Geol.*, **N 2** (in Russian).

INDEX